AF720125

Graphene Physics

Graphene Physics

Editor: Stephen Brooke

MURPHY & MOORE
www.murphy-moorepublishing.com

Published by Murphy & Moore Publishing
1 Rockefeller Plaza,
New York City, NY 10020, USA
www.murphy-moorepublishing.com

Graphene Physics
Edited by Stephen Brooke

© 2022 Murphy & Moore Publishing

International Standard Book Number: 978-1-63987-269-5 (Hardback)

This book contains information obtained from authentic and highly regarded sources. Copyright for all individual chapters remain with the respective authors as indicated. All chapters are published with permission under the Creative Commons Attribution License or equivalent. A wide variety of references are listed. Permission and sources are indicated; for detailed attributions, please refer to the permissions page and list of contributors. Reasonable efforts have been made to publish reliable data and information, but the authors, editors and publisher cannot assume any responsibility for the validity of all materials or the consequences of their use.

Trademark Notice: Registered trademark of products or corporate names are used only for explanation and identification without intent to infringe.

Cataloging-in-Publication Data

Graphene physics / edited by Stephen Brooke.
p. cm.
Includes bibliographical references and index.
ISBN 978-1-63987-269-5
1. Graphene. 2. Nanostructured materials. 3. Materials science.
4. Physics. I. Brooke, Stephen.
TA455.G65 G734 2022
620.115--dc23

Contents

Preface

The allotropic form of carbon which is one carbon atom thick is known as graphene. Atoms in a graphene layer are arranged in a hexagonal lattice structure. It is a zero gap semiconductor whose conduction and valence bands meet at the Dirac points. The propagation of electron waves in graphene takes place within a single atom layer, making it sensitive to the proximity of superconductors and ferromagnetics. It also exhibits the quantum Hall effect. The Van Der Waals forces in graphene follow an inverse cubic, asymptomatic power law. It is opaque in nature and is capable of absorbing light of all visible wavelengths. The various studies that are constantly contributing towards advancing technologies and evolution of this field are examined in detail. The ever growing need of advanced technology is the reason that has fueled the research in the field of graphene physics in recent times. A number of latest researches have been included in this book to keep the readers up-to-date with the global concepts in this area of study.

Significant researches are present in this book. Intensive efforts have been employed by authors to make this book an outstanding discourse. This book contains the enlightening chapters which have been written on the basis of significant researches done by the experts.

Finally, I would also like to thank all the members involved in this book for being a team and meeting all the deadlines for the submission of their respective works. I would also like to thank my friends and family for being supportive in my efforts.

Editor

1

Electrical Conductivity of Melt Compounded Functionalized Graphene Sheets Filled Polyethyleneterephthalate Composites

Haobin Zhang[1], Shunlun He[2*], Cao Chen[1], Wenge Zheng[1*] and Qing Yan[1]
[1]Ningbo Key Laboratory of Polymer Materials, Division of Polymers and Composites, Ningbo Institute of Material Technology & Engineering, Chinese Academy of Sciences,
[2]Orinko Plastics Group,
China

1. Introduction

Graphene, monolayer of carbon atoms arranged in a honeycomb network, has recently gained revolutionary aspirations (Novoselov et al., 2005; Novoselov et al., 2007; Heersche et al., 2007; Zhang (a) et al., 2005; Stankovich et al., 2006) because of its remarkable electronic properties (Zhang (b) et al., 2005; Berger et al., 2004), unusual thermal properties (Balandin et al., 2008) and good mechanical properties (Lee et al., 2008). These extraordinary properties make it an excellent choice as inorganic fillers to substantially improve electrical, thermal and mechanical properties of composite materials (Stankovich et al., 2006; Ramanathan et al., 2008). Several effective techniques have been developed for preparing graphene nanosheets, including chemical (Stankovich et al., 2006) and mechanical exfoliation (Novoselov et al., 2004), alkali metals intercalation and expansion (Viculis et al., 2003), microwave chemical vapor deposition (Wang et al., 2009), substrate-based thermal decomposition (Berger et al., 2004), and thermal exfoliation of graphite oxide (GO) (McAllister et al., 2007). Among them, chemical reduction of exfoliated graphite oxide in the presence of a surfactant or polymer is a relatively new method to prepare electrically conductive individual graphene sheets. Stankovich (Stankovich et al., 2006) put forward a new process to produce polystyrene/graphene nanocomposites via ultrasonic exfoliation and chemical reduction of graphite oxide and molecular-level dispersion of chemically modified graphene nanosheets. In addition, the thermal exfoliation and in situ reduction method can conveniently produce graphene nanosheets for mass production. As confirmed by Aksay and co-workers (McAllister et al., 2007), in situ reduction reaction took place during the thermal exfoliation process, which converted insulating graphite oxide to conducting graphene. More importantly, the graphene resulted through thermal exfoliation still contained some oxygen-containing groups. The oxygen functionalities on the graphene nanosheets will facilitate their dispersion in polar polymers (Ramanathan et al., 2008). Effective medium approximation indicated that graphene is more effective in improving conductivity of composites than carbon nanotubes (Xie et al., 2008). The polystyrene/graphene nanocomposites prepared by chemical modification and reduction in solution exhibited a percolation threshold in electrical conductivity as low as 0.1 vol%, comparable to those observed in single-walled carbon nanotubes (SWCNT) based nanocomposites (Grossiord et al., 2005; McLachlan et al., 2005); The electrical conductivity was as high as 1 S/m at 2.5 vol% of graphene content. Kim and Macosko (Kim & Macosko, 2008) compared the effect of graphene nanosheets and graphite on conductivity of polyester-based composites, and the results revealed that graphene gave a much lower percolation threshold (0.3 vol%) than graphite (3.0 vol%). Recently, Liang et al. (Liang et al., 2009) prepared epoxy/graphene nanocomposites via in situ process, which showed a small percolation threshold of 0.52 vol% and an electromagnetic interference shielding efficiency of 21dB with 8.8 vol% of graphene over 8.2~12.4 GHz. In contrast, unfully exfoliated graphite led to a much higher threshold (> 0.6 vol%) and a much lower electrical conductivity (< 10-2 S/m) even at high loading of 6.0 vol% (Kalaitzidou et al., 2007; Weng et al., 2005; Zheng et al., 2002).

Compared to in situ polymerization and solution mixing, melt compounding using conventional compounding devices such as mixers and commercial resins is more attractive because this approach provides manufacturers many degrees of freedom with regard to the selection of polymer grades and choice of graphene content. It is believed that melt compounding would be more economical and simple than in situ polymerization processes. Actually, melt compounding has been successfully used to prepare conductive polymeric composites by using conducting fillers such as carbon nanotubes (CNTs) (Li et al., 2006), carbon black and expanded graphite (Lagreve et al., 2006; Zhang et al., 2008). It has been confirmed that CNTs can substantially increase the electrical conductivity of PET nanocomposites with low loading of CNT, but it is expensive. As for cheap carbon black, higher loading is usually required to make a polymer conductive. Due to the low price and availability of pristine graphite in large quantities, graphene nanosheets can be an ideal choice as conductive fillers in the preparation of conductive PET nanocomposites. To the best of our knowledge, few papers have been published on PET/graphene nanocomposites prepared by conventional melt compounding. In this study, Graphene nanosheets were prepared by complete oxidation of pristine graphite followed by thermal exfoliation and reduction. Subsequently, polyethylene terephthalate/graphene nanocomposites were prepared by melt compounding. The microstructure of graphene, its dispersion in PET matrix, and the electrically conductive behavior of the resulting nanocomposites were studied.

2. Experimental

2.1 Materials

PET pellets were purchased and dried in vacuum oven at 150 oC for 5 h before use. Flaky pristine graphite with a mean size of 45 μm was bought from Qingdao Huatai Lubricant Sealing S&T (China). Graphene density is assumed to be equal to the theoretical graphite density of 2.28 g/cm3 (Kelly, 1981), and PET density is 1.34 g/cm3 (Cobbs & Burton, 1953). Fuming nitric acid (63%), sulfuric acid (98%), potassium chlorate (98%) and hydrochloric acid (37%) were obtained from Sinopharm Chemical Reagent (China).

2.2 Preparation of graphene

A technique, similar to that of Aksay and co-workers (Schniepp et al., 2006), was used to prepare bulk quantities of graphene nanosheets. Graphite oxide (GO) was prepared according to Staudenmaier method (Staudenmaier, 1898). First of all, the reaction flask was purged with nitrogen and immersed in an ice bath, 40 g graphite was then added to the homogeneous mixture of concentrated nitric acid (270 ml) and sulfuric acid (525 ml) under vigorous stirring. After uniform dispersion of the graphite powder, 330 g potassium chlorate was added slowly to minimize the risk of explosion. The reaction was then allowed to last for 120 h at room temperature. After reaction, the slurry resultants were filtered and washed with excess deionized water and 5% HCl solution to remove the sulfuric ions (SO_4^{2-}), then GO aqueous solution was neutralized with potassium hydroxide solution. GO powder was separated from the solution by using GQ75 high-speed centrifuge, and dried in an air-circulating oven at 135°C for 24 h followed by another 24 h at 135°C in a vacuum oven. As confirmed by Aksay and co-workers (McAllister et al., 2007), 1050°C was adequate for thermal exfoliation and in situ reduction of GO, this temperature was also adopted in the present work to thermally exfoliate GO. The dried GO powder was quickly inserted into a muffle furnace preheated to 1050 oC and held in the furnace for 30s.

2.3 Fabrication of PET composites

PET/graphene nanocomposites were prepared by melt compounding at 285°C using a Brabender mixer. Compounding was performed with an initial screw speed of 50 rpm/min for 4 minutes; then the screw speed was raised to 100 rpm/min within 1 min and the compounding was conducted at this speed for 5 min. The samples for microscopy and electrical conductivity analysis were prepared by pressing the composites at 275°C under a pressure of 15 MPa.

2.4 Characterization

The volume conductivity of moderately conductive composites (Conductance >10^{-6} S·m) was measured with a Agilent 34401A digital multimeter and a YOKOGAWA 7651 voltage source by the two-probe method. For PET and composites with low conductivity (Conductance ≤ 10^{-6} S·m), the conductivity testing was performed using a EST121 ultrahigh resistance and micro current meter (Beijing EST Science & Technology CO. LTD) according to National Standard of China GB/1410-1989. A three-terminal fixture (two electrodes plus guard), similar to that proposed by the ASTM D257 standard for volume conductance determination of flat specimens, was selected for the measurement of low conductance. The guard electrode connected to the earth was used to eliminate the leakage current along specimen surface. In addition, the testing fixture was embedded in a metal shield box to minimize external interference. Circular plates with 7 cm in diameter and 2.5 mm in thickness were fabricated for conductivity measurement. Sample surfaces were coated with silver paste to reduce contact resistance between sample and electrodes. For specimens with low conductivity, the side contacting with the unguarded electrode was wholly painted with silver paste while the other side only partly painted, the painted area was in accordance with the cross-section of the guarded cylinder electrode. Three specimens of each sample were tested and the average value was reported as volume electrical conductivity. The resistance R can be obtained directly from the multimeter, and thus the electrical conductivity σ can be calculated by using Eq. (1)

$$\sigma = \frac{1}{\rho} = \frac{t}{RS} \tag{1}$$

where ρ is electrical resistivity, t is thickness of sample between electrodes, and S is cross-sectional area of the guarded cylinder electrode.

The microstructures of graphene and its PET nanocomposites were characterized using a Tecnai G2 F20 TEM at an accelerating voltage of 100 kV. Graphene sheets were dispersed in methanol by ultrasonication and some pieces were collected on carbon-coated 300-mesh copper grids for observation. While for PET/graphene nanocomposites, the samples were embedded in epoxy resin, and cured at 80 oC for 6 h, and then ultrathin sections of thinner than 100 nm were cryogenically cut with a diamond knife using a microtome and collected on 300-mesh copper grids. SEM observation of PET composites was performed using an FEI scanning electron microscope (Hitachi, S-4800) with an accelerating voltage of 5 kV. PET composites were freeze-fractured in liquid nitrogen and the fractured section was coated with a thin layer of platinum before observation.
The diffraction behavior of pristine graphite, GO and graphene were studied using a Bruker AXS X-ray diffractometer with CuKα radiation at a generator voltage of 40 kV and a generator current of 40 mA. The scans were carried out in the reflection mode at a scan rate of 10°/min. Laser granularity analyzer (Malven, Zeta sizer Nano ZS) was used to determine the lateral dimension of graphene in alcohol solution.

3. Results and discussion

3.1 Characterization of graphene

Figure. 1 shows the XRD spectra of pristine graphite, graphite oxide and graphene. The strong and sharp diffraction peak of pristine graphite at 26.6o completely disappeared after oxidization and instead a new peak at 13.9° appeared, indicating a complete oxidization of graphite (Wang et al., 2008), which is a prerequisite to obtain exfoliated graphene nanosheets by ultrasonication or thermal expansion. After thermal exfoliation of the completely oxidized graphite (GO), there was no apparent diffraction peak detected, which means the periodic structure of GO was eliminated and graphene nanosheets were formed.

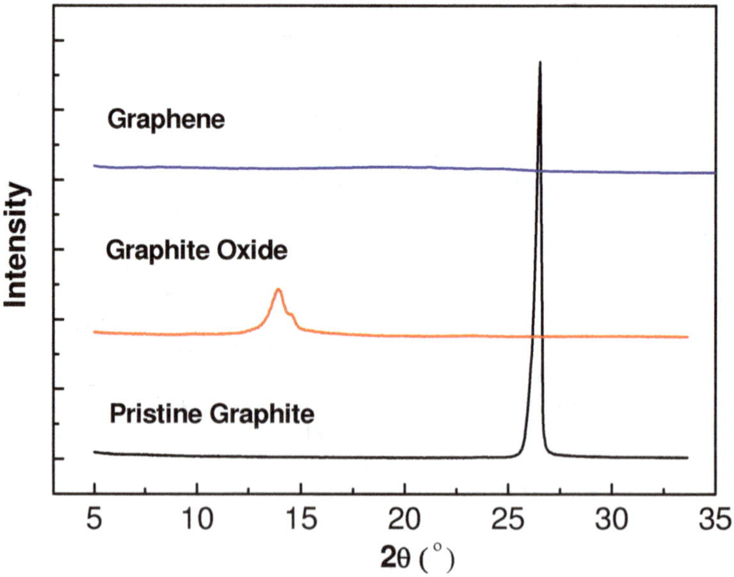

Fig. 1. X-ray diffraction patterns of pristine graphite, graphite oxide and graphene.

Figure 2 shows TEM images of the graphene under low and high magnifications. Under low magnification, graphene looks like a transparent ultrathin film with a few thin ripples within the plane (Figure. 2a), quite different from the opaque and smooth feature of pristine graphite flakes. It is known that perfect two-dimensional graphene crystal was thermodynamically unstable (Meyer et al., 2007), therefore corrugations and ripples in the two dimension were formed for thermodynamic stability (Carlsson, 2007). Thereby the transparency and rippled feature of graphene, in turn, suggests a fact that graphene sheets prepared in this study are wafer thin and consists of few-layer or even monolayer graphene. Furthermore, stacks of graphene sheets arranged with a roughly parallel structure can be clearly observed in the high-magnification TEM image from an edge-on view (Figure. 2b).

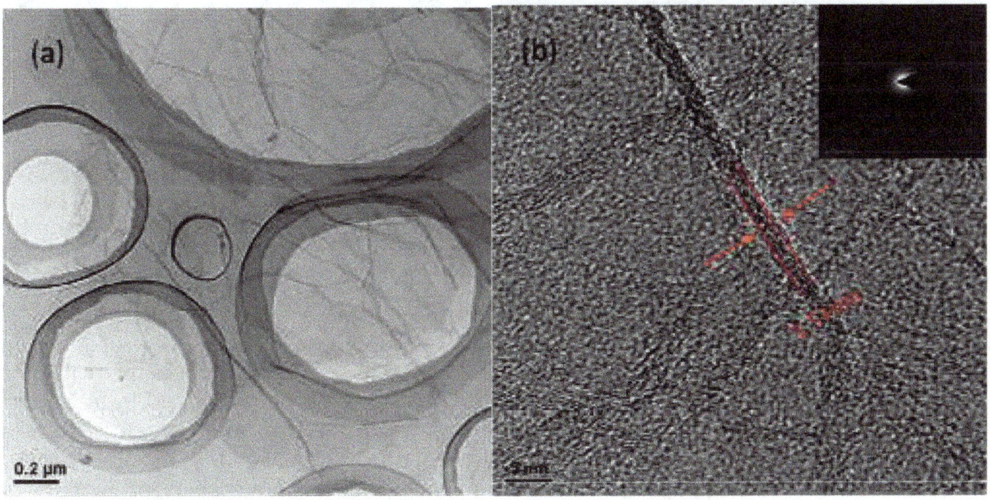

Fig. 2. TEM images of graphene under low (a) and high (b) magnifications. The scale bars are (a) 0.2 μm and (b) 5 nm. The inset is selected area electron diffraction (SAED) pattern of graphene.

The thickness of graphene sheets can thus be measured. Statistical results from many TEM images give an average thickness ~ 1.57 nm of the graphene sheets prepared in this study. In addition, the weak and diffuse selected area electron diffraction (SAED) pattern (the inset of figure. 2b) also indicates the loss of long range ordering between graphene sheets and the ultrathin feature of the graphene. However, it should be mentioned that the specific surface area of the graphene sheets measured with BET method is 555 m2/g , much smaller than the theoretical value (2630 m2/g) of single graphene sheet (Peigney et al., 2001), indicating that existence of agglomerated and overlapped parts of the graphene sheets. In spite of this, a high aspect ratio (lateral dimension to thickness) of 146 for the graphene sheets is obtained based on the results from laser granularity analyzer. In contrast, pristine graphite flake has a lower aspect ratio (~36) and smaller specific surface area (< 10 m2/g).

3.2 Microstructure of PET/graphene nanocomposites

To clearly reveal the dispersion of graphene in PET matrix, Figure 3 shows the microstructure of PET nanocomposite with 3 vol% graphene. Graphene nanosheets were

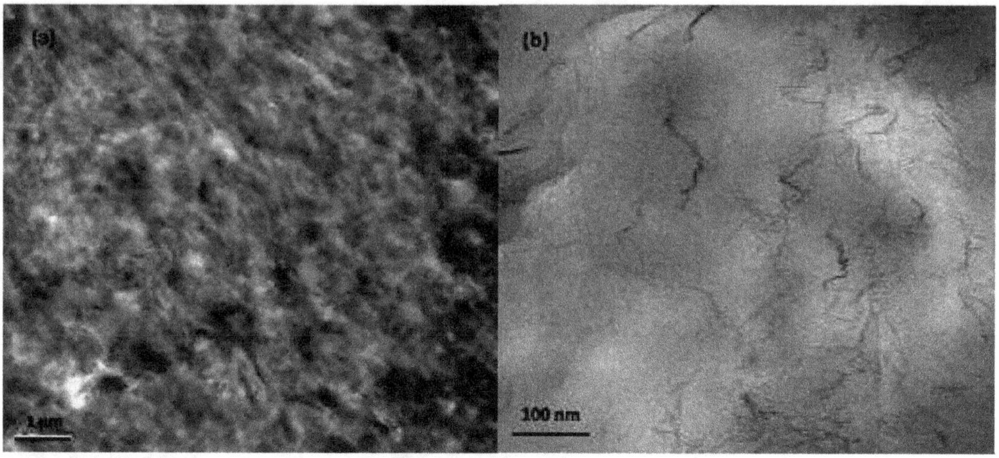

Fig. 3. Low (a) and high (b) magnification TEM micrographs of PET nanocomposite with 3 vol% graphene. The scale bars are (a) 1 μm and (b) 100 nm.

homogeneously dispersed in PET matrix and there are almost no large agglomerates observed (Figure. 3a).

The good dispersion of graphene sheets should be due to the good interaction between the oxygen and hydroxyl functional groups on the surface of graphene and the polar groups of PET. In addition, a compact continuous network throughout the matrix is also observed (Figure. 3a). As proposed by Alig et al (Alig et al., 2007), this conductive network can be considered as a network built by conductive fillers which are separated by local contact regions with polymer chains in-between, also improving electron mobility and interface polarization via tunneling (Yao et al., 2007). High magnification image (Figure. 3b) indicates that the network is composed of abundant thin stacks of a few sheets of monolayer graphene (black curly thin lines), which are bridged by the crumpled or overlapped graphene sheets (slightly thicker lines) (Figure. 3b). These wrinkled and overlapped graphene sheets can effectively link individual graphene sheets and carry high density of current, resulting in high electrical conductivity (Halperin et al., 1985).

To demonstrate the strong interaction between the oxygen and hydroxyl functional groups on the surface of graphene and the polar groups of PET, Figure 4 shows SEM image of cryo-fractured surface of PET nanocomposite with 1.2 vol% graphene. It is seen that the graphene sheets were encapsulated by PET and the thickness of the sheets increased from 1.57 to ~ 50 nm (see the inset of Fig. 4). This confirms the strong interaction between graphene and PET. Due to the large specific surface area of the graphene nanosheets, the area of interface between graphene and PET is huge, providing numerous tunneling sites for electron transport (Steinert & Dean, 2009).

3.3 Electrical properties of PET/graphene nanocomposites

Figure. 5 shows plots of electrical conductivity versus filler content for PET/graphene nanocomposites and PET/pristine graphite composites.

The PET/graphene nanocomposites exhibit a sharp transition from insulator to semiconductor with a small percolation threshold of 0.47 vol%. With a small increase of graphene content from 0.47 to 1.2 vol%, the electrical conductivity of the PET

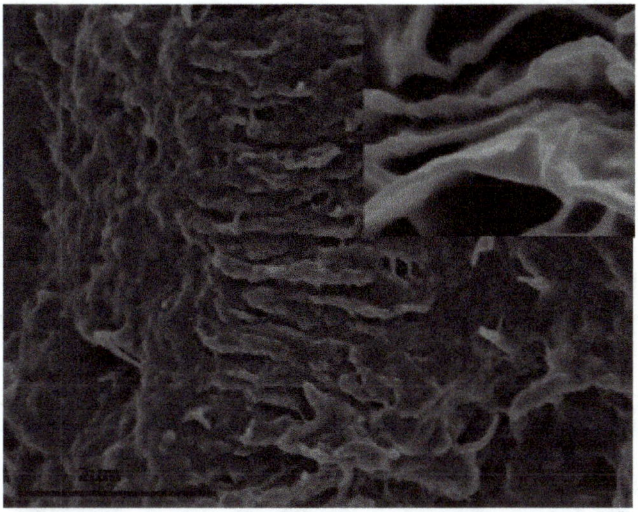

Fig. 4. SEM images of fractured surface of PET nanocomposite with 1.2 vol% of graphene. The insert is a high-magnification SEM image of the same nanocomposite.

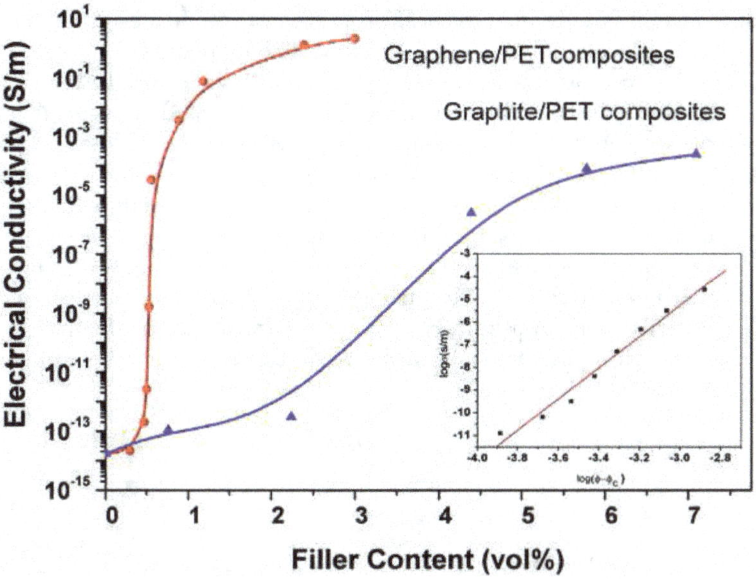

Fig. 5. Plots of electrical conductivity versus filler content for PET/graphene nanocomposites and PET/graphite composites. The inset is double-logarithmic plot of volume electrical conductivity versus (φ-φc).

nanocomposites increased quickly from 2.0×10^{-13} to 7.4×10^{-2} S/m. Actually, at 0.56 vol% of the graphene content, the conductivity the PET nanocomposite is 3.3×10^{-5} S/m, which is higher than antistatic criterion of 10^{-6} S/m. With only 3.0 vol% graphene, the conductivity of

PET nanocomposite is as high as 2.11 S/m. On the contrary, PET/graphite composites show a much higher percolation threshold of 2.4 vol% and a very broad percolation transition within a range of graphene content from 2.4 to 5.8 vol%. More interestingly, graphene is more efficient in improving electrical conductivity of PET than pristine graphite. For instance, the conductivity of PET/graphene nanocomposite with 0.92 vol% of graphene is 3.46×10^{-3} S/m, even higher than that of PET/pristine graphite composite with 7.1 vol% of graphite (2.45×10^{-4} S/m).

According to the percolation theory (Stauffer, 1985), the relationship of composite conductivity (σ) with filler volume content (φ) follows the below scaling law near the percolation threshold (φc):

$$\sigma \propto \sigma_0(\varphi - \varphi_c)^{\upsilon} \tag{2}$$

Where σo is the bulk electrical conductivity of the fillers and υ is the critical exponent describing the rapid variation of σ near φc. The percolation threshold is the critical content above which a continuous connected network is formed for the transport of electrons throughout the matrix. The inset in figure. 5 shows the double-logarithmic plot of electrical conductivity of PET/graphene nanocomposites versus (φ-φc), and an "anomalous" critical exponent (4.22) was obtained from the slope of linear fit curve, much larger than the theoretically predicted value (~ 2) for three-dimensional system (Balberg, 1987).

To predict the electrical percolation threshold in theory, filler shape must be first considered in view of its strong influence on the electrical properties of composites (Ma & Gao, 2008). However, the electrical percolation threshold cannot be directly related to filler morphology, fortunately, geometrical percolation threshold (pc) provides some insight into the former although cannot be simply identified as it. Therefore, several models were developed to explore the percolative properties of oblate objects in respect of their aspect ratio (Ambrosetti et al., 2008; Garboczi et al., 1995; Yi & Tawerghi, 2009). Recently, Ambrosetti and coworkers (Ambrosetti et al., 2008) proposed a model consisting of isotropically distributed hard oblate spheroids of identical dimensions with soft shell, more fits the system of conducting spheroids dispersed in an insulating continuum host, and the shell thickness can be interpreted as a typical tunneling length between the particles, governing the electrical connectivity of composite. The percolation threshold, pc, for infinite system was obtained by extrapolating that for finite systems through the following scaling relations (Rintoul & Torquato, 1997):

$$\Delta(L) \propto L^{-1/\delta} \tag{3}$$

$$p_c^{eff} - p_c \propto L^{-1/\delta} \tag{4}$$

where δ is the correlation length exponent, Δ represents the width of the percolation transition, pceff is the effective percolation threshold for a system of linear size L. The value of correlation exponent δ is calculated as around 0.9 from thousands of simulations for variously sized lattices, and then pc can be consequently estimated from Eq. (4) by plotting pceff against L-δ for selected values of cell size L. The simulation results, for oblate ellipsoids with various aspect ratio and penetrable shell thickness, indicated that high aspect ratio entails a lower percolation threshold and this effect is also dependant on the shell thickness, consistent with the behavior observed in our study and other composites containing oblate objects as conducting fillers. Therefore, the small φc of PET/graphene nanocomposites mainly owes to the high aspect ratios of graphene nanosheets.

Moreover, if neglecting the limiting interaction distances to a shell of constant thickness from the presence of the hard core, the model will evolve to an idealized model consisting of completely permeable ellipsoids randomly dispersed in a matrix, as Garboczi and coworkers (Garboczi et al., 1995) suggested. From this model, pc can be predicted as 1.27 α/β where β is the radius of revolution and α is the semiaxis length in the extreme oblate limit. Therefore, pc can be easily calculated as 0.87 vol% for PET/graphene nanocomposites and 3.34 vol% for PET/graphite composites if graphene and graphite considered as permeable oblate particles. These values are higher than that of φc obtained experimentally from curves of Figure 5. This difference between pc and φc may be, in part, due to the assumption that graphite and graphene were permeable, and in part, to the fact that Eq. (2) did not consider tunneling mechanism. The permeable assumption allows regions of space to be occupied by parts of more than one sheet and thereby the critical volume of fillers is overestimated. In addition, the unfavorable geometries of oblate particles also tend to reduce the interparticle connectivity and increase the percolation point. Finally, it is noticed that physically contacted network is adequate but not absolutely necessary for current flow in insulator-conductor composites, since electrons can transmit between isolated particles by tunneling through thin polymer layer (Quivy et al., 1989; Toker et al., 2003).

From above experimental results, it is can be seen that the efficiency of graphene in improving conductivity of PET is comparable to or even better than CNTs. Hu (Hu et al., 2006) reported a low electrical percolation threshold (0.9 wt%) in PET/MWCNT nanocomposites fabricated via coagulation, and the electrical conductivity reached 10-2 S/m at higher filler loading (~5 wt%). A small percolation threshold was reported by Steinert and coworkers in PET/SWCNT nanocomposite films prepared using a solution mixing and casting method. It was shown that sufficient conductivity for antistatic and electrostatic dissipation purposes was achieved at 0.5 wt% SWCNT, which was easily understood because solution casting of a film is efficient to form a conductive network due to the 2-dimentional distribution of CNT and less serious damage of CNT length. However, melt compounding like twin-screw extrusion would inevitably shorten CNT and thus reduce its aspect ratio, which is not beneficial for achieving small percolation thresholds and high conductivity. For example, CNTs-based composites prepared by melt blending with coupling agent only leads to a moderate conductivity of 0.1S/m with 2.48 vol% CNTs (Li et al., 2006). Fortunately, the electrical percolation threshold (φc = 0.47 vol %) obtained in this study is also one of the lowest values for graphite-based composites (Chen et al., 2003; Zheng & Wong, 2003) although slightly higher than that observed in polystyrene/graphene composites prepared by solution mixing (Stankovich et al., 2006). This low percolation threshold and the good conductivity are attributed to the high aspect ratio, large specific surface area and the good dispersion of graphene in PET matrix.

In addition, we found the anomalous exponent of 4.22 for the PET/graphene nanocomposites. Indeed, nonuniversal values (≥ 3) of the critical exponent were reported in many systems (Celzard et al., 1996; Carmona et al., 1984; Ezquerra et al., 1990). To explain this phenomenon, the tunneling conduction mechanism is repeatedly proposed. Balberg proposed a model based on the interparticle tunneling in percolation network. The conductance distribution function of the network, f(g), was related to the tunneling conductance, g, between two impermeable spherical particles with a radius of b and an average interparticle distance a by:

$$f(g) \propto \ln(g_0 / g) g^{(d_0/a)-1} \tag{5}$$

where g_0 is a constant and d_0 is the characteristic distance for tunneling (usually several nanometers). Percolation network and tunneling conductance are two competitive mechanisms in conductive composites, and the configuration of conductive fillers in polymer matrix determines the competition of these two mechanisms. Fillers existing as large and complicated aggregates (i.e. high structured carbon blacks) in composites are easy to cause much narrower distribution of interparticle tunneling distances than that of the distances between two adjacent particles, namely, the case of d0 > a, for small g values, the conductance function can account for the network and a universal behavior will be obtained (Balberg & Binenbaum, 1985). Whereas, smaller and less agglomerated filler structure (low structure carbon blacks) can provide a wide-enough distribution of the distances between particle surfaces, d0 < a, it can be seen that the essential part of the conductance distribution is determined and given by Eq. (6), which is well known to yield a anomalous behavior (Kogut & Straley, 1979).

$$g^{(d_0/a)-1} \equiv g^{-a} \tag{6}$$

Thus, an anomalous critical exponent of 4.0 was found in the composites filled with uniformly dispersed carbon blacks. Notice that the relationship given by Eq. (6) can be extended to general case of a ≥ 2b. For our PET/graphene nanocomposites, the evenly dispersed monolayer graphene or very small aggregates would also result in wide distribution of the interparticle distances similar to that derived from the above low structure carbon blacks in polyvinylchloride. Therefore, Eq.s (5) and (6) can be applied to interpret the nonuniversal conductivity in PET/graphene nanocomposites. Additionally, the vast nanosheet-polymer interface provides abundant potential sites for tunneling conductance, thereby an anomalous critical exponent of 4.22 was found in PET/graphene nanocomposites.

The nonuniversal behavior can also be associated with the anisotropy (Smith & Lobb, 1979) of conductive fillers and the neck-like bonds in the percolation backbone. Both experimental and theoretical results indicate that geometrical and electrical anisotropy degree is expected to influence the percolation threshold and critical exponent.

4. Conclusions

Graphene sheets were prepared by chemical oxidation of pristine graphite flakes followed by thermal exfoliation and reduction of graphite oxide. PET/graphene nanocomposites prepared by melt compounding exhibit superior electrical conductivity with a small percolation threshold of 0.47 vol%. A high electrical conductivity of 2.11 S/m of PET nanocomposite was achieved with only 3.0 vol % of graphene, which is even adequate for EMI shielding. Additionally, an anomalous critical exponent of 4.22 for PET/graphene nanocomposites can be ascribed to the tunneling conduction and conductivity anisotropy of graphene. Compared to expensive CNTs, graphene is cheaper and it is more effective in making insulating polymers electrically conductive due to its higher specific surface area than CNTs.

5. References

Alig I., Lellinger D., Dudkin S.M. & Potschke P. (2007). *Polymer*, 48(4):1020-1029.

Ambrosetti G., Johner N., Grimaldi C., Danani A. and Ryser P. (2008). *Phys Rev E*, 78(6), 061126.

Balandin A.A., Ghosh S., Bao W.Z., Calizo I., Teweldebrhan D., Miao F. & Lau C.N. (2008). *Nano Letters*, 8(3):902-907.

Balberg I. & Binenbaum N. (1985). *Phys Rev A*, 31(2):1222-1225.

Balberg I. (1987). *Phys Rev Lett*, 59(12):1305-1308.

Berger C., Song Z.M., Li T.B., Li X.B., Ogbazghi A.Y., Feng R., Dai Z.T., Marchenkov A.N., Conrad E.H., First P.N. & de Heer W.A. (2004) *JPhys Chem B*, 108(52):19912-19916.

Carmona F., Prudhon P. & Barreau F. (1984). *Solid State Commun*, 51(4):255-257.

Carlsson J.M. (2007). *Nature Materials*, 6(11):801-802.

Celzard A., Furdin G., Mareche J.F., McRae E., Dufort M. & Deleuze C. (1994). *Solid State Commun*, 92(5):377-383.

Celzard A.M.E., Mareche J.F., Furdin G., Dufort M. & Deleuze C. (1996). *JPhys Chem Solids*, 57(6-8):715-718.

Chen G.H., Weng W.G., Wu D.J. & Wu C.L. (2003). *Eur Polym J*, 39(12):2329-2335.

Cobbs W.H. & Burton R.L. (1953). *JPolym Sci*, 10(3):275-290.

Ezquerra T.A., Kulescza M., Cruz C.S. & Baltacalleja F.J. (1990). *Adv Mater*, 2(12): 597-600.

Garboczi E.J., Snyder K.A., Douglas J.F. & Thorpe M.F. (1995). *Phys Rev E*, 52(1):819-828.

Grossiord N., Loos J. & Koning C.E. (2005). *JMater Chem*, 15(24):2349-2352.

Halperin B.I., Feng S. & Sen P.N. (1985). *Phys Rev Lett*, 54(22):2391-2394.

Heersche H.B., Jarillo-Herrero P., Oostinga J.B., Vandersypen L.M.K. & Morpurgo A.F. (2007). *Nature*, 446(7131):56-59.

Hu G.J., Zhao C.G., Zhang S.M., Yang M.S. & Wang Z.G. (2006). *Polymer*, 47(1):480-488.

Kalaitzidou K., Fukushima H. & Drzal L.T. (2007). *Carbon*, 45(7):1446-1452.

Kelly B. (1981). *Physics of graphite*, London Englewood N.J: Applied science publishers, pp. 267-361.

Kim H. & Macosko C.W. (2008). *Macromolecules*, 41(9):3317-3327.

Kogut P.M. & Straley J.P. (1979). *JPhys C- Solid State Phys*, 12(11):2151-2159.

Lagreve C. Feller J.F., Linossier I. & Levesque G. (2001). *Polym Eng Sci*, 41(7): 1124-1132.

Lee C., Wei X.D., Kysar J.W. & Hone J. (2008). *Science*, 321(5887):385-388.

Li Z.F., Luo G.H., Wei F. & Huang Y. (2006). *Compos Sci Technol*, 66(7-8):1022-1029.

Liang J.J., Wang Y., Huang Y., Ma Y.F., Liu Z.F., Cai F.M., Zhang C.D., Gao H.J. & Chen Y.S. (2009). *Carbon*, 47(3):922-925.

Ma H.M. & Gao X.L. (2008). Polymer, 49(19):4230-4238.

McAllister M.J., Li J.L., Adamson D.H., Schniepp H.C., Abdala A.A., Liu J., Herrera-Alonso M., Milius D.L., Car R., Prud'homme R.K. & Aksay I.A. (2007). *Chem Mater*, 19(18): 4396-4404.

McLachlan D.S., Chiteme C., Park C., Wise K.E., Lowther S.E., Lillehei P.T., Siochi E.J. & Harrison J.S. (2005). *JPolym Sci Pol Phys*, 43(22):3273-3287.

Meyer J.C. Geim A.K., Katsnelson M.I., Novoselov K.S., Booth T.J. & Roth S. (2007). *Nature*, 446(7131):60-63.

Novoselov K.S., Geim A.K., Morozov S.V., Jiang D., Zhang Y., Dubonos S.V., Grigorieva I.V. & Firsov A.A. (2004). *Science*, 306(5696):666-669.

Novoselov K.S., Geim A.K., Morozov S.V., Jiang D., Katsnelson M.I., Grigorieva I.V., Dubonos S.V. & Firsov A.A. (2005). *Nature*, 438(7065):197-200.

Novoselov K.S., Jiang Z., Zhang Y., Morozov S.V., Stormer H.L., Zeitler U., Maan J.C., Boebinger G.S., Kim P. & Geim A.K. (2007). *Science*, 315(5817):1379-1379.

Peigney A, Laurent C, Flahaut E, Bacsa RR, & Rousset A. *Carbon*, 2001; 39(4): 507-514.

Quivy A., Deltour R., Jansen A.G.M. & Wyder P. (1989). *Phys Rev B*, 39(2):1026-1030.

Ramanathan T., Abdala A.A., Stankovich S., Dikin D.A., Herrera-Alonso M., Piner R.D., Adamson D.H., Schniepp H.C., Chen X., Ruoff R.S., Nguyen S.T., Aksay I.A., Prud'homme R.K. & Brinson L.C. (2008). *Nature Nanotechnology*, 3(6):327-331.

Rintoul M.D. & Torquato S. (1997). *JPhys A-math Gen*, 30(16):L585-L592.

Schniepp H.C., Li J.L., McAllister M.J., Sai H., Herrera-Alonso M., Adamson D.H., Prud'homme R.K., Car R., Saville D.A. & Aksay I.A. (2006). *JPhys Chem B*, 110(17): 8535-8539.

Smith L.N. & Lobb C.J. (1979). *Phys Rev B*, 20(9):3653-3658.

Stankovich S., Dikin D.A., Dommett G.H.B., Kohlhaas K.M., Zimney E.J., Stach E.A., Piner R.D., Nguyen S.T. & Ruoff R.S. (2006). *Nature*, 442(7100):282-286.

Staudenmaier L. (1898). *Ber Dtsch Chem Ges*, 31:1481-1499.

Stauffer D. (1985). *An introduction to percolation theory*, Landon and Philadephia: Taylor and Fansis, 325-355.

Steinert B.W. & Dean D.R. (2009). *Polymer*, 50(3):898-904.

Toker D., Azulay D., Shimoni N., Balberg I. & Millo O. (2003). *Phys Rev B*, 68: 041403.

Viculis L.M., Mack J.J. & Kaner R.B. (2003). Science, 299(5611):1361-1361.

Wang G.X., Yang J., Park J., Gou X.L., Wang B., Liu H. & Yao J. (2008). *J Phys Chem C*, 112(22):8192-8195.

Wang X.B., You H.J., Liu F.M., Li M.J., Wan L., Li S.Q., Li Q., Xu Y., Tian R., Yu Z.Y., Xiang D. & Cheng J. (2009). *Chem Vapor Depos*, 15(1-3):53-56.

Weng W.G., Chen G.H. & Wu D.J. (2005). *Polymer*, 46(16):6250-6257.

Xie S.H., Liu Y.Y. & Li J.Y. (2008). *Appl Phys Lett*, 92(24):243121.

Yao S.H., Dang Z.M., Jiang M.J. & Xu H.P. (2007). *Appl Phys Lett*, 91(21):042914.

Yi Y.B. & Tawerghi E. (2009). *Phys Rev E*, 79: 041134.

Zhang M., Li D.J., Wu D.F., Yan C.H., Lu P. & Qiu G.M. (2008). *JAppl Polym Sci*, 108(3):1482-1489.

Zhang Y.B., Tan Y.W., Stormer H.L. & Kim P. (2005). *Nature*, 438(7065):201-204.

Zhang Y.B., Small J.P., Amori M.E.S. & Kim P. (2005). *Phys Rev Lett*, 94(17): 178603.

Zheng W.G., Wong S.C. & Sue H.J. (2002). *Polymer*, 43(25):6767-6773.

Zheng W. & Wong S.C. (2003). *Compos Sci Technol*, 63(2):225-235.

2

Superconductivity and Electron-Phonon Coupling in Graphite Intercalation Compunds

Tonica Valla and Zhihui Pan
Condensed Matter Physics and Materials Science Department, Brookhaven National Laboratory
USA

1. Introduction

In graphite intercalation compounds (GIC), the intercalation of various atomic or molecular species in between graphene layers in graphite leads to novel properties and a very rich physics, including superconductivity (Dresselhaus & Dresselhaus, 2002). In graphite intercalated with alkaline metals, superconductivity has been known for decades (Hannay et al., 1965), but after recent discovery of relatively high T_c superconductivity in CaC_6 ($T_c =$ 11.5 K)(Emery et al., 2005; Weller et al., 2005) research in this field has been intensified. In conventional metals, the electron-phonon coupling has long been known to be the pairing interaction responsible for the superconductivity. The strength of this interaction essentially determines the superconducting transition temperature T_c. Even though the electron phonon coupling is most likely responsible for pairing in GICs (Hinks et al., 2007; Kim et al., 2006; Lamura et al., 2006), it is still not clear what electronic states, intercalant- or graphene- derived ones, and what phonons are responsible for pairing (Boeri et al., 2007; Calandra & Mauri, 2005; Mazin et al., 2007; Mazin, 2005). Due to differences in structure and composition, no clear trends have been identified that could unambiguously resolve these issues. For example, KC_8 is a superconductor and LiC_6 is not. Further, in GICs intercalated with alkaline earths, T_c ranges from zero to 11.5K, even though they share the same chemical formula MC_6, where M is an alkaline earth atom. This obviously represents a serious problem to the proposal that superconductivity originates from graphene sheets and that the only role of intercalants is to provide the charge to the graphene bands. Further, band structure calculations show that in graphite and GICs, an interlayer state exists above π^* band (Holzwarth et al., 1984; Posternak et al., 1983), prompting some researchers to propose that its partial filling and coupling to soft intercalant phonons induces superconductivity in GICs (Csányi et al., 2005; Mazin, 2005). The experimental situation is still inconclusive, with strong advocates for intercalant (Hinks et al., 2007) and graphene dominated superconductivity (Dean et al., 2010; Grüneis et al., 2009; Kim et al., 2006; Pan et al., 2010; Valla et al., 2009). Recent angle resolved photoemission spectroscopy study on CaC_6 (Valla et al., 2009) reported that the electron-phonon coupling on graphene-derived Fermi surface to graphene phonons is strong enough to explain a T_c in the range of tens of Kelvin, indicating that graphene sheets provide crucial ingredients for superconductivity in GICs. However, to test this idea, it would be important to extend similar studies to GICs with different T_c.

One manifestation of electron-phonon coupling is a renormalization of the electronic dispersion or a "kink" at the energy scale associated with the phonons. This renormalization is directly observable in a photoemission experiment, making photoemission spectroscopy an ideal probe of electron-phonon coupling. In this chapter, we present the angle resolved photoemission studies of the electronic structure and the electron-phonon coupling in the non-superconducting LiC_6 and in superconducting KC_8 ($T_c = 0.39$ K) and compare these materials with several other GICs. We find that the electron-phonon coupling on the graphene derived π^* states to the graphene derived phonons increases with the filling of π^* states in a sequence from LiC_6 to KC_8 to CaC_6, following the same trend as T_c. The positive correlation between these quantities implies that superconductivity originates in graphene sheets while the main role of intercalants is to provide the charge for filling of the graphene π^* states.

2. Photoemission and interactions

In photoelectron spectroscopy, a photon of known energy, $h\nu$, is absorbed and the outgoing electron's kinetic energy $E_K = h\nu - \phi - \varepsilon_k$ and angle are measured. These properties determine the binding energy ε_k and parallel momentum $k_{||}$ of the hole left in the occupied valence bands (Kevan, 1992). Interaction effects, including for instance electron-phonon coupling, cause the infinitely sharp line spectrum of independent electron theory, $A_0(\mathbf{k},\omega)$ = Im $G_0(\mathbf{k},\omega)$ = Im $1/(\omega - \epsilon_{\mathbf{k}0} - i\eta)$, where $\epsilon_{\mathbf{k}0}$ represents a bare band dispersion, to evolve into Im $1/[\omega - \epsilon_{\mathbf{k}0} - \Sigma(\mathbf{k},\omega)]$ where the complex self-energy $\Sigma(\mathbf{k},\omega)$ contains the effects of the many body interactions. The spectral function of the photo-hole, $A(\mathbf{k},\omega)$, then takes the form

$$A(\mathbf{k},\omega) \propto \frac{\mathrm{Im}\Sigma(\mathbf{k},\omega)}{[\omega - \epsilon_{\mathbf{k}0} - \mathrm{Re}\Sigma(\mathbf{k},\omega)]^2 + (\mathrm{Im}\Sigma(\mathbf{k},\omega))^2} \tag{1}$$

The real part, $\mathrm{Re}\Sigma(\mathbf{k},\omega)$, gives a shift in energy and associated mass enhancement, while the imaginary part $\mathrm{Im}\Sigma(\mathbf{k},\omega)$ gives the lifetime broadening of a quasiparticle. The spectral function $A(\mathbf{k},\omega)$ is a fundamental quantity that determines all the properties of a many-body system. It can be directly measured in a photoemission experiment because the intensity of photoelectrons is given by $I(\mathbf{k},\omega) = |M|^2 A(\mathbf{k},\omega) f(\omega)$ where M represents the matrix element linking the initial and final states in the photoemission process, $A(\mathbf{k},\omega)$ is the single particle spectral function given in equation (2) and $f(\omega)$ is the Fermi function which enters because the photoemission process is restricted to excitation from occupied states.

2.1 Electron-phonon coupling in photoemission

We now focus on the electron-phonon coupling. The electron-phonon coupling contribution, $\Gamma_{e-ph} = 2\mathrm{Im}\Sigma_{e-ph}$, to the total scattering rate may be calculated via the Eliashberg equation such that (Mahan, 1990)

$$\Gamma_{e-ph}(\omega,T) = 2\pi\hbar \int_{-\infty}^{\infty} d\omega' \alpha^2 F(\omega') \left[2n(\omega') + f(\omega'+\omega) + f(\omega'-\omega)\right] \tag{2}$$

where $\alpha^2 F$ is the Eliashberg function and $f(\omega)$ and $n(\omega)$ are the Fermi and Bose-Einstein distribution functions, respectively. (Grimvall, 1981) $\Gamma_{e-ph}(\omega,T)$ monotonically increases with energy over the region $|\omega| < \omega_{max}$ (for T=0), where ω_{max} is the cutoff of the phonon spectrum. The exact functional form is slightly dependent on the phonon spectrum. The temperature dependence of Γ_{e-ph} is approximately linear at higher temperatures, with slope

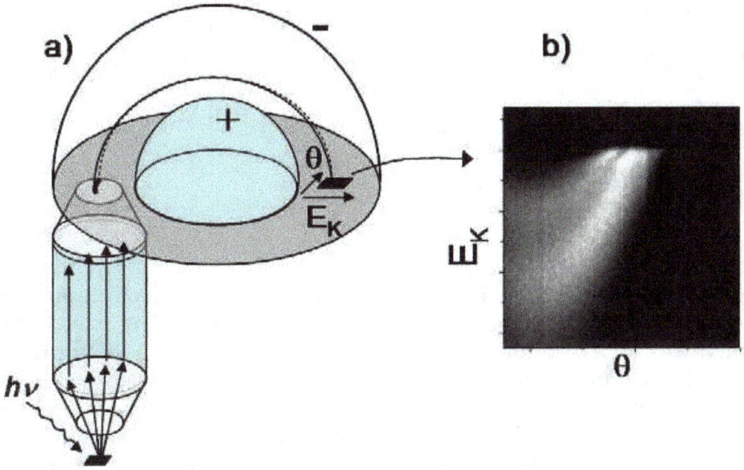

Fig. 1. Angle-resolved photoemission spectroscopy experiment. (a) Scienta hemispherical analyzer composes of a cylindrical lens, a hemispherical energy analyzer and a 2-dimensional detector (dark square). The entrance slit of the hemisphere selects a single angular line out of a finite solid angle collected by the lens. This defines a line in the inverse space over which the photoemission intensity is detected on a detector (b) Typical 2-dimensional photoemission intensity map as a function of kinetic energy E_K and angle θ from $NbSe_2$ sample (Valla et al., 2004). Bright (dark) represents high (low) photo-electron intensity.

$2\pi\lambda k_B$, where λ is the electron-phonon coupling constant given by (Grimvall, 1981)

$$\lambda = 2\int_0^\infty \frac{\alpha^2 F(\omega')}{\omega'} d\omega' \tag{3}$$

and k_B is Bolzman constant.

For most metals 10 meV $< \omega_{max} <$ 100 meV, and λ falls between 0.1 and 1.5. The electron-phonon coupling constant λ can be extracted directly from ReΣ as $\lambda = -[\partial(\mathrm{Re}\Sigma)/\partial\omega]_0$ by fitting the low energy part of ReΣ to a straight line. The scattering process is illustrated in Fig. 2 where we consider a coupling to a single phonon mode at energy Ω_0. Coupling to such a mode (at T = 0 K) will result in a step function in the scattering rate or ImΣ. The step function reflects the observation that when the photohole has enough energy to create the mode ($\omega \geq \Omega_0$), scattering from the mode opens up a decay channel, thereby limiting the lifetime. The real and imaginary parts of the self energy are related via causality through a Kramers Kronig transform. Thus the step function in ImΣ results in a cusp function for ReΣ. Such an energy dependence of Σ affects the measured spectra in two ways. Above and below the mode energy there will be a noticeable change in the quasiparticle width. Secondly, the measured dispersion will be given by $\epsilon_{\mathbf{k}0} + \mathrm{Re}\Sigma(\mathbf{k},\omega)$ and will display the mass enhancement observed immediately below the Fermi level as presented in Fig. 2(e).

Modern photoelectron spectrometers allow the simultaneous measurement of photoelectron intensities from a finite range in both energy and momentum space. A typical image is shown in Fig. 1. The ability to obtain such images has led to the development of new methodologies for the extraction of self-energies. The spectral response in Fig. 1(b) may be analysed by taking

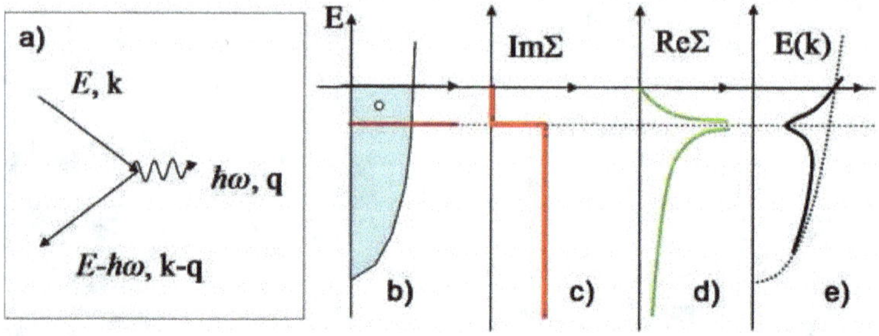

Fig. 2. Electron-phonon scattering (a) A Feynman diagram for the electron-phonon scattering. The straight lines represent the photo-hole states before and after scattering on a phonon (wiggly line). (b) The photo-hole created in photoemission (hollow circle) can emit a phonon only if it has enough energy. The red horizontal line indicates the energy of a phonon mode. (c) Imaginary part, ImΣ of the self-energy. The step function reflects the opening of a decay channel when the photohole has enough energy to create the mode ($\omega \geq \Omega_0$). (d) Real part, ReΣ of the self-energy. (e) The renormalized (observed) dispersion (thick solid line) is then obtained as a superposition of the bare dispersion (dotted line) and ReΣ.

an intensity cut at constant angle or momentum, the so called energy distribution curve or by taking an intensity cut at constant energy, a momentum distribution curve. In the limit of a momentum independent self-energy and a linear bare dispersion ($\epsilon_{k0} = v_0(k - k_F)$, where v_0 represents the bare velocity), the momentum distribution curve is a simple Lorentzian, centered at $k_m = k_F$+$[\omega_0 - \text{Re}\Sigma(\omega_0)]/v_0$ and with the full width at half maximum $\Delta k = 2\text{Im}\Sigma(\omega_0)/v_0$. The self-energy can thus be simply extracted from momentum distribution curve peaks at any binding energy, as has been discussed in several papers. (Kaminski et al., 2000; Kordyuk et al., 2005; LaShell et al., 2000; Valla et al., 2000) The same method of analysis has been employed here to study the electron-phonon coupling in GICs.

3. Photoemission from graphite intercalation compounds

The experiments were carried out on a Scienta SES-100 electron spectrometer operating in the angle resolved mode at the beamline 12.0.1 of the Advanced Light Source. The spectra were recorded at the photon energy of 50 eV with the combined instrumental energy resolution of 20-25 meV and the momentum resolution of 0.008 Å^{-1} in geometry where the polarization of light was perpendicular to the probed momentum line. The thin flake of pristine graphite was obtained by micromechanical exfoliation of HOPG on a SiO_2/Si substrate in air. From the optical contrast and from Raman data, the thickness of the flake was determined to be in the range of 7-9 graphene layers. The flake was annealed to 600 K in the ultra-high vacuum chamber before angle resolved photoemission studies. The LiC_6 and KC_8 samples were prepared by intercalating natural, single-crystal graphite flakes (Madagascan) as described in Ref. (Dresselhaus & Dresselhaus, 2002; Pruvost et al., 2004). The CaC_6 samples were prepared via immersion of a HOPG platelet or single-crystal graphite flake in a lithium/calcium alloy for 10 days. X-ray diffraction showed very high sample purity with no graphite or secondary stage phases. SQUID magnetometry revealed a sharp ($\Delta T \sim 0.3$K) superconducting transition at $\sim$ 11.6 K (onset) for CaC_6 samples. To avoid degradation, all samples were unsealed and glued to the sample holder with Ag- epoxy in an Ar filled glow box. Protected by the cured epoxy, they were then quickly transferred to the angle resolve photoemission prep-chamber, and cleaved at low temperature (15-20 K) under ultra-high vacuum conditions (2×10^{-9} Pa). All data were collected at 15-20 K.

3.1 Electronic structure and charge transfer

Fig. 3 shows the angle resolved photoemission spectra near the K point in the graphene Brillouin zone for pristine graphite, LiC_6, KC_8 and CaC_6. The upper panels (a) - (d) show the contours of photoemission intensity as a function of binding energy for a momentum line going through the K point. The intensity from a narrow interval (± 10 meV) around the Fermi level, representing the Fermi surface, is shown in the lower panels (e) - (h). The dispersing states are the graphene-derived π and π^* bands, as marked in Fig. 3(a) through (d). In KC_8 the low energy band structure is essentially graphene-like, with π and π^* bands touching at the Dirac point (Grüneis et al., 2009), shifted below the Fermi level due to the electron doping. In LiC_6 and CaC_6 a sizable gap exists between π and π^* bands. Dirac point is determined

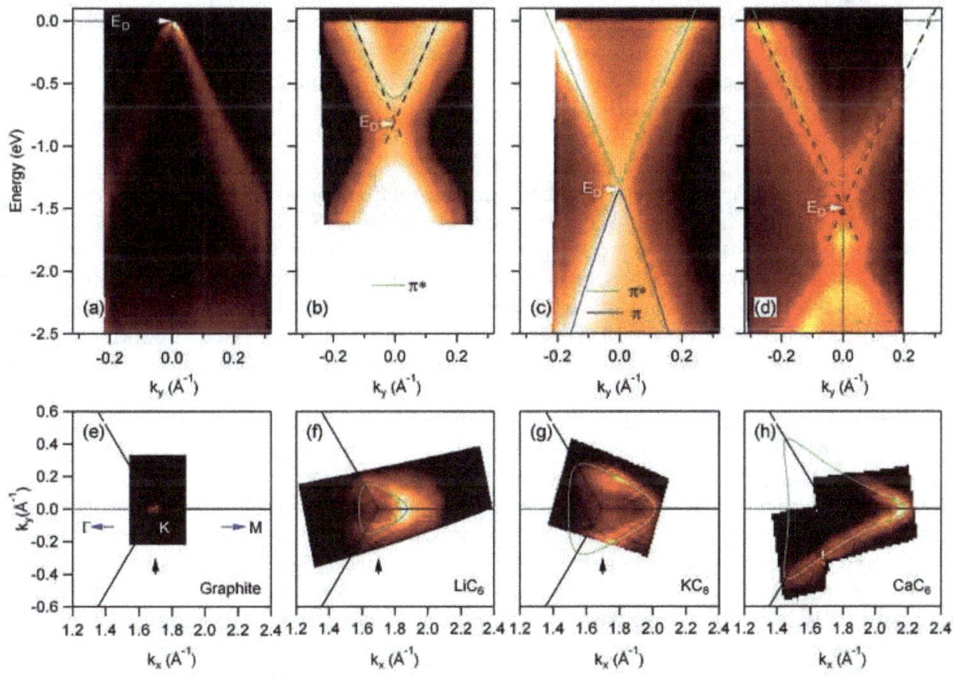

Fig. 3. Photoemission from pristine and intercalated graphite. Photoemission spectra from pristine graphite (a), LiC_6 (b), KC_8 (c) and CaC_6 (d) along the same momentum line in the graphene Brillouin Zone traversing the K point, as indicated by the black arrows in panels (e)-(h). Blue and green lines represent the π and π^* bands. White arrows indicate position of Dirac point. (e)-(h) Photoemission intensity from a narrow energy interval around the Fermi level ($\omega = \pm 10$ meV), representing the graphene-derived π^* Fermi surface, for graphite, LiC_6, KC_8 and CaC_6 respectively. The green lines represent tight-binding fits to the Fermi surface. All spectra were taken at T $\sim$ 15 K.

by extrapolating the linear part of π^* dispersion at low binding energies to the K point. The arrows indicate positions of Dirac point, (E_D), 0 eV, 0.825 eV, 1.35 eV and 1.5eV for pristine graphite, LiC_6, KC_8 and CaC_6, respectively, indicating that the filling of the π^* band increases in that sequence. In particular, it appears that the π^* band is filled more in KC_8 than in LiC_6 and that it forms a larger Fermi surface in the former material. The area enclosed by the Fermi surface (lower panels in Fig. 3) is a direct measure of doping of graphene π^* states, i.e. of the charge transferred into the graphene sheets. The Fermi surface area is determined from peak positions of momentum distribution curves at Fermi level and compared to the 3rd nearest neighbor hopping tight binding band structure. In pristine graphite, the Fermi surface is essentially a point, indicating nearly neutral graphene sheets. In the intercalated samples the Fermi surface grows from LiC_6 to KC_8 to CaC_6. Its area is 0.399 Å^{-2} in KC_8, corresponding to 0.11 electrons per graphene unit cell, or 44% of the nominal value of 0.25 for the complete charge transfer. In LiC_6, the enclosed area is 0.125 Å^{-2}, corresponding to the doping of only 0.0344 electrons per graphene unit cell (GUC). This is far below the nominal value of 1/3 electrons per graphite unit cell expected for the complete intercalant ionization. For CaC_6 the complete charge transfer would involve a donation of 2 calcium electrons to the graphene sheets, equivalent of the doping of 1/3 electrons per C atom (or $2/3e^-$ per unit cell). Instead, the measured Fermi area gives approximately 0.18 electrons per C atom.

The incomplete charge transfer into the graphene π^* states would suggest that the remaining charge occupies the so-called interlayer band. There has been a substantial body of theoretical work that depicts the intercalant-derived interlayer states and phonons as main contributors to superconductivity in GICs. However, so far there has been no clear evidence of the intercalant-derived nearly free electron band in photoemission experiments. The occupation of the interlayer state has been recently reported in CaC_6 (Sugawara et al., 2008), but the observed feature was very weak and broad. There, even the graphene-derived π^* state was very broad and did not form an enclosed contour at the Fermi level, casting doubts on these results. Our experiments always show relatively sharp π^* band that forms a well defined Fermi surface. However, in MC_6(M=Li, Ca, Ba) materials, in addition to π^* band, we always see a broad feature at slightly higher binding energy that follows the π^* band, dispersing upward from K point (Fig. 3(b)and (d)). Further from the K point, it loses intensity and its dispersion cannot be precisely traced. In CaC_6, this feature is observable over larger region of k space (Valla et al., 2009). It is possible that this is a remnant of an interlayer band, smeared out by a disorder within the intercalant layers and folded into the K point of the graphene Brillouin zone. However, our measurements do not show any evidence of the interlayer band in the region from which it should be folded to the K point - the Γ point. Of course, the missing charge has to be somewhere and even if the state is completely incoherent it still can count for the missing charge. We note that in pristine graphite, the interlayer hopping $t_\perp$ splits both π and π^* bands into the bonding and antibonding counterparts due to the AB stacking of graphene sheets. It would be tempting to assign the broad feature to a bonding π^* state, due to similarities with π^* in initial dispersion. However, unlike the pristine graphite, all the first stage GICs have the AA stacking of graphene sheets and such assignment would be incorrect. On the other hand, in LiC_6 and CaC_6 the π^* band might be split due to the AB stacking of the intercalant (if there is not too much disorder in the intercalant sites). If this was indeed the case, we counted only a portion of charge transferred to graphene layers. The alternative charge counts would be in this case 0.0616 and 0.349 electrons per C atom, for LiC_6 and for CaC_6, respectively. We point out that, if this is indeed the case, only the antibonding counterpart, which appears more coherent, plays a significant role in superconductivity, because it is its filling and its coupling to phonons that determines T_c, as will be shown in the next section. Irrespective of these issues, our experimental observation that the doping of graphene sheets is larger in KC_8 than in LiC_6 is opposite of the expected nominal doping, but is in line with the existence of superconductivity in these materials: KC_8 is a superconductor and LiC_6 is not. In the following, we identify the reason for the correlation between superconductivity and doping of the graphene sheets.

3.2 Electron-Phonon Coupling in Graphite Intercalation Compounds

It is evident from Fig. 3 that in all three intercalated materials, an anomaly or a kink in dispersion of the π^* band occurs at approximately 160 meV below the Fermi level. This is a hallmark of the interaction of the electronic states with phonons (Hengsberger et al., 1999; Valla et al., 1999) that has been attributed to a coupling to graphene in- plane high-frequency phonons(Grüneis et al., 2009; Pan et al., 2010; Valla et al., 2009). We now focus on CaC_6, a system in which these effects are the most pronounced. Fig. 4 shows the angle resolved

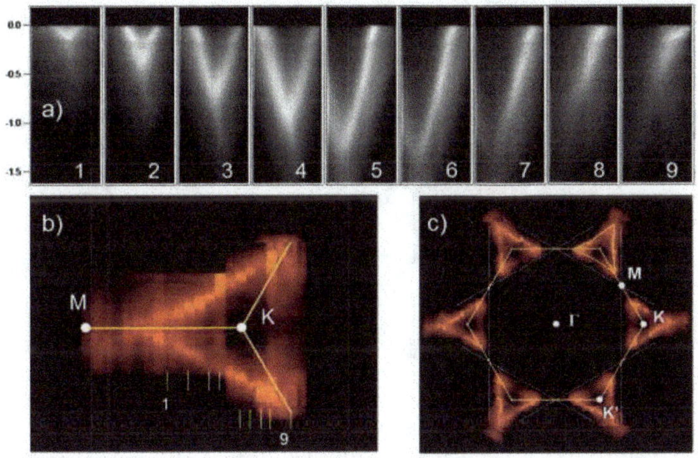

Fig. 4. Photoemission from CaC_6. (a) 1-9: photoemission spectra from CaC_6 along several momentum lines near the K point as indicated in (b). (b) Photoemission intensity from CaC_6 from a narrow energy interval around the Fermi level ($\omega = \pm 10$ meV) near the K point of the graphene Brillouin zone. Thin bars indicate momentum lines probed in (a). (c) Fermi surface of CaC_6 obtained by six fold symmetrization of intensity shown in (b). White lines correspond to the doping of 0.5 electrons per C atom, at which the van Hove singularity in the π^* band sits at the Fermi level.

photoemission spectra near the K point in the graphene Brillouin zone. The upper panel (a) shows the contours of photoemission intensity as a function of binding energy for several momentum lines in the vicinity of K point, while the intensity from a narrow interval (± 10 meV) around the Fermi level is shown in the lower contours (b and c). All the spectra are shown on the same momentum and energy scales and all were taken in the normal state, at T=15 K. The Fermi surface of CaC_6 encloses a significant area and has a concave triangular shape centred at the K point of the Brillouin zone. The charge transfer from intercalant atoms fills the graphene bands and the Dirac point has moved to 1.5 eV below the Fermi level. White lines in Fig. 4c) correspond to the doping of 0.5 electrons per C atom, at which the van Hove singularity in the π^* band sits at the Fermi level. An important observation can immediately be made: in all the spectra from Fig. 4a), there is a renormalization of the quasiparticle dispersion at $\sim$160 meV below the Fermi level.

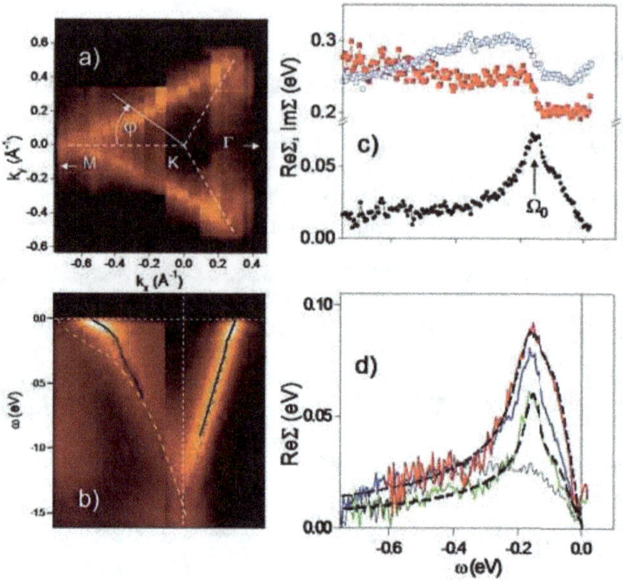

Fig. 5. Electron-phonon coupling in CaC_6. (a) Fermi surface of CaC_6. ϕ defines the polar angle around the K point, measured from the KM line. (b) Spectral intensity for $\phi = 60°$ or KΓ line (right) and $\phi = 0°$ or KM line (left). Black solid lines represent the momentum distribution curve derived dispersions. Dashed blue and yellow lines represent bands from Ref. (Calandra & Mauri, 2007), calculated for graphene and CaC_6 monolayer, respectively. (c) ReΣ (black circles) and ImΣ (red solid squares) for $\phi = 43°$ and ImΣ for $\phi = 35°$ (blue open squares). Peak (step) position in ReΣ (ImΣ) is labeled with Ω_0. (d) Real part of the self-energy for $\phi = 53°, 32°$ and $2°$ (bottom to top). Dashed lines represent the fits to the simple model self-energies as described in text. Thin black line is ReΣ of pristine graphite flake.

It is also apparent from Fig. 4 that the renormalization effects ("kinks) are not equally strong for all the points on the Fermi surface. Fig. 5 illustrates the anisotropy of the renormalization effects in CaC_6. We have extracted the dispersions and the linewidths of the quasiparticle states by fitting the spectral intensities at constant energies, or momentum distribution curves (Valla et al., 1999), with Lorentzian distributions. The real and imaginary components of the self-energy are then derived in the usual manner from the peak positions k_m and widths Δk of the momentum distribution curve peaks at each ω, as described, using the expressions

$$\mathrm{Re}\Sigma(\omega) = (k_m - k_0)v_0(\omega), \mathrm{Im}\Sigma(\omega) = \frac{\Delta k}{2}v_0(\omega) \tag{4}$$

where v_0 represents the bare band velocity at ω (LaShell et al., 2000; Valla et al., 2000). The starting approximation for the bare band was a tight binding dispersion which was then refined until the obtained ReΣ and ImΣ satisfied Kramers-Kronig transformations (Kordyuk et al., 2005). We note that the choice of the bare dispersion may play a significant role in determining the exact shape and magnitude of the self-energy (Valla et al., 2000). Therefore, the self-consistency check through the Kramers-Kronig criterion is important in getting the correct results. We note that the bands calculated for a CaC_6 monolayer (dashed blue line in Fig. 5(b) (Calandra & Mauri, 2007) do not realistically represent the bare dispersions in our case. Instead, we have found that the bare dispersions calculated within 3rd nearest neighbor tight binding approximation and appropriately filled, work well for the whole Fermi surface. The results for ReΣ and ImΣ corresponding to the azimuthal angle $\phi = 43°$ and ImΣ for $\phi = 35°$ are shown in Fig. 5(c). Fig. 5(d) shows ReΣ for different momentum lines corresponding, from bottom to top, to azimuthal angles $\phi = 53°$, $32°$ and $2°$. All the self-energies have a structure typical for the interaction with a well defined bosonic mode: the imaginary part has a sharp step-like feature, while the real part has a peak at the energy of the mode, Ω_0. Therefore, the mode energy can be read directly from the measured self-energy. In our case, the dominant structure occurs at ~160 meV, with an additional feature at ~75 meV. These features can be naturally attributed to the interaction with in-plane and out-of-plane phonons of the graphene sheets. We note that the finite energy resolution also affects the measured ReΣ near the Fermi level, within the resolution range, but with the tendency to underestimate the measured electron-phonon coupling (Valla, 2006). We have excluded the affected interval $|\omega| < 20$ meV from the considerations and any fine structure, related to a possible coupling to the intercalant modes, is out of our detection limits. At higher energies, we can model the measured self-energy at any ϕ with an Eliashberg function, $\alpha^2F(\omega)$ (see equation 2), consisting of two peaks: one fixed at 75 meV and the other ranging from 155 to 165 meV. The relative contribution of these two peaks increases from 1:5 to almost 1:1 on moving from $\phi = 60°$ to $\phi = 0$. The modelled ReΣ are shown for these two limiting cases. Our model assumes a constant density of states (DOS), whereas a linear DOS would be a better approximation since the scattering is expected to reflect the density of final states. This is visible in Fig. 5(c), where ImΣ quickly acquires a slope as one moves from ΓK to KM. Fitting of measured ReΣ in the energy range $20 \leq |\omega| \leq 300$ meV to the model gives the energy of the high frequency mode Ω_0. In addition, the coupling strength, λ, can be extracted from the low energy slope of ReΣ, $\lambda = -[\partial(\text{Re}\Sigma)/\partial\omega]_0$ by fitting the low energy part of ReΣ to a straight line. From Fig. 5(d) it is clear that the characteristic energy Ω_0 does not vary a lot, whereas the coupling strength λ displays a significant anisotropy. In Fig. 6(e-f), we plot both quantities as functions of azimuthal angle ϕ. The coupling constant ranges from 0.38 (for $\phi \approx 60°$) to 0.89 (for $\phi \approx 0$), with the momentum averaged value of 0.53. The coupling constant in pristine graphite flake is much smaller ($\lambda = 0.16$) and does not show a significant anisotropy, in agreement with Leem et al. (Leem et al., 2008). The measured Fermi velocity on the graphene-derived Fermi surface is also very anisotropic: $v_F(\phi = 0) \approx 1.1$ eVÅ, $v_F(\phi = 60°) \approx 4.1$eVÅ, with the momentum average $\langle v_F(k) \rangle \approx 2.5$ eVÅ. If the graphene electrons were responsible for superconducting properties, the in-plane coherence length, $\epsilon_{ab} = v_F/(\pi\Delta)$, would be in the range 200Å$< \epsilon_{ab} <$ 800 Å, if we take $\Delta \approx 1.6$ meV for the superconducting gap (Bergeal et al., 2006). $\langle v_F(k) \rangle$ gives $\epsilon_{ab} \approx$500 Å, in reasonable agreement with $\epsilon_{ab} \approx$350 Å reported in scanning tunneling microscopy (Park et al., 2007), specific heat (Kim et al., 2006) and magnetization (Emery et al., 2005; Lamura et al., 2006) studies. We also note that peculiar linearity in the upper critical filed, H_{C2}, at low temperatures, could also be explained by anisotropic Fermi velocity on the graphene-derived Fermi surface (Mazin et al., 2007). In contrast, the highly isotropic intercalant Fermi surface would not produce linear H_{C2}. Recent theoretical calculations suggest that in CaC_6, the crucial role is played by calcium, i.e. electrons on the intercalant-derived Fermi surface are strongly coupled to the intercalant vibrations (Boeri et al., 2007; Calandra & Mauri, 2005; Csányi et al., 2005; Mazin, 2005). Those calculations show nearly one order of magnitude weaker coupling of the graphene-derived electrons to the in-plane graphene phonons, with essentially no anisotropy (Park et al., 2008). The large calcium isotope effect also implies the dominant role of soft Ca vibrations in superconductivity (Hinks et al., 2007). Our results, on the contrary, suggest that graphene sheets play the dominant role because the obtained electron-phonon coupling on graphene-derived Fermi surface with the in-plane and out-of-plane graphene phonons seems to be more than sufficient to give $T_c \sim 11.6$ K in CaC_6.

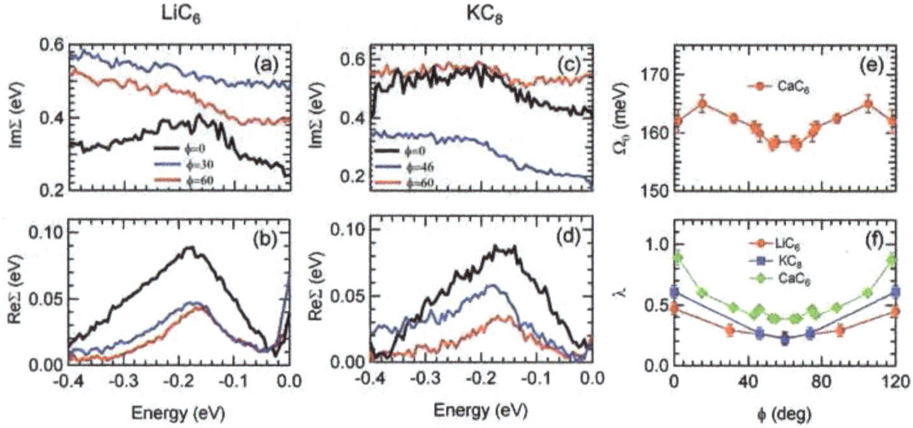

Fig. 6. Self-energy and electron-phonon coupling in graphite intercalation componds. $\mathrm{Im}\Sigma(\omega)$ (a) and $\mathrm{Re}\Sigma(\omega)$ (b) for LiC_6 and KC_8 ((c) and (d)) for several different points at the Fermi surface as indicated by the polar angle ϕ. (e) Characteristic energy Ω_0 of the maximum in $\mathrm{Re}\Sigma(\omega)$ of CaC_6 as a function of ϕ. (f) The electron phonon coupling constant λ in LiC_6, KC_8 and CaC_6 , extracted from $\mathrm{Re}\Sigma$ as a function of ϕ.

By extending similar analysis to a whole series of graphite intercalation compounds, including LiC_6, KC_8, KC_{24} (Camacho et al., n.d.; Pan et al., 2010), and BaC_6, we have been able to track certain trends and to get a better insight into the mechanism of superconductivity in the intercalated graphite. Fig. 6 shows the self-energies and the extracted electron-phonon coupling for the two most interesting members of the series: LiC_6, KC_8. The superconducting transition temperature, T_c, in these two materials is inverted from the expected trend in which the more charge nominally transferred to graphene sheets would mean the higher T_c. However, the actual charge transfer to graphene sheets measured in photoemission, is different from the nominal one dictated by the chemical formula (see Fig. 3) and, as previously discussed, it is actually larger in KC_8 than in LiC_6. Following the same procedure as for CaC_6, we have extracted the electron-phonon coupling from the self-energies. The coupling in both materials shows some anisotropy [Fig. 6(f)], with the maximum along the KM direction and the minimum along the ΓK direction, similar, but significantly weaker than in CaC_6(Valla et al., 2009) and what was recently reported for KC_8 (Grüneis et al., 2009). The most important observation, however, is that the momentum averaged $\langle \lambda_k \rangle$ is stronger in KC_8 than in LiC_6. The coupling constant and its anisotropy both increase from LiC_6 to KC_8 to CaC_6, exactly in the same sequence as the actual filling of the graphene π^* band and in previously established sequence for T_c. Strengthening of the electron-phonon coupling with the filling of the π^* band has also been observed in the epitaxial graphene (McChesney et al., 2007). This is not surprising, because the density of states near the Fermi level increases with the filling of the π^* band. Also, a larger Fermi surface makes an electron-phonon scattering process more probable as the phase space available for the scattering events grows. In the pristine graphite, the Fermi surface is nearly a point and the coupling is strongly suppressed (Leem et al.,

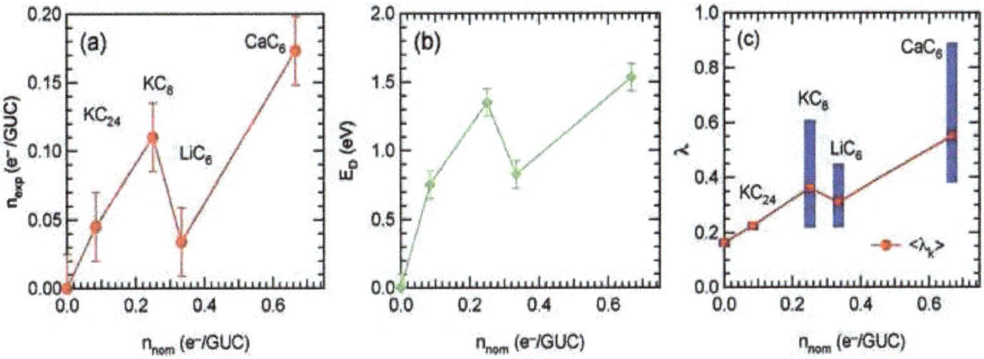

Fig. 7. Electron-phonon coupling and the filling of graphene states. (a) Measured charge transfer from the intercalant to the graphite π^* states. (b) Binding energy of Dirac point (chemical potential shift). (c) electron-phonon coupling, λ. The range of each vertical bar indicates the anisotropy with the maximum along the K-M line and minimum along the K-Γ line. Red circles represent the momentum averaged values, $\langle \lambda_k \rangle$.

2008; Valla et al., 2009). To better illustrate the positive correlation between the filling of the graphene π^* band and the coupling strength λ in different GICs, we plot these quantities in Fig. 7 as functions of nominal chemical composition for several different materials. The actual (measured) charge transfer (in electrons/GUC), obtained fom the Fermi surface area, is shown in Fig. 7(a). The increase in size of the Fermi surface is consistent with the energy of Dirac point [panel (b)] as the chemical potential, μ, shifts from the pristine graphite to CaC_6. It is interesting to note that $\mu \propto \sqrt{n_{exp}}$ still holds, regardless of the shape of Fermi surface and number of Fermi sheets in these five different materials. The coupling constant λ [panel (c)] follows the same trend as the filling of the π^* band. Note that this is the very trend of superconducting transition temperatures (T_c) in these materials, plotted in Fig. 8(c). This correlation suggests that the graphene π^* states and their coupling to graphene in-plane phonons is crucial for superconductivity in graphite intercalation compounds. The only role that the intercalants seem to play is to provide the charge for filling of the π^* states. As an additional test of this idea we have calculated the superconducting transition temperature, T_c, from McMillan's formula (McMillan, 1968), using the measured coupling constants $\langle \lambda_k \rangle$ and Coulomb pseudo-potential $\mu^* = 0.14$. The calculated T_c is compared with the experimental one in Fig. 8(b-c). It is clear not only that the calculated T_c follows the same trend, but even the absolute values are very close to the measured ones. This reinforces our conclusion that the graphene electronic states and graphene-derived phonons are crucial for superconductivity in GICs. Note that the threshold-like behavior near $\langle \lambda_k \rangle = 0.3$ places LiC_6 on one side and KC_8 on another side of a steep increase in T_c. We note that superconductivity in LiC_3 and LiC_2, materials in which more Li is pushed in under pressure, supports our picture where the electron-phonon coupling and superconductivity strengthen with the filling of graphene π^* states. The increase in T_c from 0.39 K for stoichiometric KC_8 to 0.55K in material with excess K is also in line with this picture. A further test would be a systematic photoemission study on alkaline-earth GICs (Ca, Sr, Ba) where T_c decreases with the atomic mass of alkaline-earth intercalant.

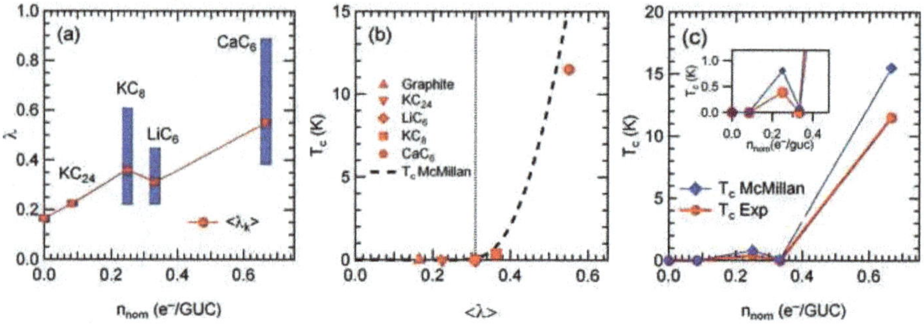

Fig. 8. Electron-phonon coupling and superconducting critical temperature (T_c) in graphite intercalation compunds. (a) Measured λ for several different GICs. The range of each vertical bar indicates the anisotropy with the maximum along the KM line and minimum along the KΓ line. The red circles and line represent the momentum averaged values $\langle \lambda_k \rangle$. (b) T_c as a function of $\langle \lambda_k \rangle$ calculated from measured $\langle \lambda_k \rangle$ using McMillan's formula (McMillan, 1968). The dashed line represents the calculated values while symbols represent experimental T_c values. (c) Comparison of calculated (blue diamond) and measured (red circle) values as functions of nominal chemical composition (electrons per graphene unit cell (GUC)) for measured GICs. Zero corresponds to the pristine graphite. The inset zooms in the low T_c values.

3.3 Enhancement of electron-phonon coupling via dynamical nesting

A huge discrepancy between the calculated and measured coupling, where the later one is almost one order of magnitude stronger, naturally provokes the question: Where is such strong and anisotropic coupling coming from? Similar anisotropy was recently reported for electron-phonon coupling in K and Ca doped graphene on SiC (McChesney et al., 2007). There, the proximity of van Hove singularity to the Fermi level was suggested to be an origin of the enhanced coupling. Even though the claim on very high anisotropy was subsequently retracted due to the erroneous procedure of extracting the coupling near the KM lines, the measured coupling on the unaffected portion of the Fermi surface was still significantly stronger than the calculated one (Park et al., 2008). In the case of CaC_6 we believe that we have found the likely mechanism of enhancement of the coupling: very favourable dynamical nesting conditions exist on the Fermi surface that may strongly enhance the inter-valley coupling.

This is illustrated in Fig. 9. In order to satisfy the energy conservation rule in an electron-phonon scattering event, the photohole created at $\sim$160 meV has to be scattered to the Fermi level after emission of a 160 meV phonon. To explore the allowed momenta in such a process, we plot the states at $\omega = -160$ meV at one corner of the Brillouin zone and the states at the Fermi level ($\omega = 0$) in another corner for two highly doped GICs: KC_8 (panel (a)) and CaC_6 (panel (b)). A wavevector connecting any pair of points on different contours represents an allowed scattering process. Note that in the case of intra-valley scattering, both contours would have to be centred at the same K point. However, from the measured phonon spectra in graphite (Maultzsch et al., 2004), and recent Raman experiments on CaC_6 (Hlinka et al., 2007), it is clear that there are no $q \approx 0$ modes at $\sim$160 meV and we have to consider inter-valley scattering.

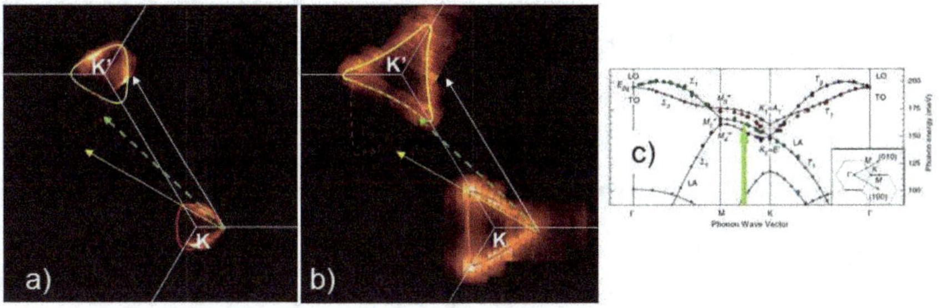

Fig. 9. Dynamical nesting of the Fermi surface in GICs. Contour of photoemission intensity from KC_8 (a) and CaC_6 (b) at $\omega = 0$ meV (yellow contour) around K′ point and at the "kink" energy $\omega = -\Omega_0 = -160$ meV (red contour), in the vicinity of of K point of the Brillouin zone. White (yellow) arrow represents the ΓK (ΓM) wave vector, while the dashed green arrow represents the wave vector that "nests" the two contours. Evidently, the nesting is much more efficient in CaC_6 than in KC_8. (c) Phonon dispersions in pristine graphite (Maultzsch et al., 2004). The position of dynamical nesting wave vector for CaC_6 from(b) is marked with green arrow. Its length corresponds to the exchanged energy.

From Fig. 9, it appears that in the case of CaC_6, some wave-vectors are particularly efficient in connecting the two contours. As the $\omega = -160$ meV contour is slightly convex and the Fermi surface is slightly concave with essentially the same curvature, the green vector in Fig 9(b) effectively nests the whole side, or $\sim$ 1/3 of the length of the two contours. The tips of contours can be nested by two such vectors, which might explain a portion of the observed anisotropy in λ. The additional factor characterizing the tips is a gain in coupling to lower frequency phonons at $\omega \approx 75$ meV, as can be seen in Fig. 5d. We note that the observed "nesting" is not static: it involves emission/absorption of a high frequency phonon and would not produce the charge density wave instability. The dynamical nesting and resulting enhancement of the electron-phonon coupling is obviously very sensitive to the filling of graphene-derived bands. For lower doping levels, the nesting efficiency is quickly lost as both contours become convex, as is evident in the case of KC_8 (Fig. 9(a)). At even higher fillings, possibly realized under pressure (Gauzzi et al., 2007; Smith et al., 2006), the nesting conditions would likely improve further, but at some point, the coupling to lower frequency phonons near the KM lines might become too strong and eventually drive the system into the charge-density wave state.

4. Summary and outlook

In conclusion, we have performed angle resolved photoemission spectroscopy studies of the electronic structure in pristine graphite and several alkali and alkaline-earth GICs. We have measured the strength of the electron-phonon interaction on the graphene-derived Fermi surface to carbon derived phonons in these materials and found that it follows a universal trend where the coupling strength and superconductivity monotonically increase with transfer of electrons into the graphene-derived π^* states. This correlation between the filling of graphene π^* states, the strength of the electron-phonon coupling and superconductivity suggests that both graphene-derived electrons and graphene-derived phonons are crucial for superconductivity in GICs. We have also identified a possible mechanism of enhancement of the electron-phonon coupling in these materials that stems from the particular Fermi surface geometry. For highly doped systems (CaC_6), a concave triangular shape of the Fermi surface perfectly matches the convex energy contour at the energy of the exchanged phonon, effectivley enhancing the electron-phonon coupling. Such mechanism is expected to be very sensitive on the occupation of the π^* band and might be used to fine-tune the coupling in order to reach a maximal T_c. Perhaps this is what is going on under pressure in some superconducting GICs (Gauzzi et al., 2007; Smith et al., 2006).

We note that similar results were obtained for graphene monolayers, epitaxially grown on SiC(0001) and Ir(111) and chemically doped by adsorbed (or subsurface) alkali and alkaline-earth atoms (Bianchi et al., 2010; McChesney et al., 2007; 2010). Due to the reduced dimensionality, we probably cannot expect these systems to be superconducting, but the ability to contionusly fill the π^* states and to fine tune the electronic structure in these films offers the opportunity of measuring the electron-phonon coupling in much finer steps of doping. Such studies might help in understanding the role of the Fermi surface topology in shaping the properties of graphite intercalation compounds and open the new avenues of designed synthesis of new intercalated materials with targeted properties.

5. Acknowledgments

We would like to acknowledge the many contributions of our collaborators on the work described here. These include T. Berlijn, J. Camacho, M. Dean, M. Ellerby, A. V. Fedorov, C. A. Howard, M. Khodas, W. Ku, M. Strongin, M. H. Upton, A.C. Walters and I. Zaliznyak. The research work described in this paper was supported by the Department of Energy under Contract No. DE-AC02-98CH10886.

6. References

Bergeal, N., Dubost, V., Noat, Y., Sacks, W., Roditchev, D., Emery, N., Hérold, C., Marêché, J.-F., Lagrange, P. & Loupias, G. (2006). Scanning tunneling spectroscopy on the novel superconductor CaC_6, *Phys. Rev. Lett.* 97(7): 077003.

Bianchi, M., Rienks, E. D. L., Lizzit, S., Baraldi, A., Balog, R., Hornekær, L. & Hofmann, P. (2010). Electron-phonon coupling in potassium-doped graphene: Angle-resolved photoemission spectroscopy, *Phys. Rev. B* 81(4): 041403.

Boeri, L., Bachelet, G. B., Giantomassi, M. & Andersen, O. K. (2007). Electron-phonon interaction in graphite intercalation compounds, *Phys. Rev. B* 76(6): 064510.

Calandra, M. & Mauri, F. (2005). Theoretical explanation of superconductivity in C_6Ca, *Phys. Rev. Lett.* 95(23): 237002.

Calandra, M. & Mauri, F. (2007). Electronic structure of heavily doped graphene: The role of foreign atom states, *Phys. Rev. B* 76(16): 161406.

Camacho, J., Upton, M., Pan, Z.-H., Howard, C., Ellerby, M. & Valla, T. (n.d.). unpublished.

Csányi, G., Littlewood, P. B., Nevidomskyy, A. H., Pickard, C. J. & Simons, B. D. (2005). The role of the interlayer state in the electronic structure of superconducting graphite intercalated compounds, *Nature Physics* 1(1): 42–45.
URL: *http://www.nature.com/doifinder/10.1038/nphys119*

Dean, M. P. M., Howard, C. A., Saxena, S. S. & Ellerby, M. (2010). Nonadiabatic phonons within the doped graphene layers of XC_6 compounds, *Phys. Rev. B* 81(4): 045405.

Dresselhaus, M. S. & Dresselhaus, G. (2002). Intercalation compounds of graphite, *Advances in Physics* 51(1): 1–186.
URL: *http://www.informaworld.com*

Emery, N., Hérold, C., d'Astuto, M., Garcia, V., Bellin, C., Marêché, J. F., Lagrange, P. & Loupias, G. (2005). Superconductivity of bulk CaC_6, *Phys. Rev. Lett.* 95(8): 087003.

Gauzzi, A., Takashima, S., Takeshita, N., Terakura, C., Takagi, H., Emery, N., Hérold, C., Lagrange, P. & Loupias, G. (2007). Enhancement of superconductivity and evidence of structural instability in intercalated graphite CaC_6 under high pressure, *Phys. Rev. Lett.* 98(6): 067002.

Grimvall, G. (1981). *The Electron-Phonon Interaction in Metals*, North-Holland, New York.

Grüneis, A., Attaccalite, C., Rubio, A., Vyalikh, D. V., Molodtsov, S. L., Fink, J., Follath, R., Eberhardt, W., Büchner, B. & Pichler, T. (2009). Angle-resolved photoemission study of the graphite intercalation compound KC_8 : A key to graphene, *Phys. Rev. B* 80(7): 075431.

Hannay, N. B., Geballe, T. H., Matthias, B. T., Andres, K., Schmidt, P. & MacNair, D. (1965). Superconductivity in graphitic compounds, *Phys. Rev. Lett.* 14(7): 225–226.

Hengsberger, M., Purdie, D., Segovia, P., Garnier, M. & Baer, Y. (1999). Photoemission study of a strongly coupled electron-phonon system, *Phys. Rev. Lett.* 83(3): 592–595.

Hinks, D. G., Rosenmann, D., Claus, H., Bailey, M. S. & Jorgensen, J. D. (2007). Large Ca isotope effect in the CaC_6 superconductor, *Phys. Rev. B* 75(1): 014509.

Hlinka, J., Gregora, I., Pokorny, J., Hérold, C., Emery, N., Marêché, J. F. & Lagrange, P. (2007). Lattice dynamics of CaC_6 by Raman spectroscopy, *Phys. Rev. B* 76(14): 144512.

Holzwarth, N. A. W., Louie, S. G. & Rabii, S. (1984). Interlayer states in graphite and in alkali-metal-graphite intercalation compounds, *Phys. Rev. B* 30(4): 2219–2222.

Kaminski, A., Mesot, J., Fretwell, H., Campuzano, J. C., Norman, M. R., Randeria, M., Ding, H., Sato, T., Takahashi, T., Mochiku, T., Kadowaki, K. & Hoechst, H. (2000). Quasiparticles in the superconducting state of $Bi_2Sr_2CaCu_2O_{8+\delta}$, *Phys. Rev. Lett.* 84(8): 1788–1791.

Kevan, E. (1992). *Angle-Resolved Photoemission*, Elsevier, Amsterdam.

Kim, J. S., Kremer, R. K., Boeri, L. & Razavi, F. S. (2006). Specific heat of the Ca-intercalated graphite superconductor CaC_6, *Phys. Rev. Lett.* 96(21): 217002.

Kordyuk, A. A., Borisenko, S. V., Koitzsch, A., Fink, J., Knupfer, M. & Berger, H. (2005). Bare electron dispersion from experiment: Self-consistent self-energy analysis of photoemission data, *Phys. Rev. B* 71(21): 214513.

Lamura, G., Aurino, M., Cifariello, G., Di Gennaro, E., Andreone, A., Emery, N., Hérold, C., Marêché, J.-F. & Lagrange, P. (2006). Experimental evidence of *s*-wave superconductivity in bulk CaC_6, *Phys. Rev. Lett.* 96(10): 107008.

LaShell, S., Jensen, E. & Balasubramanian, T. (2000). Nonquasiparticle structure in the photoemission spectra from the Be(0001) surface and determination of the electron self energy, *Phys. Rev. B* 61(3): 2371–2374.

Leem, C. S., Kim, B. J., Kim, C., Park, S. R., Ohta, T., Bostwick, A., Rotenberg, E., Kim, H. D., Kim, M. K., Choi, H. J. & Kim, C. (2008). Effect of linear density of states on the quasiparticle dynamics and small electron-phonon coupling in graphite, *Phys. Rev. Lett.* 100(1): 016802.

Mahan, G. (1990). *Many Particle Physics*, Plenum Press, New York.

Maultzsch, J., Reich, S., Thomsen, C., Requardt, H. & Ordejón, P. (2004). Phonon dispersion in graphite, *Phys. Rev. Lett.* 92(7): 075501.

Mazin, I., Boeri, L., Dolgov, O., Golubov, A., Bachelet, G., Giantomassi, M. & Andersen, O. (2007). Unresolved problems in superconductivity of CaC_6, *Physica C: Superconductivity* 460-462: 116–120.
URL: *http://dx.doi.org/10.1016/j.physc.2007.03.276*

Mazin, I. I. (2005). Intercalant-driven superconductivity in YbC_6 and CaC_6, *Phys. Rev. Lett.* 95(22): 227001.

McChesney, J. L., Bostwick, A., Ohta, T., Emtsev, K. V., Seyller, T., Horn, K. & Rotenberg, E. (2007). Massive enhancement of electron-phonon coupling in doped graphene by an electronic singularity, p. 8.
URL: *http://arxiv.org/abs/0705.3264*

McChesney, J. L., Bostwick, A., Ohta, T., Seyller, T., Horn, K., González, J. & Rotenberg, E. (2010). Extended van hove singularity and superconducting instability in doped graphene, *Phys. Rev. Lett.* 104(13): 136803.

McMillan, W. L. (1968). Transition temperature of strong-coupled superconductors, *Phys. Rev.* 167(2): 331–344.

Pan, Z.-H., Camacho, J., Upton, M. H., Fedorov, A. V., Walters, A. C., Howard, C. A, Ellerby, M. & Valla, T. (2010). Why is KC8 a superconductor and LiC6 is not? URL: http://arxiv.org/abs/1003.3903.

Park, C.-H., Giustino, F., Cohen, M. L. & Louie, S. G. (2007). Velocity renormalization and carrier lifetime in graphene from the electron-phonon interaction, *Phys. Rev. Lett.* 99(8): 086804.

Park, C.-H., Giustino, F., McChesney, J. L., Bostwick, A., Ohta, T., Rotenberg, E., Cohen, M. L. & Louie, S. G. (2008). Van hove singularity and apparent anisotropy in the electron-phonon interaction in graphene, *Phys. Rev. B* 77(11): 113410.

Posternak, M., Baldereschi, A., Freeman, A. J., Wimmer, E. & Weinert, M. (1983). Prediction of electronic interlayer states in graphite and reinterpretation of alkali bands in graphite intercalation compounds, *Phys. Rev. Lett.* 50(10): 761–764.

Pruvost, S., Herold, C., Herold, A. & Lagrange, P. (2004). Co-intercalation into graphite of lithium and sodium with an alkaline earth metal, *Carbon* 42(8-9): 1825–1831.
URL: *http://linkinghub.elsevier.com/retrieve/pii/S0008622304002106*

Smith, R. P., Kusmartseva, A., Ko, Y. T. C., Saxena, S. S., Akrap, A., Forró, L., Laad, M., Weller, T. E., Ellerby, M. & Skipper, N. T. (2006). Pressure dependence of the superconducting transition temperature in C_6Yb and C_6Ca , *Phys. Rev. B* 74(2): 024505.

Sugawara, K., Sato, T. & Takahashi, T. (2008). Fermi-surface-dependent superconducting gap in C_6Ca, *Nature Physics* 5(1): 40–43.
URL: *http://www.nature.com/doifinder/10.1038/nphys1128*

Valla, T. (2006). Comment on "Multiple bosonic mode coupling in the electron self-energy of $(La_{2-x}Sr_x)CuO_4$", *Phys. Rev. Lett.* 96(11): 119701.

Valla, T., Camacho, J., Pan, Z.-H., Fedorov, A. V., Walters, A. C., Howard, C. A. & Ellerby, M. (2009). Anisotropic electron-phonon coupling and dynamical nesting on the graphene sheets in superconducting CaC_6 using angle-resolved photoemission spectroscopy, *Phys. Rev. Lett.* 102(10): 107007.

Valla, T., Fedorov, A. V., Johnson, P. D., Glans, P.-A., McGuinness, C., Smith, K. E., Andrei, E. Y. & Berger, H. (2004). Quasiparticle spectra, charge-density waves, superconductivity, and electron-phonon coupling in 2H-$NbSe_2$, *Phys. Rev. Lett.* 92(8): 086401.

Valla, T., Fedorov, A. V., Johnson, P. D. & Hulbert, S. L. (1999). Many-body effects in angle-resolved photoemission: quasiparticle energy and lifetime of a Mo(110) surface state, *Phys. Rev. Lett.* 83(10): 2085–2088.

Valla, T., Fedorov, A. V., Johnson, P. D., Xue, J., Smith, K. E. & DiSalvo, F. J. (2000). Charge-density-wave-induced modifications to the quasiparticle self-energy in 2H-$TaSe_2$, *Phys. Rev. Lett.* 85(22): 4759–4762.

Weller, T. E., Ellerby, M., Saxena, S. S., Smith, R. P. & Skipper, N. T. (2005). Superconductivity in the intercalated graphite compounds C_6Yb and C_6Ca, *Nature Physics* 1(1): 39–41.
URL: *http://dx.doi.org/10.1038/nphys0010*

3

Electronic and Magnetic Properties of the Graphene-Ferromagnet Interfaces: Theory vs. Experiment

Elena Voloshina[1] and Yuriy Dedkov[2]
[1]*Physikalische und Theoretische Chemie, Freie Universität Berlin*
[2]*Fritz-Haber-Institut der Max-Planck-Gesellschaft, Berlin*
Germany

1. Introduction

Graphene is a two-dimensional sheet of carbon atoms arranged in a honeycomb lattice with two crystallographically equivalent atoms (C1 and C2) in the unit cell [Geim (2009); Geim & Novoselov (2007); Neto et al. (2009)] [Fig. 1(a)]. The sp^2 hybridization between one $2s$ orbital and two $2p$ orbitals leads to a trigonal planar structure with a formation of a σ bonds between carbon atoms that are separated by 1.42 Å. These corresponding σ bands have a filled shell and, hence, form a deep valence band. The half-filled $2p_z$ orbitals, which are perpendicular to the planar structure, form the bonding (π) and antibonding (π^*) bands in the electronic structure of graphene [Fig. 1(b)]. The π and π^* bands touch in a single point exactly at the Fermi energy (E_F) at the corner of the hexagonal graphene's Brillouin zone (K-points). Close to this so-called Dirac point (E_D) the bands display a linear dispersion and form perfect Dirac cones [Neto et al. (2009)] [Fig. 1(c)]. Thus, undoped graphene is a semimetal ("zero-gap semiconductor"). The linear dispersion of the bands mimics the physics of quasiparticles with zero mass, so-called Dirac fermions [Geim (2009); Geim & Novoselov (2007); Neto et al. (2009)].

The unique "zero-gap" electronic structure of graphene leads to some limitations for application of this material in real electronic devices. In order, for example, to prepare a practical transistor, one has to perform a "doping" of a graphene layer inducing electrons or holes in the electronic structure [Fig. 1(d)]. There are several ways of the modification of the electronic structure of graphene with the aim of doping. Among them are: (i) incorporation in its structure of nitrogen [Wang et al. (2009)] and/or boron [Dutta & Pati (2008)] or transition-metal atoms [Mao et al. (2008)]; (ii) using different substrates that influence the electronic structure of the graphene layer, e. g. metallic substrates for the graphene growth [Wintterlin & Bocquet (2009)]; (iii) intercalation of different materials underneath graphene grown on different substrates [Dedkov et al. (2008a; 2001); Enderlein et al. (2010); Gierz et al. (2008; 2010); Shikin et al. (2000)]; (iv) deposition of atoms or molecules on top [Boukhvalov & Katsnelson (2009); Coletti et al. (2010)]; etc.

The exceptional transport properties of graphene [Geim & Novoselov (2007)] make it a promising material for applications in microelectronics [Morozov et al. (2008); Novoselov et al. (2005)] and sensing [Schedin et al. (2007)]. This has recently led to a revival of interest

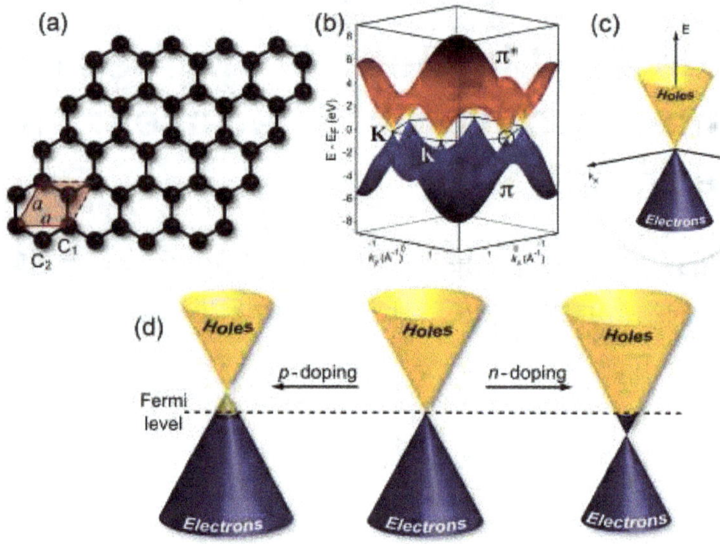

Fig. 1. (a) Crystal structure of the graphene layer where carbon atoms are arranged in the honeycomb lattice. Unit cell of graphene with lattice constant a has two carbon atoms per unit cell, C1 and C2. (b) Electronic dispersion of π and π^* states in the honeycomb lattice of free-standing graphene obtained in the framework of tight-binding approach. These branches have linear dispersion (c) in the vicinity of the K points of the Brillouin zone of graphene. (d) Position of the Dirac point and the Fermi level as a function of doping for the free-standing graphene.

in graphene on transition metal surfaces [Coraux et al. (2008; 2009); Dedkov et al. (2008a;b); Gamo et al. (1997); Hu et al. (1987); Land et al. (1992); Martoccia et al. (2008); N'Diaye et al. (2006); Sasaki et al. (2000); Sutter et al. (2009); Wang et al. (2008)], as large area epitaxial graphene layers of exceptional quality can be grown, which might be an alternative to micromechanical cleavage for producing macroscopic graphene films [Bae et al. (2010); Kim et al. (2029)].

The electronic interaction of graphene with a metal is both of fundamental and technological interest in view of possible device applications. Recent theoretical calculations by V. M. Karpan and co-workers [Karpan et al. (2007; 2008)] for graphene/metal interfaces imply the possibility of an ideal spin-filtering in the current-perpendicular-to-the-plane configuration (CPP) for the ferromagnet/graphene/ferromagnet sandwich-like structures. The close-packed surfaces of Co and Ni were considered as ferromagnetic (FM) electrodes, which perfectly coincide with graphene from the crystallographic point of view. The spin-filtering effect originates form the unique overlapping of the electronic structures of the graphene monolayer and close-packed surfaces of ferromagnetic Ni and Co. As discussed earlier, graphene is a semimetal with electronic density in the vicinity of E_F at corners (K points) of the hexagonal Brillouin zone of graphene. If the Fermi surface projections of ferromagnetic metals, *fcc* Ni or Co, on the (111) close-packed plane are considered, then in both cases one finds only minority electron density around the K points of the surface Brillouin zone. In this case, it is expected that the preferential transport of only minority electrons and perfect spin-filtering will appear in a FM/graphene/FM stack [Karpan et al. (2007; 2008)], because graphene has electronic states only around K. The interaction between graphene and ferromagnetic material will however change the electronic properties of the interface partially quenching the spin-filtering effect in the sandwich-like structure, but a sizable effect can still be detected by choosing the proper combination of FM materials [Yazyev & Pasquarello (2009)] and this effect is predicted to increase strongly when multilayers of graphene are used [Karpan et al. (2007; 2008)].

Besides spin-filtering, graphene might be the best material for the realization of spintronic devices. Such systems usually require the effective injection of the spin-polarized electrons in the conductive channel which can be made from graphene [Tombros et al. (2007)]. However, prior to being able to implement graphene/ferromagnet systems in any kind of spintronic unit, a study of their electronic, magnetic, and interfacial properties has to be performed.

In the present chapter the electronic and magnetic structures of the graphene/FM interface will be discussed. From the theoretical side the different interfaces will be analyzed: graphene/Ni(111), graphene/Co(0001), graphene/1 ML Fe/Ni(111). Here the electronic structure of the interface as well as effect of induced magnetism in the graphene layer are discussed. As for experiment, the crystallographic structure, morphology, electronic and magnetic properties of a graphene/ferromagnet interface are considered for the case of the Ni(111) close-packed surface. Electronic structure studies of this interface reveal the existence of interface states, which originate from the strong hybridization of the graphene π and Ni $3d$ valence-band states leading to the appearance of the induced magnetic moment on the carbon atoms in the graphene layer that is confirmed by both x-ray magnetic circular dichroism (XMCD) and spin-resolved photoemission (PES). Detailed analysis of the Fermi surface of the graphene/Ni(111) system is performed via comparison of experimental and theoretical data for this interface.

2. Methods of investigations of graphene on ferromagnetic surfaces

2.1 Methods of calculation of the electronic and magnetic structure

Binding between graphene and metallic surface can be either chemical or physical in nature. Chemical binding typically implies a strong interaction through a charge sharing between the substrate and the adsorbate, yielding modification of their electronic structures. Physisorption, on the other hand, arises due to classical electrostatic or dispersion (van der Waals) interactions. The latter ones are the long-range correlation effects, which are not captured in the density functional theory (DFT) calculations because of the local character of standard functionals. In such cases, local density approximation (LDA) is known to give rather good results due to error cancellation. This is the reason of using for the graphene-containing systems the LDA, instead of generalized gradient approximation (GGA). Clearly, this is not a proper solution of the problem, and a better way would be to resort to methods beyond the DFT approach, either phenomenological or quantum-chemical post-Hartree-Fock correlation methods (see, e. g. [Paulus & Rosciszewski (2009); Pisani et al. (2008)]). An alternative approach is an inclusion of the dispersion correction to the total energy obtained with standard DFT approximation explicitly by hand with DFT-D method, that is atom pair-wise sum over C_6R^{-6} potentials (see, e. g. [Grimme (2004)]). High accuracy is accessible with a new exchange-correlation functional named van der Waals-density functional (vdW-DF), recently developed by Dion *et al.* [Dion et al. (2004)]. This functional is parameter free and therefore is considered *ab initio* in true spirit of DFT as it depends only on the total electron density.

The latter approach is shown to be working well for graphene/metal interface, where weak interaction between graphene and substrate is observed [Brako et al. (2010)].
In the previous theoretical studies on graphene/FM interface, the strong interaction for FM = Ni(111) or Co(0001) was found [Bertoni et al. (2005); Karpan et al. (2007; 2008)]. While existence of a non-magnetic configuration of graphene/Ni(111) interface, where graphene is weakly bound with substrate, was theoretically predicted, this structure was shown to be energetically unfavourable. The purpose of the present work is a comparison between theory and experiment, therefore the most stable configurations will be in the focus of the study. For them, standard GGA functionals are supposed to give reasonable result.
In our DFT studies, the electronic and structural properties of the graphene-substrate system are obtained using the Perdew-Burke-Ernzerhof (PBE) functional [Perdew et al. (1996)]. For solving the resulting Kohn-Sham equation we have used the Vienna Ab Initio Simulation Package (VASP) [Kresse & Furthmuller (1996a;b)] with the projector augmented wave basis sets [Blochl (1994)]. The plane-wave kinetic energy cutoff is set to 500 eV. The supercell used to model the graphene-metal interface is constructed from a slab of 13 layers of metal atoms with a graphene sheet adsorbed at both sides and a vacuum region of approximately 14 Å. When optimizing the geometry, the positions (z-coordinates) of the carbon atoms as well as those of the top two layers of metal atoms are allowed to relax. In the total energy calculations and during the structural relaxations the k-meshes for sampling the supercell Brillouin zone are chosen to be as dense as 24×24 and 12×12, respectively.

2.2 Experimental methods of investigation

The presented experimental studies of the graphene/Ni(111) interface were performed in different experimental stations in identical experimental conditions allowing for the reproducible sample quality in different experiments. In all experiments the W(110) single crystal was used as a substrate. Prior to preparation of the graphene/Ni(111) system the well-established cleaning procedure of the tungsten substrate was applied [Dedkov et al. (2008)]: several cycles of oxygen treatment with subsequent flashes to 2300°C. A well-ordered Ni(111) surface was prepared by thermal deposition of Ni films with a thickness of more than 200 Å on to a clean W(110) substrate and subsequent annealing at 300°C. An ordered graphene overlayer was prepared via thermal decomposition of propene (C_3H_6) according to the recipe described elsewhere [Dedkov et al. (2008;a;b; 2001); Nagashima et al. (1994)]. The quality, homogeneity, and cleanliness of the prepared graphene/Ni(111) system was verified by means of low-energy electron diffraction (LEED), scanning tunneling microscopy (STM), and core-level as well as valence-band photoemission.
STM experiments were carried out in an ultra-high vacuum (UHV) system (base pressure 8×10^{-11} mbar) equipped with an Omicron variable temperature scanning tunneling microscope. All STM measurements were performed in the constant-current-mode at room temperature using electrochemically etched polycrystalline tungsten tips cleaned in UHV by flash-annealing. The sign of the bias voltage corresponds to the voltage at the sample. Tunneling parameters are given separately for each STM image: U_T for tunneling voltage and I_T for tunneling current.
Near-edge absorption spectroscopy (NEXAFS) and XMCD spectra were collected at the D1011 beamline of the MAX-lab Synchrotron Facility (Lund, Sweden) at both Ni $L_{2,3}$ and C K absorption edges in partial (repulsive potential $U = -100$ V) and total electron yield modes (PEY and TEY, respectively) with an energy resolution of 80 meV. Magnetic dichroism spectra were obtained with circularly polarized light (degree of polarization is $P = 0.75$)

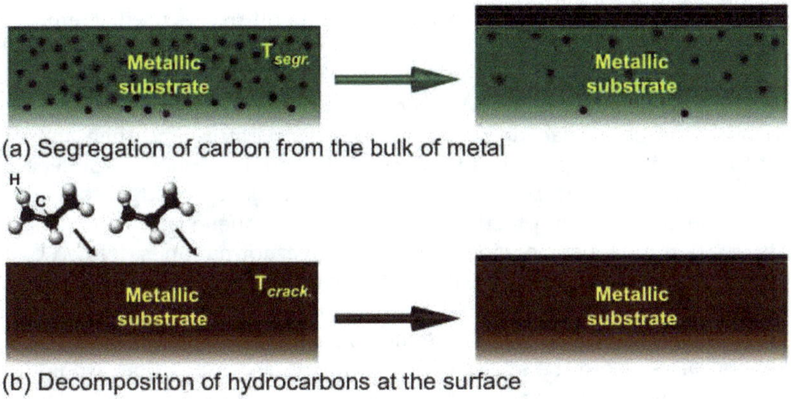

Fig. 2. Two ways of the graphene preparation on metal surfaces: (a) Segregation of bulk-dissolved carbon atoms to the surface at high temperature T_{segr}; (b) Decomposition (cracking) of hydrocarbon molecules at the surface of transition metals at high temperature T_{crack}.

in the remanence magnetic state of the graphene/Ni(111) system after applying of an external magnetic field of 500 Oe along the $< 1\bar{1}0 >$ easy magnetization axis of the Ni(111) film. All absorption measurements were performed at 300 K. The base pressure during the measurements did not exceed 1×10^{-10} mbar.

Angle-resolved PES (ARPES) experiments were performed at the UE56/2-PGM-1 beam-line at BESSY (Berlin, Germany). The photoemission intensity data sets $I(E, k_x, k_y)$ were collected with a PHOIBOS 100 energy analyzer (SPECS) while the graphene/Ni(111)/W(110) sample was placed on a 6-axes manipulator (3 translation and 3 rotation axes). The temperature of the sample during the measurements was kept at 80 K or 300 K. The energy/angular resolution was set to 80 meV/0.2°. In case of the spin-resolved PES experiments the mini-Mott-spin-detector (SPECS) was used instead of the 2D CCD detector. The spin-resolved spectra were collected in the remanent magnetic state of the graphene/Ni(111) system (see above) in normal emission geometry. The effective Sherman function was estimated to be $S_{eff} = 0.1$ and instrumental asymmetry was accounted via measuring of spin-resolved spectra for two opposite directions of the sample magnetization. The base pressure during all photoemission measurements was below 7×10^{-11} mbar.

3. Methods of graphene preparation on metallic surfaces

There are two common methods of the graphene preparation on metallic surfaces: (i) Segregation of the carbon atoms at the surface of bulk metallic sample, which was previously doped with carbon [Fig. 2(a)] and (ii) Decomposition of the carbon-containing molecules on the surface of transition metal [Fig. 2(b)]. In the first method the transition metal sample with some amount of carbon impurities or the bulk crystal, which was preliminary loaded by carbon (via keeping the sample at elevated temperature in the atmosphere of CO or hydrocarbons) is annealed at higher temperatures. This procedure leads to the segregation of carbon atoms at the surface of metal. The careful control of the temperature and the cooling rate of the sample allows to obtain the different thicknesses of the grown graphene layer: mono- vs. multilayer growth. In the second method the decomposition (cracking) of carbon-containing molecules at the particular temperatures is used. Here, mostly light hydrocarbon molecules, like ethylene or propene, are used, but the decomposition of CO, acetylene, and of heavy hydrocarbon molecules, like cyclohexane, *n*-heptane, benzene, and toluene, also was successfully applied [Wintterlin & Bocquet (2009)]. In this method molecules could be adsorbed on the surface of metal at room temperature and then annealing of the sample leads to the molecule decomposition and hydrogen desorption; or molecules could be directly adsorbed on the hot sample. The recent experiments demonstrate that both methods, segregation and decomposition, practically lead to the graphene layers of the same quality. In case of segregation the kinetic of single graphene layer formation is defined by the careful controlling of the annealing temperature (but multilayers of graphene could be also prepared). In the second method the restriction to the formation of only single-layered graphene is defined by the fact that in this case the chemical reaction on the catalitically active metallic surface takes the place. Here the speed of hydrocarbon decomposition drops down by the several order of magnitude as soon the first grapehen monolayer is formed [Nagashima et al. (1994)]. List of metal surfaces, which were used for the preparation of graphene layers is compiled in Ref.[Wintterlin & Bocquet (2009)], where methods of preparation (segregation vs. decomposition), main experimental method, and corresponding reference are presented for different metallic surfaces (lattice-matched and lattice-mismatched).

4. Structure and bonding of graphene on ferromagnetic surfaces: Theory

The surface lattice constants of Ni(111), Co(0001), and the 1 ML Fe/Ni(111) system match the in-plane lattice constants of the graphene layer almost perfectly (compare: $a_{\mathrm{Ni}}/\sqrt{2} = 2.49$ Å, $a_{\mathrm{Co}} = 2.51$ Å, $a_{1\,\mathrm{ML\,Fe/Ni(111)}} = 2.52$ Å, and $a_{\mathrm{graphene}} = 2.46$ Å), that leads to the perfect crystallographic order at the interface between considered FM substrates and graphene. Several possible atomic configurations are usually considered for the interface between graphene and close-packed FM surface. Three "high-symmetry" structures, which preserve $3m$ symmetry, are known as $fcc - hcp$, $top - hcp$, and $top - fcc$. The latter one is schematically shown in Fig. 3: the C atoms are arranged above the Ni atoms of the first and third (fcc) layers. In the $top - hcp$ configuration, the C atoms are placed directly above the Ni atoms of the first layer (top site) and the second layer (hcp site). In the $fcc - hcp$ configuration, the C atoms are placed above fcc and hcp sites. Three additional configurations were considered recently, which were called $bridge - top$, $bridge - fcc$, and $bridge - hcp$. In these structures, one of the carbon atoms is not placed in top, fcc or hcp sites but in-between [Fuentes-Cabrera et al. (2008)].

In the present studies the electronic and magnetic structures of the graphene/Ni(111), graphene/Co(0001), and graphene/1 ML Fe/Ni(111) systems were calculated for the slab consisting of the graphene layer and 13 layers of FM material (in case of the latter system, 1 ML Fe was placed on the both sides of the 11-layers Ni slab in fcc stacking). As a first step the energy minimization for different structural arrangements of carbon atoms on FM surface was carried out. The results of these calculations are compiled in Table 1.

It has been found that the lowest energy configuration for FM = Ni or Co corresponds to a $top - fcc$ geometry in which one carbon atom is positioned above a surface FM atom (C-*top*) while the second carbon atom is situated above a third layer FM atom (C-fcc). This is in agreement with other previously published first-principles calculations [Bertoni et al. (2005); Eom et al. (2009); Karpan et al. (2007)] as well as with experiments for graphene on the Ni(111) surface [Gamo et al. (1997)]. In contrary to the latter result, where LEED $I - V$-curve analysis has been used in order to determine the interface arrangement, in experiments of C. Klink *et al.*

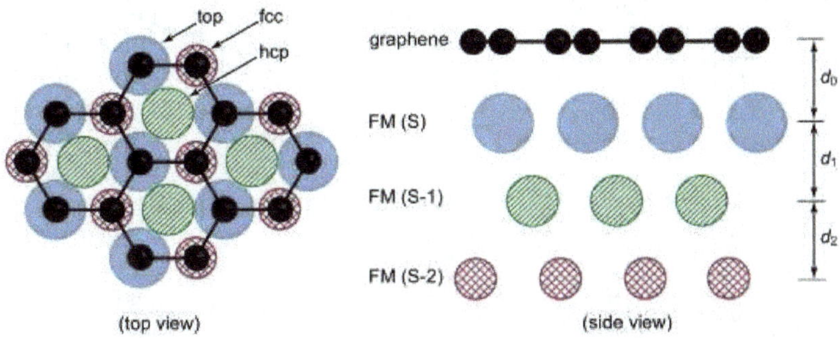

Fig. 3. Top- and side-view of a ball model of one of a possible arrangements of carbon atoms in a graphene layer on the lattice matched (111) surface. The different positions of carbon atoms in the graphene layer above FM interface atom (S) and on the hollow places above FM atoms in the second (S-1) as well as third layer (S-2) are marked as *top*, *hcp*, and *fcc*, respectively. d_{0-2} are distances between layers in the system.

[Klink et al. (1995)], involving combined STM and LEED studies, the $fcc - hcp$ configuration is shown to be the most stable one. The same conclusion was made by R. Rosei *et al.* [Rosei et al. (1983)], when using the extended-energy-loss fine-structure spectroscopy method. The calculations including three additional "low symmetry" configurations showed that within DFT employing GGA-PBE, none of the structures is stable at the experimentally relevant temperatures; with local-density approximation (LDA), the $bridge - top$ configuration was found to be the most energetically favorable one [Fuentes-Cabrera et al. (2008)].

The situation is different in the case of the graphene/1 ML Fe/Ni(111) system where the lowest-energy configuration was found to be $top - hcp$ with one carbon atom staying above Fe interface atom and the second one in the hollow *hcp* cite of the structure. The available experimental data [Dedkov et al. (2008a)] confirms the three-fold symmetry of this system, but cannot give any additional information about the stacking of the graphene and metallic layers. Structural studies by means of, e. g. LEED $I - V$-curve-analysis or x-ray diffraction, are necessary for this system.

The presented results (Table 1) demonstrate the reduction of the magnetic moment of the surface atoms of FM material upon adsorption of graphene on top. This reduction originates from the strong interaction between graphene π states and the $3d$ states of the FM interface layer. Such interaction leads to the noticeable modification of the graphene and metal valence band states, which will be discussed further in the text.

The electronic structure of a graphene sheet depends on its interaction strength with the ferromagnetic substrate. Clearly, the stronger bonding between graphene and underlaying metal, the shorter equilibrium distance d_0 between graphene and FM interface layer. As expected, this calculated distance is nearly the same for $top - fcc$ and $top - hcp$ arrangements, where one carbon atom of the graphene unit cell sits on top of a metal atom. But the energy difference found between these two models shows that FM atoms of the second and third layers also interact with the graphene layer. For the less strongly bound $fcc - hcp$ configurations of graphene on FM, the equilibrium separation is rather large, comparable to the layer spacing of 3.35 Å in bulk graphite (note: in our calculations, d_0 is slightly larger than can be expected due to the inability to describe dispersion interactions by PBE functional), indicating physisorption, and the characteristic band structure of an isolated graphene sheet is clearly recognizable.

	FM	Structure of the graphene/FM interface		
	Ni(111)	*top-fcc*	*top-hcp*	*fcc-hcp*
d_0 (Å)		2.135/2.133	2.145/2.146	3.540/3.540
d_1 (Å)	2.005	2.020	2.020	2.020
d_2 (Å)	2.031	2.017	2.015	2.036
ΔE (eV)		0.000	0.049	0.054
m_{FM} (μ_B)	−0.030/0.710	−0.025/0.543	−0.026/0.514	−0.029/0.669
m_{C} (μ_B)		−0.019/0.031	−0.019/0.027	0.000/0.000
	Co(0001)	*top-fcc*	*top-hcp*	*fcc-hcp*
d_0 (Å)		2.155/2.137	2.155/2.138	3.639/3.639
d_1 (Å)	1.954	1.963	1.956	1.952
d_2 (Å)	2.046	2.035	2.043	2.046
ΔE (eV)		0.000	0.017	0.090
m_{FM} (μ_B)	−0.053/1.784	−0.060/1.600	−0.059/1.583	−0.055/1.771
m_{C} (μ_B)		−0.043/0.038	−0.048/0.044	0.000/0.000
	1ML Fe/Ni(111)	*top-fcc*	*top-hcp*	*fcc-hcp*
d_0 (Å)		2.117/2.092	2.114/2.089	3.487/3.487
d_1 (Å)	2.039	2.029	2.019	2.028
d_2 (Å)	2.061	2.044	2.057	2.062
ΔE (eV)		0.021	0.000	0.104
m_{FM} (μ_B)	−0.028/2.622	−0.034/2.486	−0.035/2.469	−0.022/2.616
m_{C} (μ_B)		−0.048/0.040	−0.050/0.039	0.000/0.000

Table 1. Results for the atomic structure of the three graphene/FM interface models and for the clean FM surface: d_0 is the distance between the graphene overlayer and the interface FM layer (the two values for the two nonequivalent carbon atoms are indicated); d_1 is the distance between the interface FM layer and the second FM layer; d_2 is the distance between the second and third FM layers; ΔE is the energy difference between the energy calculated for the different slabs and the energy calculated for the most stable geometry; m_{FM} is the interface/surface FM spin magnetic moment (the two values for the *sp* and *d* magnetizations are indicated); m_{C} is the interface carbon spin magnetic moment (the two values for the two nonequivalent carbon atoms are indicated).

We consider the case of the graphene/Ni(111) as a representative one for the description of general features of the electronic structure of the graphene/FM interface with the analysis of the region around the Fermi level further in the text. The majority and minority spin band structures of the graphene/Ni(111) system (13 ML of Ni in the slab) with the $top - fcc$ arrangement of the carbon atoms are shown in Fig. 4. Due to the strong interaction between carbon layer and FM substrate, noticeable modification of the both graphene and FM band structure is observed. First of all, this is a nonrigid downward shift of the occupied graphene bands compare to those in pure graphene: at the Γ point, the π band is shifted by 2.25 eV while the σ band is only shifted by 1.15 eV. The unoccupied σ^* band is also shifted by 1.01eV. The interaction between graphene and metal layers implies a hybridisation of the graphene π bands with the nickel (cobalt, iron) 3*d* bands (and, in the second turn, with the FM 4*s* and 4*p* bands).

Strong hybridization of the graphene π and FM substrate 3*d* valence band states results in three occupied and two unoccupied hybridized states, with a strong contribution of the carbon p_z orbitals (see Fig. 5 for details). Four of them are clearly visible near the K point: *HS*1,

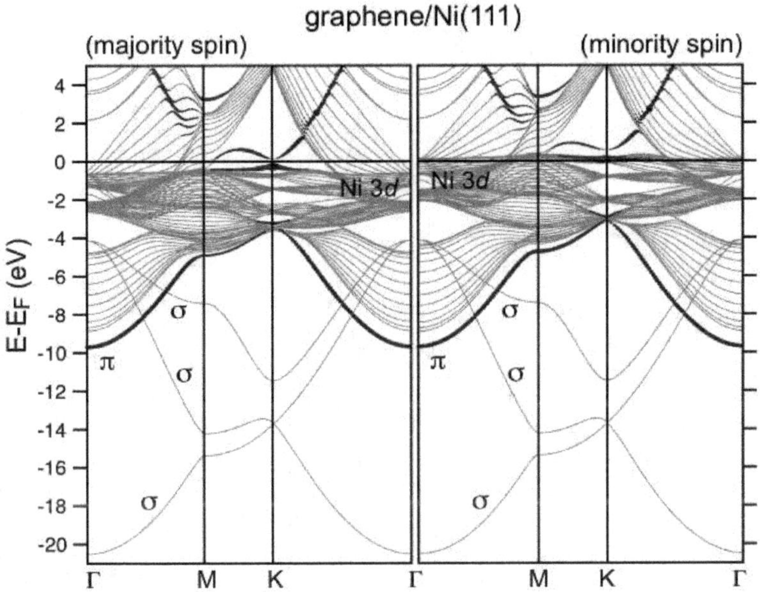

Fig. 4. Calculated majority and minority spin band structures for a slab terminated by graphene/Ni(111) interface for a most energetically favorable $top-fcc$ configuration. For the blue (darker) lines, the carbon p_z character is used as a weighting factor.

$HS2$, $HS3$, and $HS4$, respectively (Tab. 2). The states $HS2$ and $HS4$ can be attributed to the carbon atoms just above the interface nickel atoms, while the states $HS1$ and $HS3$ only involve the carbon atoms occupying the fcc hollow sites (Fig. 5). A 0.40 eV gap opening between the graphene occupied π and unoccupied π^* bands at the K point for the graphene/Ni(111) system is related to the $HS3$ and $HS4$. One more hybrid state $HS5$, observed near the M point of the Brillouin zone, results from a hybridisation of the interface nickel p_x, p_y, and d orbitals with carbon p_z orbitals (Fig. 5).

The resembling situation is observed for the graphene/Co(0001) and graphene/1 ML Fe/Ni(111) interfaces where similar hybridized states can be clearly identified in the calculated electronic structure (Fig. 5). The results of the analysis are summarized in Table 2. In the row Ni(111), Co(0001), 1 ML Fe/Ni(111) one can clearly identify increasing the exchange splitting between spin-up and spin-down counter-parts for every interface state, that can be assigned to increasing the magnetic moment of the underlying FM substrate. The interesting case is observed for graphene/1 ML Fe/Ni(111) where two Fe 3d quantum-well states of the minority character exist in the energy gap above the Fermi level around the K point [M. Weser, E.N.Voloshina, K.Horn, Yu.S.Dedkov (submitted)]. These states are strongly hybridized with C p_z states of C-top and C-hcp atoms that leads to increasing the magnetic moments of carbon atoms compared to those in the graphene/Ni(111) system (Table 1).

The calculated magnetic moment of the FM substrate for three systems, which are considered here are presented in Table 1. In all cases the deposition of the graphene layer on FM surface leads to the reduction of the magnetic moment of the interface FM atoms and appearing the induced magnetism of carbon atoms of graphene. For example, in the case of the most stable $top-fcc$ arrangement of the graphene/Ni(111) interface the magnetic moment of

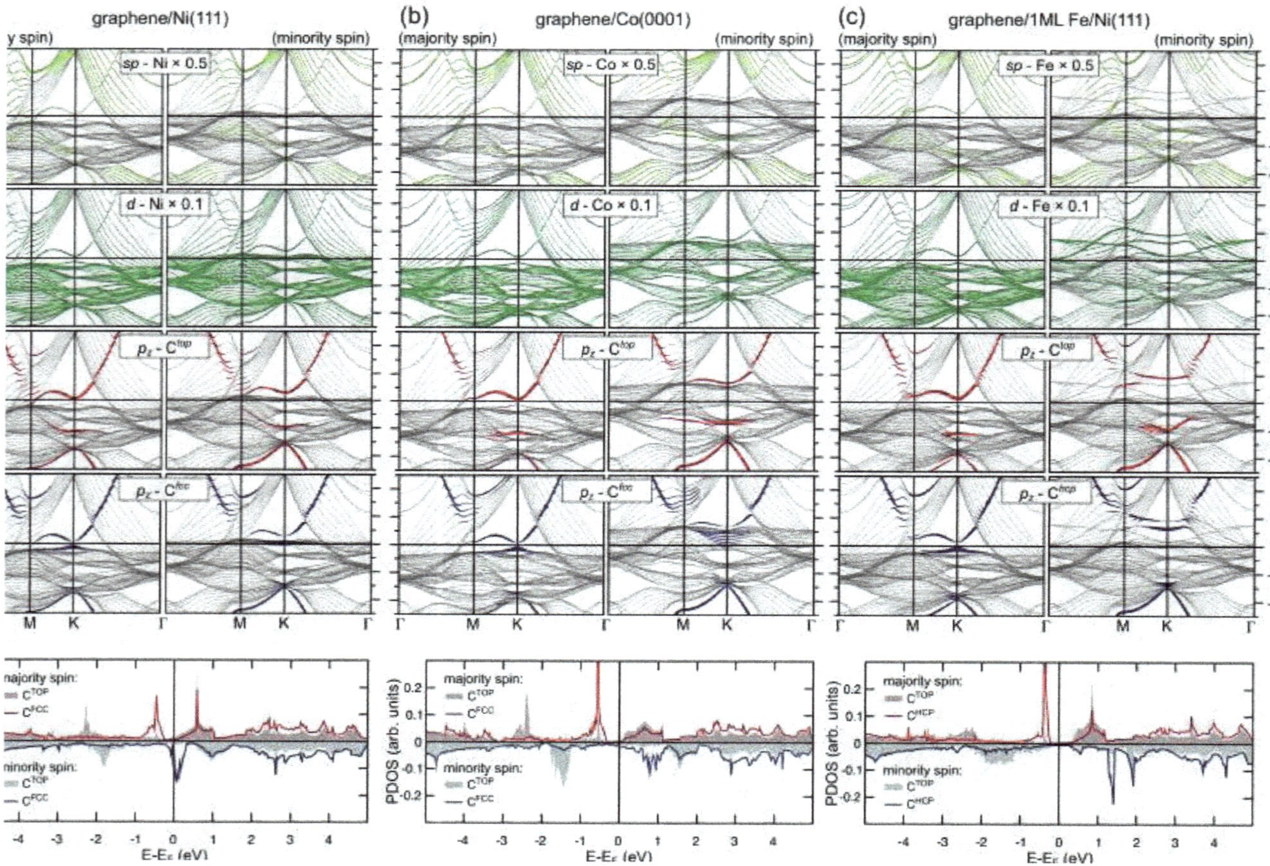

Fig. 5. Calculated majority and minority spin band structures for a slab terminated by: (a) graphene/Ni(111) interfaces ($top - fcc$ arrangement); (b) graphene/Co(0001) interfaces ($top - fcc$ arrangement); (c) graphene/1 ML Fe/Ni(111) interfaces ($top - hcp$ arrangement). The radius of the circles is proportional to the partial density of states at the interface between FM or C atoms (see legend). The corresponding calculated majority and minority partial p_z density of states of graphene is shown for every example in the lower panel of the figure.

State	FM = Ni(111)		FM = Co(0001)		FM = 1ML Fe/Ni(111)	
	spin ↑	spin ↓	spin ↑	spin ↓	spin ↑	spin ↓
*HS*5	−3.20	−3.32	−3.51	−4.01	−3.56	−4.05
*HS*4	−0.12	−0.55	−0.12	−1.27	−0.28	−1.72
*HS*3	0.28	−0.16	0.27	−0.95	0.26	−1.19
*HS*2	2.28	1.93	2.26	1.63	2.32	1.99
*HS*1	3.27	2.93	3.44	2.81	3.75	2.77

Table 2. Binding energies (in eV) (positive sign for states below the Fermi level and negative sign for states above the Fermi level) of the interface hybrid states (*HS*) extracted from the calculated band structures of the graphene/FM system at the K and M points (see text for details).

Ni interface atoms shows a reduction by about 20 % for 3*d* electrons. Our DFT calculations predict an induced magnetic moment of the carbon atoms in the graphene layer, which is equal to $-0.019\mu_B$ and $0.031\,\mu_B$ depending on the position of the carbon atom on the Ni(111) surface, due to the different behavior of the majority and minority spin interface states. The same behavior is also observed for the graphene/Co(0001) and graphene/1 ML Fe/Ni(111) interfaces where significantly large induced magnetic moments of carbon can be observed (Table 1).

5. Electronic and magnetic structure of graphene on ferromagnetic substrates: experiment

5.1 Growth and surface structure of graphene on Ni(111)

The quality of the samples at each preparation step was checked with LEED and STM. Since Ni(111) and graphene have the nearly similar lattice parameters (the lattice mismatch is of only 1.3 %), graphene forms the hexagonal (1×1) structure. Fig. 6(a) shows an overview of a graphene domain on Ni(111) after thermal decomposition of propene. The graphene layer is continuous and exhibits a highly ordered crystallographic structure without any visible defects even over large areas. Fig. 6(b) represents a magnified topographic image of the graphene lattice together with a typical LEED pattern of monolayer graphene on Ni(111) [inset of Fig. 6(b)]. A higher magnification STM image of the graphene surface is shown in Fig. 6(c) with the graphene hexagonal unit cell marked in the image.

So far several possible atomic configurations were considered for the graphene/Ni(111) intrface. Three "high-symmetry" structures are known as $hcp - fcc$, $top - hcp$, and $top - fcc$ and they are discussed in the previous section. In our case, graphene terraces have a peak-to-peak roughness of 0.2 Å and show a honeycomb structure with a lattice constant of 2.4 ± 0.1 Å [Fig. 6(b,c)], which agrees well with the expected 2.46 Å lattice spacing of graphene. STM images show that in the honeycomb unit cell carbon atoms corresponding to different sites appear with a different contrast, which can be attributed to the differences in the local stacking of the graphene sheet and the Ni(111) substrate. Therefore we interpret our STM images in the following way: Fig. 6 shows a single layer graphene, where carbon atoms most possibly occupy positions corresponding to one of the two non-equivalent configurations – $top - fcc$ or $top - hcp$. However, it turns to be impossible to directly identify which of the sites are occupied.

Additionally, some different orientations of the graphene relative to the Ni(111) substrate could be observed [Dedkov & Fonin (2010); Dedkov et al. (2008b)]. These observations show

Fig. 6. High-quality graphene/Ni(111) system. (a) Large scale constant current STM image of the graphene/Ni(111) surface. Tunneling parameters: $U_T = 0.002\,V$; $I_T = 48\,nA$. (b) Magnified STM image of the perfect graphene lattice. The inset shows a LEED image obtained at 63 eV. (c) High magnification STM image showing atomic structure of the graphene monolayer. Tunneling parameters: $U_T = 0.002\,V$; $I_T = 48\,nA$. Graphene hexagonal unit cell is marked in (c).

that although the interaction between nickel and graphene is relatively strong, different adsorption geometries are locally possible.

5.2 Bonding and magnetism at the graphene/Ni(111) interface

In order to address the average spatial orientation of selected molecular orbitals (for example π or σ) at the graphene/Ni(111) interface, we vary the sample orientation with respect to the wave vector of the linearly polarized x-ray light and monitor the NEXAFS intensity. The observed changes of the NEXAFS lineshape at the C K edge in the graphene/Ni(111) system represent a nice example demonstrating the so-called *search-light*-like effect [Stöhr & Samant (1999)]. In such an experiment, the absorption intensity associated with a specific molecular orbital final state has a maximum if the electric field vector is aligned parallel to the direction of maximum charge or hole density, i. e. along a molecular orbital, and the intensity vanishes if the electric field vector is perpendicular to the orbital axis. The detailed description of the angular dependence of NEXAFS intensities can be found elsewhere [Stöhr (1999); Stöhr & Samant (1999)].

Figure 7 shows NEXAFS spectra of the graphene/Ni(111) system recorded at the C K absorption edge as a function of the angle, θ, between the direction of the incident linearly polarized x-ray light and the sample surface, e. g. between the electrical field vector of light and the sample surface normal (see the insets of Fig.7). The reference NEXAFS spectrum of the highly-oriented pyrolitic graphite (HOPG) crystal measured at $\theta = 30^\circ$ is shown in the upper part of the figure. The spectral features in the two broad regions $283 - 289$ eV and $289 - 295$ eV can be ascribed to C $1s \rightarrow \pi^*$ and C $1s \rightarrow \sigma^*$ transitions of core electrons into unoccupied states (π^*, σ_1^*, and σ_2^*), respectively. Upon the comparison of the NEXAFS C $1s \rightarrow \pi^*, \sigma^*$ spectrum of the graphene/Ni(111) system with the reference graphite spectrum, considerable changes in the spectral shapes are observed, which can be attributed to a strong chemisorption. A broadening of the π^* and σ^* resonances gives an evidence for a strong orbital mixing at the graphene/Ni interface, indicating a strong delocalization of the corresponding core-excited state. The characteristic double-peak structure at 285.5 and 287.1 eV can be qualitatively understood already with the ground-state DFT calculations (see present calculations discussed in the previous section and Ref. [Bertoni et al. (2005)]), while the dynamics of core-hole screening has to be considered for correct reproducing of the spectral profile [Rusz et al. (2010)]. In the ground-state approximation the first π^* resonance can be roughly associated with the unoccupied C $2p$ density of states located at the C atoms on top of Ni atoms, while the second π^* peak is mainly due to the C $2p$ density of states on the C atoms located at fcc hollow sites. However, in a real experiment individual contributions are considerably mixed and further affected by the core hole and its dynamical screening. The σ^* resonances are also influenced by the interaction with the substrate: they are visibly broadened and shifted by 0.6 eV to lower energies. The broadening is a result of the increased screening by the substrate electrons. The reduction of the $\pi^* - \sigma^*$ separation reflects the reduced anisotropy of the potential for outgoing electrons due to the slight ripple in the graphene layer on Ni(111) accompanied by a softening of the C-C bonds. A comparison of the present NEXAFS results for graphene on Ni(111) with those recently obtained for graphene/Rh and graphene/Ru [Preobrajenski et al. (2008)] indicates the existing of a very strong covalent interfacial bonding between carbon atoms in the graphene layer and Ni atoms of the substrate.

In the following we would like to compare our NEXAFS results with the recently calculated C K-edge electron energy-loss spectra (EELS) for the graphene/Ni(111) interface [Bertoni et al. (2005)]. In this case, experimental NEXAFS spectra taken at $\theta = 10^\circ$ and $\theta = 90^\circ$ correspond to the calculated EELS spectra for the scattering vector $\mathbf{q}$ perpendicular and parallel to the graphene layer, respectively. The calculated EELS spectra are found to agree well with the experimental NEXAFS data: (i) the spectra show the same angle (scattering vector) dependence and (ii) the experimentally observed NEXAFS features are well reproduced in the calculated EELS spectra. For example, two peaks in the NEXAFS spectra in the $1s \rightarrow \pi^*$ spectral region at 285.5 eV and 287.1 eV of photon energy can be assigned to the double-peak structure in the calculated EELS spectrum at 0.8 eV and 3.0 eV above the Fermi level [Bertoni et al. (2005)]. According to the theoretical calculations [Bertoni et al. (2005)], the first sharp feature in the NEXAFS spectrum is due to the transition of the electron from the $1s$ core level into the interface hybridized state HS_4 above the Fermi level (around the K point in the hexagonal Brillouin zone), which originates from the C p_z–Ni $3d$ hybridization and corresponds to the antibonding orbital between a carbon atom C-top and an interface Ni atom. The second peak in the NEXAFS spectrum is due to the dipole transition of an electron from the $1s$ core level into the interface hybrid state HS_5 above the Fermi level (around the M point in the hexagonal Brillouin zone), which originates from C p_z–Ni $p_x, p_y, 3d$ hybridization and

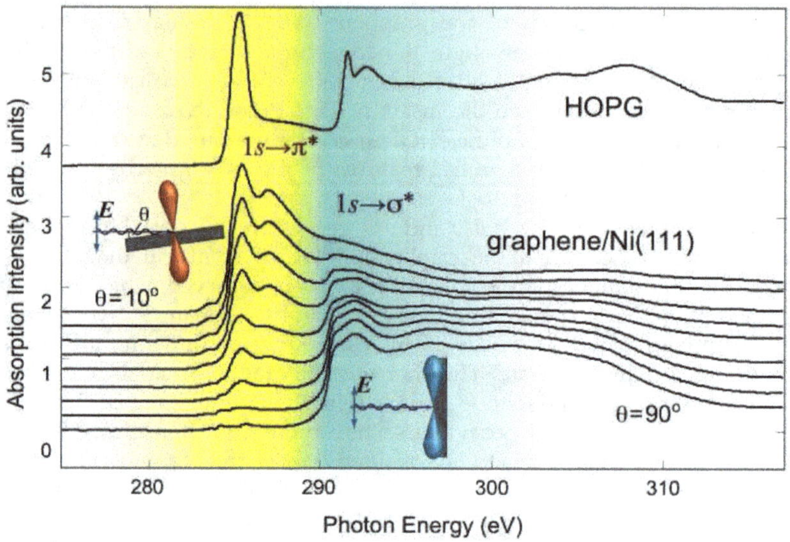

Fig. 7. Angular dependence of the absorption signal at the C K edge of the graphene/Ni(111) system measured as a function of angle, θ, between polarization vector of incoming linearly polarized light and the surface normal of the sample. Spectra were collected in the partial electron yield mode and angle was changed with a step of 10° from top to bottom. The reference spectrum of HOPG is shown in the upper part of the panel for comparison.

corresponds to a bonding orbital between C-*top* and C-*fcc* atoms, involving a Ni interface atom (see discussion about interface hybrid states in the previous section). In case of the NEXAFS C 1$s \rightarrow \sigma^*$, the theory also correctly describes the shape of the absorption spectra [Bertoni et al. (2005)].

The strong hybridization between graphene π and Ni 3d valence band states at the graphene/Ni(111) interface leads to the partial charge transfer of the spin-polarized electrons from Ni onto C with the appearance of an induced effective magnetic moment of carbon atoms (Section 4), which can be detected in an experiment sensitive to the magnetic state of particular element, like XMCD. Figure 8 shows NEXAFS spectra of the graphene/Ni(111) system obtained for two opposite magnetization directions with respect to the polarization of the incident X-ray beam (upper panels) together with the resulting XMCD signal (lower panels). The spectra collected at the Ni $L_{2,3}$ edge in TEY mode and at the C K edge in PEY mode are presented in the left-hand panel and in the right-hand panel, respectively. The Ni $L_{2,3}$ XMCD spectrum (white line as well as fine structure behind the absorption edge) is in perfect agreement with previously published spectroscopic data [Dhesi et al. (1999); Nesvizhskii et al. (2000); Srivastava et al. (1998)]. The quantitative analysis of the absorption spectra obtained on a magnetic sample with circularly polarized light can be performed with the help of the so-called sum rules for spin- and orbital-magnetic moments [Carra et al. (1993); Thole et al. (1992)]. The intensities of L_3 and L_2 absorption lines and their differences for the parallel and anti-parallel orientations of the projection of photon spin on the sample magnetization direction are quantitatively related by the sum rules to the number of 3d holes in the valence band of the ferromagnetic material and the size of the spin and orbital magnetic moments.

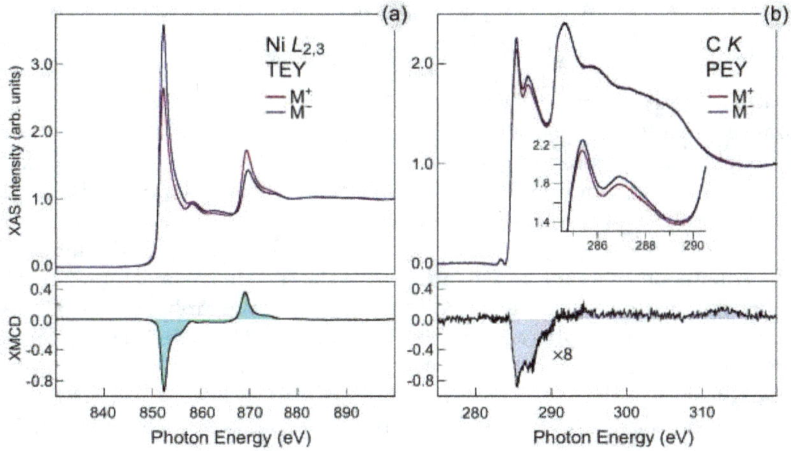

Fig. 8. XMCD spectra of the graphene/Ni(111) system: absorption spectra measured with circularly polarized light for two opposite orientations of the sample magnetization are shown in the upper part for the Ni $L_{2,3}$- (a) and C K-edges (b). The corresponding differences reflecting the strength of the dichroic signal are shown in the lower part of the respective figures.

From the experimental Ni $L_{2,3}$ TEY NEXAFS data [Fig. 8(a)] the bulk values of Ni $3d$ magnetic moments were derived, using a number of $3d$ holes in the valence band of Ni $n_h = 1.45$ [Sorg et al. (2004)] and polarization of light $P = 0.75$. At room temperature and in TEY mode (more bulk sensitive), Ni provides a spin moment of $3d$ electrons of $\mu_S = 0.69\,\mu_B$ and an orbital moment of $\mu_L = 0.07\,\mu_B$, respectively. These values coincide with the previously published experimental results [Baberschke (1996); Srivastava et al. (1998)]. The experimentally obtained spin-magnetic moment is very close to the calculated bulk value of $\mu_S = 0.67\,\mu_B$ for the graphene/Ni(111) system (Section 4). For the most energetically favorable configuration of carbon atoms on Ni(111), $top - fcc$, the calculations predict a reduction of the spin-magnetic moments of Ni interface atoms by 16 % down to $0.52\,\mu_B$ (Section 4). The experimental data collected at the Ni $L_{2,3}$ absorption edge in the PEY mode (more surface/interface sensitive) also shows a slight reduction of the spin moment to $\mu_S = 0.63\,\mu_B$. However, the observed decreasing is not so pronounced as yielded by the theoretical calculation, which can be explained by the large contribution of the bulk-derived signal to the XMCD spectra.

The most important and interesting results of these XMCD experiments on the graphene/Ni(111) system is the observation of the relatively large dichroic contrast at the C K absorption edge [Fig. 8(b)]. In order to magnify the measured magnetic contrast at the $1s \to \pi^*$ absorption edge, these XMCD spectra were collected in the PEY mode with the circularly polarized light at an angle of $\theta = 20^\circ$. We note that the observed differences in the NEXAFS spectra collected at this angle visible in Figs. 7 and 8 are due to the different polarization of light: in Fig. 7(a) the data are obtained with the linearly polarized x-rays, i. e. the strong angular dependence of the absorption signal is due to the different graphene's orbital orientation; whereas the data in Fig. 8 are taken with the circularly polarized light, i. e. both $1s \to \pi^*$ and $1s \to \sigma^*$ transitions are nearly equivalently possible. The C K XMCD spectrum reveals that the major magnetic response stems from transitions of the $1s$ electron onto the π^*-states, while transitions onto the σ^*-states yield very weak (if any) magnetic signal. These results indicate that only the C $2p_z$ orbitals of the graphene layer are magnetically polarized due to the hybridization with the Ni $3d$ band. The sharp structure at the $1s \rightarrow \pi^*$ absorption edge originates from hybridized C p_z–Ni $3d$ and C p_z–Ni p_x, p_y $3d$ states (see earlier discussion and Section 4).

The appearance of XMCD signal at the C K-edge shows that indeed the Ni film induces a magnetic moment in the graphene overlayer. However, at the C K absorption edge, the electron transitions occur from the non-spin-orbit split $1s$ initial states to the $2p$ final states and thus, in the analysis of the dichroism effect at the K edge one equation in the selection rules is missed. This means that the XMCD signal at K edges provides the information only on the orbital magnetic moment μ_{orb} [Carra et al. (1993); Huang et al. (2002); Thole et al. (1992)]. The partial charge transfer from Ni to C atoms in the graphene/Ni(111) system was calculated for the 22-atom (graphene)/Ni cluster [Yamamoto et al. (1992)] yielding $0.205e^-$ per C atom in graphene, which leads to the $2p$-orbital occupation number of $n_p = 2.205e^-$. Using the C K NEXAFS spectra the procedure described in work [Huang et al. (2002)], the orbital magnetic moment of $\mu_{orb} = 1.8 \pm 0.6 \times 10^{-3}\,\mu_B$ per C atom was extracted. The relatively large uncertainty arises mainly from the estimation of the number of C $2p$ holes, the background subtraction of NEXAFS spectra, and from the error for the degree of circular polarization of light.

The theoretical calculations presented in Section 4 also give the values for the spin magnetic moment of $-0.019\,\mu_B$ and $0.031\,\mu_B$ for C-*top* and C-fcc atoms, respectively. However, the magnetic splitting of the majority and minority parts of the interface hybrid states HS_3 and HS_4 was found between 0.44 and 0.43 eV, respectively, which may yield higher values for the magnetic moment. Due to the impossibility to directly extract the value of the spin magnetic moment form the K edge XMCD spectra, we apply a simple comparison with the magnetic measurements on similar systems in oder to estimate the average μ_S value for the carbon atoms at the graphene/Ni(111) interface. For the Fe/C multilayers clear magnetic signals of carbon were obtained by using the resonant magnetic reflectivity technique [Mertins et al. (2004)]. Hysteresis loop recorded at C K absorption edge gave a clear proof of ferromagnetism of carbon atoms at room temperature with a measured spin magnetic moment of $\mu_S = 0.05\,\mu_B$ induced by adjacent Fe atoms. The observed ferromagnetism of carbon in the Fe/C multilayered system was related to the hybridization between the Fe $3d$ orbitals and the C p_z orbitals, which are normal to the graphene-type layered sp^2-coordination. The second comparison can be performed with carbon nanotubes on ferromagnetic Co substrate [Céspedes et al. (2004)]. Carbon nanotubes were shown to become magnetized when they are in contact with magnetic material. Spin-polarized charge transfer at the interface between a flat ferromagnetic metal substrate and a multiwalled carbon nanotube leads to a spin transfer of about $0.1\,\mu_B$ per contact carbon atom. Additionally, a comparison of the XMCD spectra obtained at the C K edge in graphene/Ni(111) (present work) and at the O K edge in O/Ni(100) [Sorg et al. (2006)], reveals the approximately same magnitude of the XMCD signal. For the O/Ni(100) system, where the induced spin-magnetic moment of $0.053\,\mu_B$ on oxygen atom was calculated, the theoretically simulated NEXAFS and XMCD spectra agree well with the experimental data. Considering these analogous systems, we estimate the induced magnetic moment for graphene on Ni(111) to be in the range of 0.05-0.1 μ_B per carbon atom.

The experimentally observed effective magnetic moment of carbon atoms of the graphene layer on Ni(111) was recently confirmed by spin-resolved photoemission measurements [Dedkov & Fonin (2010)]. These data present the clear spin-contrast in the valence band region and a spin polarization value of about -60% at the Fermi level for the pure Ni(111) surface. The presence of graphene on Ni(111) strongly modifies the valence band spectrum of Ni indicating the strong interaction between valence band states of graphene and Ni. In the graphene/Ni(111) system the spin polarization of Ni 3*d* states at E_F is strongly reduced to about -25%. The significant modifications of the spin-resolved structure of Ni 3*d* states as well as the reduction of the spin polarization at E_F is considered as an indication of decreasing the magnetic moment of Ni atoms at the graphene/Ni(111) interface. These spin-resolved photoemission measurements demonstrate the clear spin contrast for the graphene π states with the maximum spin-polarization of $P = -12 \pm 2\%$ at room temperature. The two spin-resolved π components are clearly resolved giving the exchange splitting of about 34 ± 9 meV for these states, which agrees well with the value extracted from theoretical calculations (Section 4).

5.3 Electronic properties of graphene on Ni(111)

Fig. 9(a) shows a series of angle-resolved photoemission spectra measured along the $\Gamma - \mathrm{K}$ direction of the hexagonal Brillouin zone of the graphene/Ni(111) system. This series is extracted from the 3D sets of data of photoemission intensity $I(E_B, k_x, k_y)$, where E_B is the binding energy and k_x, k_y are the orthogonal components of the in-plane wave vector. For the graphene/Ni(111) system the K and M points of the Brillouin zone are reached at 1.7 Å^{-1} and at 1.4 Å^{-1}, respectively. The presented photoemission data are in very good agreement with previously published results [Dedkov et al. (2008b; 2001); Grüneis & Vyalikh (2008); Nagashima et al. (1994); Shikin et al. (2000)]. In Fig. 9(a) one can clearly discriminate dispersions of graphene π- and σ-derived states in the region below 2 eV of the binding energy (BE) as well as Ni 3*d*-derived states near E_F. The binding energy of the graphene π states in the center of the Brillouin zone (in the Γ point) equals to 10.1 eV which is approximately by 2.4 eV larger than the binding energy of these states in pure graphite. The shift to larger binding energy is different for σ and π valence band graphene-derived states. This behavior can be explained by the different hybridization strength between these states and Ni 3*d* valence band states, which is larger for the out-of-plane oriented π states compared with the one for the in-plane oriented σ states of the graphene layer as discussed in Section 4. The binding energy difference of ≈ 2.4 eV for the π states and ≈ 1 eV for the σ states between graphite and graphene on Ni(111) is in good agreement with previously reported experimental and theoretical values [Bertoni et al. (2005); Dedkov et al. (2008b)]. The effect of hybridization between Ni 3*d* and graphene π states can be clearly demonstarted in the region around the K point of the Brillouin zone: (i) one of the Ni 3*d* bands at 1.50 eV changes its binding energy by ≈ 150 meV to larger BE when approaching the K point; (ii) a hybridization shoulder is visible in photoemission spectra, which disperses from approximately 1.6 eV to the binding energy of the graphene π states at the K point [see also Fig. 9(b,c) for a detailed view]. The strong hybridization observed in PES spectra underlines the fact that the π states might become spin-polarized and might gain a non zero-magnetic moment due to charge transfer from the Ni atoms to the carbon atoms.

Considering the electronic band structure of the graphene/Ni(111), the region around the K point delivers the most interesting and important information with respect to the possible spin-filtering effects in the graphene/ferromagnet or ferromagnet/graphene/ferromagnet sandwich-like structures. This part of the electronic structure measured with two different photon energies ($h\nu = 70$ eV and 100 eV) is shown in Fig. 9(b,c) as color maps (upper panels) together with the corresponding intensity profiles directly at the K point (lower panels). Firstly, the spectral function of the graphene layer on Ni(111) is characterized by the absence of

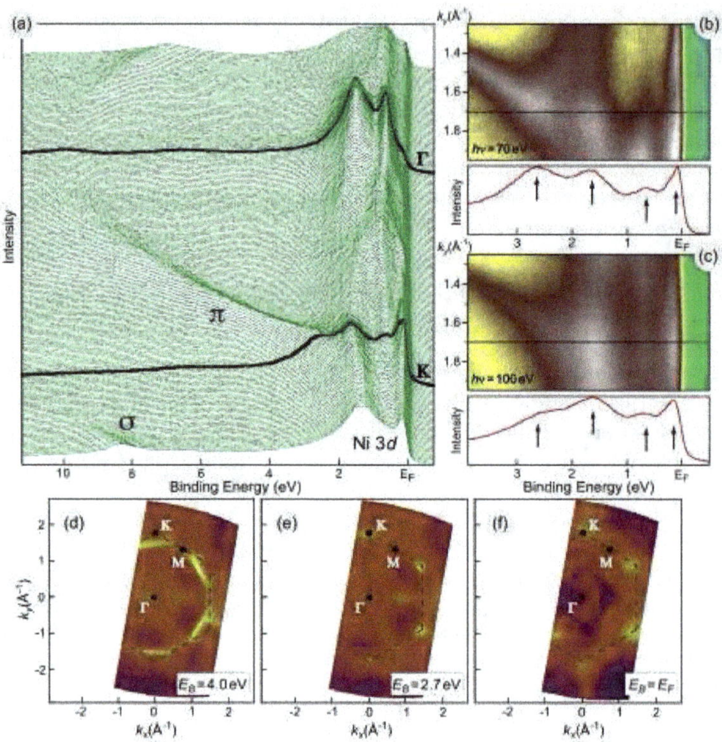

Fig. 9. (a) Series of the angle-resolved photoemission spectra of the graphene/Ni(111) system recorded along the $\Gamma - \mathrm{K}$ direction of the hexagonal Brillouin zone at 100 eV of photon energy. Spectra corresponding to Γ and K points are marked by thick black lines. (b,c) Regions around the K point of photoemission intensity of the graphene/Ni(111) system measured at 70 eV and 100 eV of photon energy. The corresponding intensity profiles at the K point are shown in the lower panels with arrows indicating the main photoemission features discussed in the text. (d-f) Constant energy cuts of the 3D data sets in the energy-wave vector space at 4 eV, 2.7 eV of BE, and at E_F.

well-ordered structure of the graphene π-bands in the vicinity of the Fermi level and secondly, the Dirac-cone is not preserved. Both observations can be attributed to a strong interaction between graphene layer and metallic substrate leading to a strong hybridization between the graphene π and the Ni 3*d* valence band states. In the vicinity of the K point a number of photoemission peaks can be clearly distinguished: (i) a sharp peak about the Fermi level at $0.1 - 0.2$ eV BE, (ii) a graphene π-states-related peak at 2.65 eV BE, (iii) two peaks at 0.7 eV and 1.65 eV BE.

In the following we perform a detailed analysis of the experimentally obtained electronic structure relying on electronic structure calculations for the graphene/Ni(111) system discussed in Section 4. These calculations predict the existence of three interface hybrid states below the Fermi level originating from the strong hybridization between the Ni 3*d* and the graphene π states and corresponding to: (HS_1) bonding between C-fcc and interface Ni atoms; (HS_2) bonding between C-top and interface Ni atoms; (HS_3) antibonding between

C-fcc and interface Ni atoms. From the analysis of the experimental data in the region around the K point of the hexagonal Brillouin zone we could distinguish a number of flat bands, which are clearly separated from each other.
The interpretation of the experimentally observed photoemission features around the Fermi level could be performed as presented in the following. The photoemission peak close to the Fermi level (0.1-0.2 eV BE) could be considered as a combination of the interface hybrid state HS_3 (both spins) with a large contribution of the graphene π-character and the Ni $3d(\downarrow)$-band. The second peak at 0.7 eV BE could be assigned to the combination of the Ni $3d(\uparrow)$- and Ni $3d(\downarrow)$-bands present calculations. The feature at 1.65 eV could be considered as a combination of Ni $3d(\uparrow)$-band and $HS_2(\downarrow)$-state with a large graphene π-character. The last photoemission peak (2.65 eV BE) could be assigned to the interface hybrid state $HS_2(\uparrow)$ with large contribution of the graphene π-character.
In order to check the theoretical predictions concerning the spin-dependent electronic transport properties of the ideal graphene/Ni(111) interface, we perform a careful analysis of the constant energy photoemission maps close to the Fermi level. We would like to admit that such an analysis can be rather complicated due to the fact that Ni $3d$ bands, which dominate the photoemission intensity around the Fermi level, are very flat in the vicinity of E_F. Figure 9(d-f) shows the constant energy cuts of the 3D data set obtained at $h\nu$=100 eV for the graphene/Ni(111) system. These energy cuts were taken at (a) 4 eV and (b) 2.7 eV of BE as well as at (c) the Fermi level. The energy cut at $E_B = 4$ eV shows characteristic photoemission intensity patterns of the graphene layer, which reflect the symmetry of the system. Below the Dirac point (crossing of straight dispersion lines of π states in free-standing graphene) the graphene π bands are visible in the first Brillouin zone whereas no bands can be seen in the second one. Additionally several energy bands are present in the middle of the Brillouin zone, which also show hexagonal symmetry. These bands originate from the hybridization of the Ni and graphene valence band states. The constant energy cut taken in the region of the minimal binding energy of the graphene π states ($E_B = 2.7$ eV) is shown in Fig. 9(e). In the case of graphene/Ni(111) the Dirac point is not preserved due to the strong hybridization of Ni $3d$ and graphene π states around the K point. This can also be directly recognized at this energy cut where graphene π states produce broad intensity spots instead of sharp points in the wave-vector space. As in the previous case, we observe a number of valence band states in the middle part of the Brillouin zone, which again could be assigned to the hybridization-derived states.

The most interesting and important information in view on the spin-dependent transport properties of the graphene/Ni(111) system can be extracted form the constant energy cut obtained at the Fermi energy, which is presented in Fig. 9(f). Already the analysis of Fig. 9(a-c) shows that the photoemission intensity is increased around the K point and along the K – M direction of the hexagonal Brillouin zone, that correlates with the increased photoemission intensity observed in the energy cut shown in Fig. 9(f) for the Fermi energy. Additionally, a number of arcs surrounding the K points and weak (but distinguished) diamond-shape regions of increased intensity is clearly visible in the middle part and around the M points of the Brillouin zone, respectively. Upon the comparison of the obtained photoemission results for the graphene/Ni(111) system [Fig. 9(a-f)] with the band structure calculations for this system (Section 4), we find very good agreement between theory and experiment. Particularly, the region around the Fermi level for the ideal graphene/Ni(111) system is well reproduced in the experiment, confirming the main predictions of the theory. Unfortunately, at the present stage of the experiment, we can not specify the spin-character of energy bands, which should be the subject of future spin- and angle-resolved photoemission investigations.

6. Conclusions and outlook

In conclusion, the electronic structure and magnetic properties of the graphene/ferromagnet interface were investigated via combination of theoretical and experimental methods: DFT calculations, NEXAFS, XMCD, and mapping of the band structure by means of ARPES. From the theoretical side the different interfaces were analyzed: graphene/Ni(111), graphene/Co(0001), and graphene/1 ML Fe/Ni(111). Here the electronic structure of the interface as well as the effect of induced magnetism in graphene layer were discussed. In all cases a strong modification of the electronic structure of the graphene layer and FM substrate upon graphene adsorption were detected in theoretical calculations and confirmed by spectroscopic methods. This modification is due to the considerable hybridization of the graphene π and FM $3d$ valence band states accompanied by the partial charge transfer of spin-polarized electrons from FM onto C atoms leading to the appearance of the effective magnetic moment in the graphene layer. The presence of an effective magnetic moment on carbon atoms of the graphene layer was unambigously proven by XMCD and spin-resolved photoemission measurements for the graphene/Ni(111) system. The experimentally obtained electronic structure of occupied and unoccupied states of graphene/Ni(111) was compared with band structure calculations allowing the clear assignment of spectral features in the NEXAFS and ARPES data. The good agreement between theory and experiment was also found upon the analysis of the Fermi energy cuts, that give us an opportunity to confirm the main statements of the theoretical works predicting the spin-filtering effects of the graphene/Ni(111) interface. However, the clear assignment of the spectroscopic valence band features have to be performed in future spin-resolved photoemission experiments.

7. Acknowledgements

We would like to thank K. Horn, M. Weser, S. Böttcher, M. Fonin, and B. Paulus. We are also grateful to P. Kelly and L. Calmels for fruitful discussions. This work has been supported by the European Science Foundation (ESF) under the EUROCORES Programme EuroGRAPHENE (Project "SpinGraph"). E. V. appreciate the support from the German Research Foundation (DFG) through the Collaborative Research Center (SFB) 765 "Multivalency as chemical organisation and action principle: New architectures, functions and applications". Y. D. acknowledges the financial support by the DFG under project DE 1679/2-1. We would like to acknowledge our co-workers in Max-lab (Lund) and BESSY (Berlin) synchrotron facilities for the technical assistance during experiment. We appreciate the support from the HLRN (High Performance Computing Network of Northern Germany) in Berlin.

8. References

Baberschke, K. (1996). The magnetism of nickel monolayers, *Appl. Phys. A* 62(5): 417–427.

Bae, S., Kim, H., Lee, Y., Xu, X., Park, J.-S., Zheng, Y., Balakrishnan, J., Lei, T., Kim, H. R., Song, Y. I., Kim, Y.-J., Kim, K. S., Ozyilmaz, B., Ahn, J.-H., Hong, B. H. & Iijima, S. (2010). Roll-to-roll production of 30-inch graphene films for transparent electrodes, *Nature Nanotech.* 5(8): 574–578.

Bertoni, G., Calmels, L., Altibelli, A. & Serin, V. (2005). First-principles calculation of the electronic structure and eels spectra at the graphene/Ni(111) interface, *Phys. Rev. B* 71(7): 075402 1–8.

Blochl, P. (1994). Projector augmented-wave method, *Phys. Rev. B* 50(24): 17953–17979.

Boukhvalov, D. W. & Katsnelson, M. I. (2009). Chemical functionalization of graphene, *J. Phys.: Condens. Matter.* 21(34): 344205 1–12.

Brako, R., Sokcevic, D., Lazic, P. & Atodiresei, N. (2010). Graphene on Ir(111) surface: From van der Waals to strong bonding, *arXiv:1006.1280v1 [cond-mat.mes-hall]*. URL: *http://arxiv.org/abs/1006.1280*.

Carra, P., Thole, B., Altarelli, M. & Wang, X. (1993). X-ray circular-dichroism and local magnetic-fields, *Phys. Rev. Lett.* 70(5): 694–697.

Céspedes, O., Ferreira, M. S., Sanvito, S., Kociak, M. & Coey, J. M. D. (2004). Contact induced magnetism in carbon nanotubes, *J. Phys.: Condens. Matter.* 16: L155–L161.

Coletti, C., Riedl, C., Lee, D. S., Krauss, B., Patthey, L., von Klitzing, K., Smet, J. H. & Starke, U. (2010). Charge neutrality and band-gap tuning of epitaxial graphene on SiC by molecular doping, *Phys. Rev. B* 81(23): 235401 1–8.

Coraux, J., N'Diaye, A. T., Busse, C. & Michely, T. (2008). Structural coherency of graphene on Ir(111), *Nano Lett.* 8(2): 565–570.

Coraux, J., N'Diaye, A. T., Engler, M., Busse, C., Wall, D., Buckanie, N., zu Heringdorf, F.-J. M., van Gastel, R., Poelsema, B. & Michely, T. (2009). Growth of graphene on Ir(111), *New J. Phys.* 11(2): 023006 1–22.

Dedkov, Y. S. & Fonin, M. (2010). Electronic and magnetic properties of the graphene-ferromagnet interface, *New J. Phys. (accepted)* .

Dedkov, Y. S., Fonin, M. & Laubschat, C. (2008). A possible source of spin-polarized electrons: The inert graphene/Ni(111) system, *Appl. Phys. Lett.* 92(5): 052506 1–3.

Dedkov, Y. S., Fonin, M., Rüdiger, U. & Laubschat, C. (2008a). Graphene-protected iron layer on Ni(111), *Appl. Phys. Lett.* 93(2): 022509 1–3.

Dedkov, Y. S., Fonin, M., Rüdiger, U. & Laubschat, C. (2008b). Rashba effect in the graphene/Ni(111) system, *Phys. Rev. Lett.* 100(10): 107602 1–4.

Dedkov, Y. S., Shikin, A. M., Adamchuk, V. K., Molodtsov, S. L., Laubschat, C., Bauer, A. & Kaindl, G. (2001). Intercalation of copper underneath a monolayer of graphite on Ni(111), *Phys. Rev. B* 64(3): 035405 1–6.

Dedkov, Y., Vyalikh, D., Holder, M., Weser, M., Molodtsov, S., Laubschat, C., Kucherenko, Y. & Fonin, M. (2008). Dispersion of 4f impurity states in photoemission spectra of Yb/W(110), *Phys. Rev. B* 78(15): 153404 1–4.

Dhesi, S., Dürr, H., van der Laan, G., Dudzik, E. & Brookes, N. (1999). Electronic and magnetic structure of thin Ni films on Co/Cu(001), *Phys. Rev. B* 60(18): 12852–12860.

Dion, M., Rydberg, H., Schroder, E., Langreth, D. & Lundqvist, B. (2004). Van der Waals density functional for general geometries, *Phys. Rev. Lett.* 92(24): 246401 1–4.

Dutta, S. & Pati, S. K. (2008). Half-metallicity in undoped and boron doped graphene nanoribbons in the presence of semilocal exchange-correlation interactions, *J. Phys. Chem. B* 112(5): 1333–1335.

Enderlein, C., Kim, Y. S., Bostwick, A., Rotenberg, E. & Horn, K. (2010). The formation of an energy gap in graphene on ruthenium by controlling the interface, *New J. Phys.* 12: 033014 1–9.

Eom, D., Prezzi, D., Rim, K. T., Zhou, H., Lefenfeld, M., Xiao, S., Nuckolls, C., Hybertsen, M. S., Heinz, T. F. & Flynn, G. W. (2009). Structure and electronic properties of graphene nanoislands on Co(0001), *Nano. Lett.* 9(8): 2844–2848.

Fuentes-Cabrera, M., Baskes, M. I., Melechko, A. V. & Simpson, M. L. (2008). Bridge structure for the graphene/Ni(111) system: A first principles study, *Phys. Rev. B* 77(3): 1–5.

Gamo, Y., Nagashima, A., Wakabayashi, M., Terai, M. & Oshima, C. (1997). Atomic structure of monolayer graphite formed on Ni(111), *Surf. Sci.* 374(1-3): 61–64.

Geim, A. (2009). Graphene: Status and prospects, *Science* 324(5934): 1530–1534.

Geim, A. K. & Novoselov, K. S. (2007). The rise of graphene, *Nature Mater.* 6: 183–191.

Gierz, I., Riedl, C., Starke, U., Ast, C. & Kern, K. (2008). Atomic hole doping of graphene, *Nano Lett.* 8(12): 4603–4607.

Gierz, I., Suzuki, T., Weitz, R. T., Lee, D. S., Krauss, B., Riedl, C., Starke, U., Hoechst, H., Smet, J. H., Ast, C. R. & Kern, K. (2010). Electronic decoupling of an epitaxial graphene monolayer by gold intercalation, *Phys. Rev. B* 81(23): 235408 1–6.

Grimme, S. (2004). Accurate description of van der waals complexes by density functional theory including empirical corrections, *J. Comput. Chem.* 25(12): 1463–1473.

Grüneis, A. & Vyalikh, D. (2008). Tunable hybridization between electronic states of graphene and a metal surface, *Phys. Rev. B* 77(19): 193401 1–4.

Hu, Z., Ogletree, D., van Hove, M. & Somorjai, G. (1987). LEED theory for incommensurate overlayers - application to graphite on Pt(111), *Surf. Sci.* 180(2-3): 433–459.

Huang, D., Jeng, H., Chang, C., Guo, G., Chen, J., Wu, W., Chung, S., Shyu, S., Wu, C., Lin, H. & Chen, C. (2002). Orbital magnetic moments of oxygen and chromium in CrO_2, *Phys. Rev. B* 66(17): 174440 1–5.

Karpan, V. M., Giovannetti, G., Khomyakov, P. A., Talanana, M., Starikov, A. A., Zwierzycki, M., van den Brink, J., Brocks, G. & Kelly, P. J. (2007). Graphite and graphene as perfect spin filters, *Phys. Rev. Lett.* 99(17): 176602 1–4.

Karpan, V. M., Khomyakov, P. A., Starikov, A. A., Giovannetti, G., Zwierzycki, M., Talanana, M., Brocks, G., Brink, J. V. D. & Kelly, P. J. (2008). Theoretical prediction of perfect spin filtering at interfaces between close-packed surfaces of Ni or Co and graphite or graphene, *Phys. Rev. B* 78(19): 1–11.

Kim, K. S., Zhao, Y., Jang, H., Lee, S. Y., Kim, J. M., Kim, K. S., Ahn, J.-H., Kim, P., Choi, J.-Y. & Hong, B. H. (2029). Large-scale pattern growth of graphene films for stretchable transparent electrodes, *Nature* 457(7230): 706–710.

Klink, C., Stensgaard, I., Besenbacher, F. & Lægsgaard, E. (1995). An STM study of carbon-induced structures on Ni(111) - evidence for a carbidic-phase clock reconstruction, *Surf. Sci.* 342(1-3): 250–260.

Kresse, G. & Furthmuller, J. (1996a). Efficiency of ab-initio total energy calculations for metals and semiconductors using a plane-wave basis set, *Comp. Mater. Sci.* 6(1): 15–50.

Kresse, G. & Furthmuller, J. (1996b). Efficient iterative schemes for ab initio total-energy calculations using a plane-wave basis set, *Phys. Rev. B* 54(16): 11169–11186.

Land, T., Michely, T., Behm, R., Hemminger, J. & Comsa, G. (1992). STM investigation of single layer graphite structures produced on Pt(111) by hydrocarbon decomposition, *Surf. Sci.* 264(3): 261–270.

Mao, Y., Yuan, J. & Zhong, J. (2008). Density functional calculation of transition metal adatom adsorption on graphene, *J. Phys.: Condens. Matter.* 20(11): 115209 1–6.

Martoccia, D., Willmott, P. R., Brugger, T., Bjorck, M., Gunther, S., Schleputz, C. M., Cervellino, A., Pauli, S. A., Patterson, B. D., Marchini, S., Wintterlin, J., Moritz, W. & Greber, T. (2008). Graphene on Ru(0001): A 25×25 supercell, *Phys. Rev. Lett.* 101(12): 126102 1–4.

Mertins, H., Valencia, S., Gudat, W., Oppeneer, P., Zaharko, O. & Grimmer, H. (2004). Direct observation of local ferromagnetism on carbon in C/Fe multilayers, *Europhys. Lett.* 66: 743–748.

Morozov, S. V., Novoselov, K. S., Katsnelson, M. I., Schedin, F., Elias, D. C., Jaszczak, J. A. & Geim, A. K. (2008). Giant intrinsic carrier mobilities in graphene and its bilayer, *Phys. Rev. Lett.* 100(1): 016602 1–4.

Nagashima, A., Itoh, H., Ichinokawa, T., Oshima, C. & Otani, S. (1994). Change in the electronic states of graphite overlayers depending on thickness, *Phys. Rev. B* 50(7): 4756–4763.

Nagashima, A., Tejima, N. & Oshima, C. (1994). Electronic states of the pristine and alkali-metal-intercalated monolayer graphite/Ni(111) systems, *Phys. Rev. B* 50(23): 17487–17495.

N'Diaye, A., Bleikamp, S., Feibelman, P. & Michely, T. (2006). Two-dimensional Ir cluster lattice on a graphene moiré on Ir(111), *Phys. Rev. Lett.* 97(21): 215501 1–4.

Nesvizhskii, A., Ankudinov, A., Rehr, J. & Baberschke, K. (2000). Interpretation of x-ray magnetic circular dichroism and x-ray absorption near-edge structure in Ni, *Phys. Rev. B* 62(23): 15295–15298.

Neto, A. C., Guinea, F., Peres, N., Novoselov, K. & Geim, A. (2009). The electronic properties of graphene, *Rev. Mod. Phys.* 81(1): 109–162.

Novoselov, K., Geim, A., Morozov, S., Jiang, D., Katsnelson, M., Grigorieva, I., Dubonos, S. & Firsov, A. (2005). Two-dimensional gas of massless dirac fermions in graphene, *Nature* 438(7065): 197–200.

Paulus, B. & Rosciszewski, K. (2009). Application of the method of increments to the adsorption of h2s on graphene, *Int. J. Quantum. Chem.* 109(13): 3055–3062.

Perdew, J., Burke, K. & Ernzerhof, M. (1996). Generalized gradient approximation made simple, *Phys. Rev. Lett.* 77(18): 3865–3868.

Pisani, C., Maschio, L., Casassa, S., Halo, M., Schuetz, M. & Usvyat, D. (2008). Periodic local MP2 method for the study of electronic correlation in crystals: Theory and preliminary applications, *J. Comput. Chem.* 29(13): 2113–2124.

Preobrajenski, A. B., Ng, M. L., Vinogradov, A. S. & Mårtensson, N. (2008). Controlling graphene corrugation on lattice-mismatched substrates, *Phys. Rev. B* 78(7): 073401 1–4.

Rosei, R., Decresceni, M., Sette, F., Quaresima, C., Savoia, A. & Perfetti, P. (1983). Structure of graphitic carbon on Ni(111) - a surface extended-energy-loss fine-structure study, *Phys. Rev. B* 28(2): 1161–1164.

Rusz, J., Preobrajenski, A. B., Ng, M. L., Vinogradov, N. A., Mårtensson, N., Wessely, O., Sanyal, B. & Eriksson, O. (2010). Dynamical effects in x-ray absorption spectra of graphene and monolayered h-BN on Ni(111), *Phys. Rev. B* 81(7): 073402 1–4.

Sasaki, M., Yamada, Y., Ogiwara, Y., Yagyu, S. & Yamamoto, S. (2000). Moire contrast in the local tunneling barrier height images of monolayer graphite on Pt(111), *Phys. Rev. B* 61(23): 15653–15656.

Schedin, F., Geim, A. K., Morozov, S. V., Hill, E. W., Blake, P., Katsnelson, M. I. & Novoselov, K. S. (2007). Detection of individual gas molecules adsorbed on graphene, *Nature Mater.* 6(9): 652–655.

Shikin, A., Prudnikova, G., Adamchuk, V., Moresco, F. & Rieder, K. (2000). Surface intercalation of gold underneath a graphite monolayer on Ni(111) studied by angle-resolved photoemission and high-resolution electron-energy-loss spectroscopy, *Phys. Rev. B* 62(19): 13202–13208.

Sorg, C., Ponpandian, N., Bernien, M., Baberschke, K., Wende, H. & Wu, R. Q. (2006). Induced magnetism of oxygen in surfactant-grown Fe, Co, and Ni monolayers, *Phys. Rev. B* 73(6): 064409 1–7.

Sorg, C., Ponpandian, N., Scherz, A., Wende, H., Nunthel, R., Gleitsmann, T. & Baberschke, K. (2004). The magnetism of ultrathin Ni films grown with O surfactant, *Surf. Sci.* 565(2-3): 197–205.

Srivastava, P., Wilhelm, F., Ney, A., Farle, M., Wende, H., Haack, N., Ceballos, G. & Baberschke, K. (1998). Magnetic moments and curie temperatures of Ni and Co thin films and coupled trilayers, *Phys. Rev. B* 58(9): 5701–5706.

Stöhr, J. (1999). Exploring the microscopic origin of magnetic anisotropies with x-ray magnetic circular dichroism (xmcd) spectroscopy, *J. Magn. Magn. Mat.* 200(1): 470–497.

Stöhr, J. & Samant, M. (1999). Liquid crystal alignment by rubbed polymer surfaces: a microscopic bond orientation model, *J. Electr. Spectr. Rel. Phenom.* 98-99: 189–207.

Sutter, P., Hybertsen, M. S., Sadowski, J. T. & Sutter, E. (2009). Electronic structure of few-layer epitaxial graphene on Ru(0001), *Nano Lett.* 9(7): 2654–60.

Thole, B., Carra, P., Sette, F. & van der Laan, G. (1992). X-ray circular-dichroism as a probe of orbital magnetization, *Phys. Rev. Lett.* 68(12): 1943–1946.

Tombros, N., Jozsa, C., Popinciuc, M., Jonkman, H. T. & van Wees, B. J. (2007). Electronic spin transport and spin precession in single graphene layers at room temperature, *Nature* 448(7153): 571–574.

Wang, B., Bocquet, M. L., Marchini, S., Guenther, S. & Wintterlin, J. (2008). Chemical origin of a graphene moire overlayer on Ru(0001), *Phys. Chem. Chem. Phys.* 10(24): 3530–3534.

Wang, X., Li, X., Zhang, L., Yoon, Y., Weber, P. K., Wang, H., Guo, J. & Dai, H. (2009). N-doping of graphene through electrothermal reactions with ammonia, *Science* 324(5928): 768–771.

Wintterlin, J. & Bocquet, M. L. (2009). Graphene on metal surfaces, *Surf. Sci.* 603(10-12): 1841–1852.

Yamamoto, K., Fukishima, M., Osaka, T. & Oshima, C. (1992). Charge-transfer mechanism for the (monolayer graphite)-Ni(111) system, *Phys. Rev. B* 45(19): 11358–11361.

Yazyev, O. V. & Pasquarello, A. (2009). Magnetoresistive junctions based on epitaxial graphene and hexagonal boron nitride, *Phys. Rev. B* 80(3): 035408 1–5.

4

Graphene Transistors

Kristóf Tahy, Tian Fang, Pei Zhao, Aniruddha Konar, Chuanxin Lian, Huili (Grace) Xing, Michelle Kelly and Debdeep Jena
University of Notre Dame, USA

1. Introduction

The recent discovery of graphene (Novoselov et al., 2004), a single atomic sheet of graphite, has ignited intense research activities to explore the electronic properties of this novel two-dimensional (2D) electronic system. Charge transport in graphene differs from that in conventional 2D electronic systems as a consequence of the conical energy dispersion relation near the charge neutrality (Dirac) point in the electronic band structure (Zhang et al., 2005). Field-effect mobility as high as 15 000 $cm^2/V{\cdot}s$ and a Fermi velocity of ~10^8 cm/s have been demonstrated at room temperature (Geim & Novoselov 2007). These properties make graphene a possible candidate for electronic devices in the future.

The major benefit of graphene frequently quoted is superior electron/hole mobility compared to other semiconductors, but as Fig. 1 shows, this is not experimentally the case *yet* when compared to narrow bandgap III-V semiconductors for comparable carrier densities. In addition, the lack of a bandgap limits the usage of two dimensional graphene for digital switching, where high on/off ratios are necessary. However, several potential advantages may be listed: the perfect 2D confinement of carriers, electron/hole symmetry originating from a conical bandstructure, and the possibility of opening bandgaps lithographically by fabricating graphene nanoribbons (GNRs). In Section 2, high field characteristics of 2D exfoliated graphene are reported on both short-channel and long-channel back gated field effect transistors (FETs). We will elaborate on the problem of metal contact formation and high field transport of 2D graphene. The comparison of fabricated devices based on exfoliated, epitaxial and chemical vapour deposited (CVD) graphene will follow. The large area graphene opens the possibility to make 2D graphene devices for RF/analog amplification (Lin et al., 2010). Superior frequency performance of such devices makes it a promising application of 2D graphene FETs.

The real advantage of graphene, however, can be highlighted from a novel proposed device architecture that is yet to be demonstrated (Zhang et al., 2008). If GNRs are made by lithographic patterning, and are either chemically or electrostatically doped into GNR p-n junctions, then the planar form yields to a tunnelling field-effect transistor (TFET). Towards the realization of the GNR TFET we designed a single but tuneable junction structure to demonstrate the operational principle. We discuss these GNR p-n junction transistors in Section 3 along with the developed analytical and device models used to explain the measured transistor characteristics.

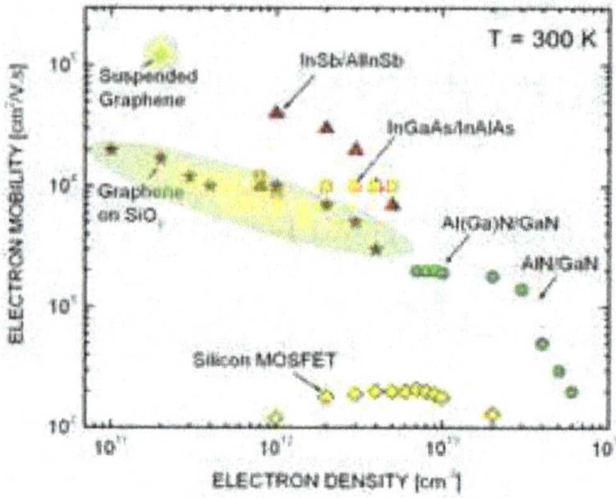

Fig. 1. 300 K electron mobility vs. 2DEG density for various semiconductor materials.

2. 2D graphene transistors

2.1 Device fabrication and characteristics of 2D exfoliated graphene transistors

Graphene offers advantages for interconnects due to high current carrying capacity despite its single layer nature (Murali et al., 2009) but more often graphene is being considered as a possible channel material for transistor technology. So far, graphene has been studied mostly at low biases, but for the application in practical devices, it is essential to investigate the high-field transport properties. High current drives were measured by pushing the devices up to breakdown. The saturation current density for many samples has been measured to be in the 1.5 – 2 A/mm range. Gate modulation of the drain current is found to depend strongly on the channel length. For comparable 2D carrier concentrations, saturation currents in 2D graphene are found to be higher than Si MOSFETs and III-V Nitride HEMTs - a feature that is attractive for various applications.

We first discuss 2D graphene FETs fabricated from exfoliated graphene and later we compare the results to graphene devices made of catalyzed CVD or epitaxial graphene. Exfoliated graphene flakes on heavily n-type doped silicon wafers with t_{ox} = 300 nm thick thermal oxide from Graphene Industries (for preparation of samples, see: Novoselov et al., 2004) were used for the experiments. Single layer graphene flakes were identified by Raman spectroscopy (inset of Fig. 2 (a)). The wafers were backside-metalized after oxide removal in HF to form back-gate contacts. E-beam evaporated Cr/Au (2/200 nm) was used to define the drain and source contacts, which were patterned by e-beam lithography. The source-drain separations ranged from L = 250 nm - 8 μm. After metal deposition and lift-off, the samples were annealed in forming gas at ~650 K for ~2 hours to remove the e-beam resist residue. The graphene flakes were then patterned by O_2 plasma reactive ion etch with PMMA masks to widths ranging from W = 1 - 10 μm. The SEM image (inset of Fig. 2 (b)) shows a typical FET. High current annealing was performed to drive off impurities for some FETs to recover their intrinsic performance. The devices were measured using a semiconductor parameter analyzer in ambient environment and in vacuum, at room temperature and at 77 K. The drain currents of the FETs were first measured at a low bias of V_{DS} = 20 mV, while the gate voltage was varied over a wide range. The gate leakage current was many orders of magnitude lower than the drain current.

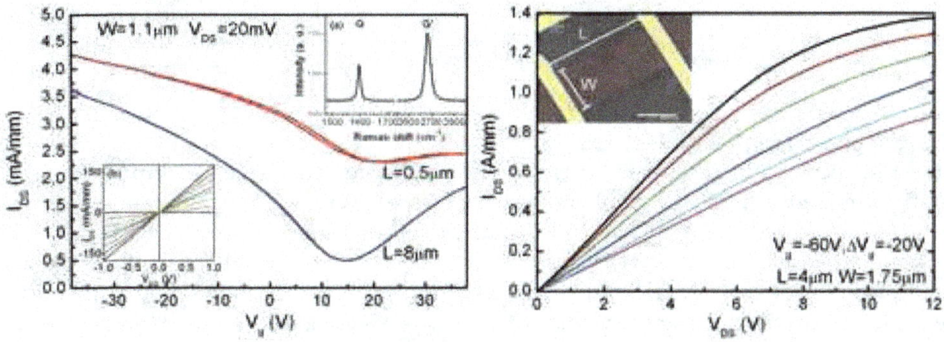

(a) (b)

Fig. 2. (a) I_{DS} as a function of V_G at V_{DS} = 20 mV for short-channel (0.5 μm) and long-channel (8 μm) back-gated graphene FETs. Inset (a) Raman spectrum of single layer graphene. Inset (b) Low field I-V curves of long-channel FETs with V_G = -35 V, ΔV_G = +10 V. (b) I-V curves of a long-channel FETs. Inset is a SEM image of this FET. The scale bar is 1.5 μm.

Fig. 2 (a) shows the representative transfer characteristics of a long- and a short-channel FET. Over the same range of gate overdrives, the long channel FET was observed to exhibit higher gate modulation (~8x) than the short-channel FETs (~2x). The field-effect mobilities were calculated to be ~2000 - 4000 $cm^2/V{\cdot}s$ for long-channel FET and ~200 $cm^2/V{\cdot}s$ for short-channel FET respectively. These characteristics remained similar at different pressures, as well as at lower temperature.

Recently, saturation of drain current at high source-drain biases has been reported, and theory of optical phonon emission limited saturation current has been proposed (Meric et al., 2008; Luo et al., 2008). In particular, back- and top-gated 2D graphene FETs showed saturation current densities in the 0.5 - 0.7 A/mm range as reported by Meric et al., 2008. In Fig. 2 (b), the high bias characteristics of a back-gated FET (W/L = 1.75/4 μm) are shown. A saturation current of 1.38 A/mm is measured for this device. Saturation current densities in the 1.5 - 2.0 A/mm range were measured on a number of devices, in agreement with another earlier report for back-gated devices (Luo et al., 2008). This is ~2x higher than the reported saturation current densities for dual-gate devices, indicating that the top-gate high-κ oxide lowers the current drive through graphene. The effective 2D carrier densities were in the $2 - 4{\cdot}10^{12}$ cm^{-2} range, as estimated from the effective gate voltage. The ~0.5 - 0.7 A/mm saturation current reported earlier has been attributed to substrate-induced optical phonon scattering. However the ~2x higher current drives measured here on SiO_2 substrates indicate that the lower current drives are due to the top gate high-κ oxide rather than the SiO_2 substrate. This conjecture will be discussed in detail in the next Section.

Such current drives can be obtained in Si MOSFETs and III-V HEMTs, but only with a) high 2DEG densities (>10^{13} cm^{-2}), and/or b) very short channel lengths taking advantage of ballistic transport. Our measurements show that graphene FETs are able to achieve these high current densities without these criteria, implying that even higher current drives are possible when they are met in the future. For a rough comparison, a carbon nanotube with diameter ~2 nm and a saturation current ~25 μA has an effective current per the circumferential width ~4 A/mm (Chen & Fuhrer, 2005), much higher than either graphene or other semiconductors. At high drain biases, the current saturates at more than 1 A/mm, but with weak gate modulation. Graphene carries over the high current carrying capacity of carbon nanotubes, but the absence of an energy bandgap hampers the modulation of this high current density.

2.2 Graphene on high-κ dielectrics

In the last section, we discussed carrier mobility and current drive of graphene on SiO_2. Low-field mobility values for graphene on SiO_2 was 4000 cm²/V·s at room temperature and the highest reported value at $>10^{13}$ cm^{-2} carrier concentration is around 9000 cm²/V·s in literature (Novoselov et al., 2004). It is widely believed that ionized charged impurities at the graphene-oxide interface (Hwang et al., 2007) limits the carrier mobility in graphene. As electron-ionized charged impurity interaction $v_{cou} \approx \kappa^{-2}$, where κ is the average dielectric constant of graphene's environment, it seems that use of very high-κ substrate instead of SiO_2 can improve the carrier mobility of graphene at low fields. But use of high-κ dielectrics for graphene transistors comes with an extra scattering mechanism—the remote interface-phonon or surface optical phonon (SO) scattering. Most of the high-κ materials are polar in nature because of large difference in electronegativity of constituent atoms. At room temperature, the vibrating dipoles of the high-κ dielectric material create an electric field, which decays exponentially away from the graphene-dielectric interface. Due to the fact that graphene is an atomically thin (0.3 - 0.4 nm) membrane, the decaying polarization field (also known as surface-phonon field) can scatter the carriers in the graphene channel and limit the low-field mobility. The surface phonon energies can be calculated by solving Maxwell's equation in the dielectric continuum model.

	SiO_2	AlN	Al_2O_3	HfO_2	ZrO_2	SiC
κ^0_{ox}	3.9	9.14	12.53	22.0	24.0	9.7
κ^∞_{ox}	2.5	4.8	3.2	5.03	4.0	6.5
$\omega_{TO,1}$	55.6	81.4	48.18	12.4	16.67	-
$\omega_{TO,2}$	138.1	88.5	71.41	48.35	57.7	-
$\omega_{SO,1}$	59.98	83.60	55.01	19.42	25.02	116.0
$\omega_{SO,2}$	146.51	104.96	94.29	52.87	70.80	167.58

Table 1. Surface phonon energies of graphene/dielectric/air system for various dielectric materials.

In Table 1, we present the calculated values of surface-phonon energies for commonly used dielectrics in device-physics. Most binary dielectric compounds exhibit two surface-phonon modes corresponding to two bulk transverse phonon (TO) modes. It should be noted that, high-κ dielectric materials such as HfO_2 and ZrO_2 exhibit very low- energy SO modes. The surface-phonon modes are optical-like (dispersionless) as shown in Fig. 3 and cause inelastic scattering of carrier electrons in graphene. The electron-SO phonon scattering rates are calculated using second order perturbation theory and transport coefficient such as electron mobility have been calculated using the Boltzmann transport equation under the relaxation time approximation. When ionized impurity scattering is combined with SO-phonon scattering, a comprehensive picture of charge transport emerges. At low temperature, when SO-phonons are frozen, indeed electron mobility can be enhanced by a factor of 3-4 by using ZrO_2 as the gate dielectric, but at room temperature, this dielectric enhancement is washed out by strong electron-SO phonon scattering. Fig. 3 (b) provides a relative assessment of electron mobility on different dielectrics. The curve represents the expected mobility enhancement by high-κ dielectrics, whereas the dots represent the effective mobility after inclusion of SO-phonon scattering rates. It is seen that there is not much room for improving carrier mobility with commonly available high-κ dielectrics at room temperature. In the future, high-κ dielectrics with high SO-phonon energies should be sought to achieve high mobility and speed in graphene.

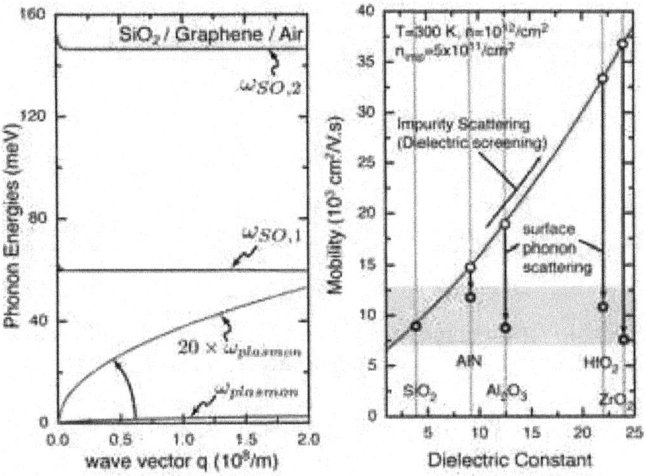

(a) (b)

Fig. 3. (a) Dispersion of SO phonon and (b) mobility of graphene on common dielectrics.

2.3 Short channel and contact effects

The dependence of high bias drain currents and their modulation efficiency with the back-gate is observed to depend strongly on the channel length of graphene. In Fig. 4, the FET characteristics are shown for four different channel lengths, ranging from 0.5 to 8 μm. Two features are observed - a) for the same channel length, the gate modulation at higher drain biases is lower, as is expected from the saturation of the current, and b) for the same bias voltages (or fields), the gate modulation efficiency decreases sharply as the channel length reduces. As shown in Fig. 4, the gate modulation reduces from ~7 to ~1.4 as the source-drain separation is scaled from 8 μm to 0.5 μm.

Due to the absence of a bandgap in graphene, charge exchange is expected at the metal-graphene source/drain contacts owing to the work function differences. Considering the work function of Cr (W_{Cr} ~4.5 eV), and that of pristine graphene (W_{Gr} ~4.5 eV) one would expect them to form a flat-band (or 'neutral') contact without any charge transfer. However, recent calculations have shown that the necessary condition for the formation of a neutral contact with a metal (work function W_M) in intimate contact with graphene is $W_M - W_{Gr}$ ~0.9eV (Giovannetti et al., 2008). Therefore, we expect that the graphene region adjacent to the Cr contact pad is effectively doped with excess carriers. The excess charge region extends over an effective Debye length from each contact resulting in source/drain extension regions. Quantitative modelling of this interesting phenomenon will be presented in detail in Section

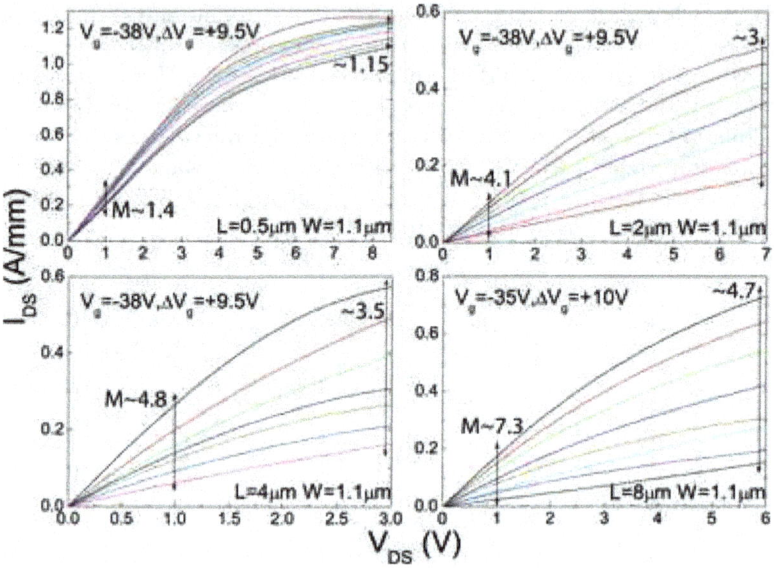

Fig. 4. Current density as a function of drain biases for FETs with different channel lengths varied from 0.5 μm to 8 μm. Clear loss of gate modulation is observed with the shrinking of the graphene channel length.

2.4. When the source/drain separation is smaller than twice the Debye length, the channel conductivity is controlled by the source/drain contacts, and the back-gate gradually loses the ability to modulate the current, as seen in the FET characteristics in Figs. 2 & 4.

2.4 Influence of metal/graphene coupling on 2D graphene transistor

Recently, many experimental studies point out that metal contacts could play a critical role on device performance of graphene transistors (Lee et al., 2008; Mueller et al., 2009). When graphene is covered by a metal electrode, the Fermi level of the underlying graphene will be shifted according to density functional theory calculations (Giovannetti et al., 2008) and have been observed in experiments (Huard et al., 2008; Nagashio et al., 2009). Here, we explore the effects of metal contacts on the operation and scalability of 2D graphene FETs using detailed numerical device simulations based on the non-equilibrium Green's function (NEGF) formalism in the ballistic limit. Our treatment of metal/graphene contacts captures (1) the doping effect due to the shift of the Fermi level at graphene contacts; (2) the Metal-Induced-States (MIS) in the graphene channel.

The modelled device is shown in the inset of Fig. 5 (left). The channel is assumed to be uniform graphene with width of 150 nm. At w = 150 nm, the current density is similar to analytical 2D results, which implies an effective 2D approximation. The contact regions are assumed to be semi-infinitely long. The top gate insulator is t_{ox} = 1.5 nm thick with a dielectric constant ε_{ox} = 20. The bottom gate dielectric SiO2 is 50 nm thick, and the Si substrate is connected to the ground. The simulated area is only the channel (dashed rectangle) with fixed boundary conditions at the source and the drain. The major challenge in experiments is the aforementioned degradation of graphene channel mobility due to the top high-κ gate dielectric; in this theoretical evaluation, a ballistic model has been assumed to describe the charge transport in the graphene channel for simplicity.

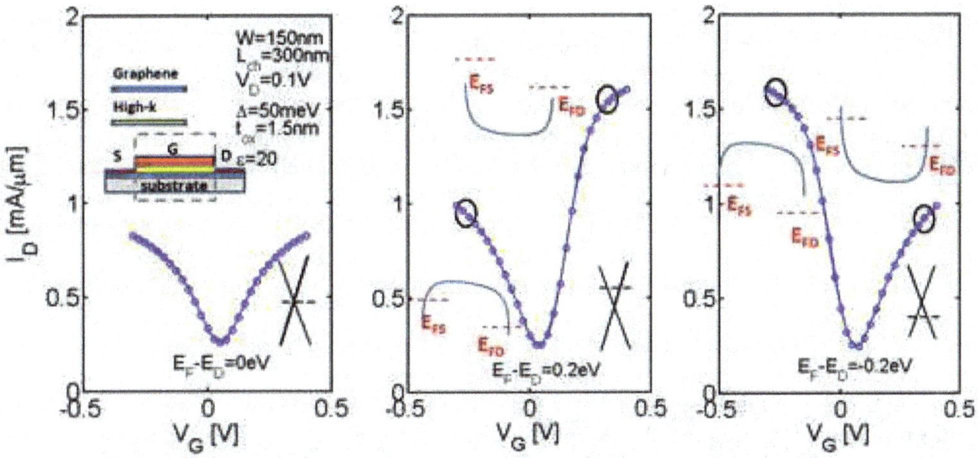

Fig. 5. (inset) Structure of the modelled device. (left) Without any shift of Fermi level in the contact region, symmetric transfer characteristics can be seen. When the Fermi level of contact regions of graphene is shifted due to the metal by E_F-E_D = 0.2 eV (center) and by E_F-E_D = -0.2 eV (right), asymmetric transfer characteristics are observed.

The channel transport is solved using a mode-space based non-equilibrium Green's function formalism. A mode space approach significantly reduces the computational cost while still maintaining the accuracy, as already demonstrated in simulations of silicon (Venugopal et al., 2002), carbon nanotube (Guo et al., 2004), and GNR (Zhao & Guo, 2009) FETs. In our calculation, we assume that the potential variation along the channel width is negligible. Based on this approximation, the electrostatics is solved in 2D, and the mode space method yields accurate results. The channel potential is computed using the 2D electrostatic solver, which also capture the shift of Fermi level at graphene contact. The details of the electrostatic solution procedure are discussed in Zhang et al., 2010. The effect of contacts on the channel is captured through the contact self-energy $\Sigma_{S,D} = \tau g_S \tau$, where τ is the coupling matrix between the contact and channel, and g_S is the surface Green's function of the contact, $g_S = [(E+i\eta+i\Delta)I-H_{contact}]^{-1}$ (Datta, 2005), where η is typically an infinitesimal quantity, and Δ describes the metal/graphene coupling (Golizadeh-Mojarad & Datta 2009; Nemec et al., 2008). By iterating between the NEGF transport solver to find the charge density and the 2D electrostatic solver to find the self-consistent potential, a self-consistent solution is obtained. The source/drain current is then computed from $I_{DS} = (2e/h)\int T(E)(f_S(E)-f_D(E))dE$, where T(E) is the source/drain transmission coefficient.

Fig. 5 shows the calculated transfer characteristics at V_{DS}=0.1 V with different Fermi level shifts of the contact regions of graphene. The metal/graphene coupling strength is Δ = 50 meV. The effects of different Δ's are discussed later. If the contact does not introduce any Fermi level shift as in Fig. 5 (left) E_F-E_D = 0 eV, the transfer curve is symmetric. The minimum conduction point locates at $V_{GS} = V_{DS}/2$. Fig. 5 (right) shows a typical case (Au contact), when the value of the Fermi level shift is E_F-E_D = -0.2 eV (Giovannetti et al., 2008). At negative V_G, carriers can directly go through the channel without experiencing any barriers (see schematic potential profile of Dirac point in Fig. 5 (right)). However at positive V_G, electron transport through the channel will be suppressed by the barrier. Thus the positive current branch is reduced compared to the negative current branch. The complete transfer curve shows a clearly asymmetric behaviour. We point out that the minimum conduction point gate voltage is also shifted due to the barriers at contacts and I_{on} also increases with contact doping effect. When E_F-E_D = -0.2 eV as shown in Fig. 5 (center), an effective n-type doping is introduced by the metal contacts. A similar asymmetric behaviour with positive current branch larger than negative current branch is expected in that case. We point out that even though previous studies have described an asymmetric transfer characteristics in 2D FETs due to electrostatically doped graphene p-n junctions (Low et al., 2009), metal/graphene contacts themselves may also lead to asymmetric conduction as experiments have indicated (Huard et al., 2008; Nagashio et al., 2009).

Fig. 6 shows a comparison of the effect of different coupling strengths Δ at large drain bias (V_{DS}) voltages. The rigorous explanation of the values used for Δ is related to the metal/graphene hybrid system, which is beyond the focus of this study. For a comparison, we define three Δ values here, Δ = 0 eV for the intrinsic graphene case, Δ = 8 meV for weak metal-graphene coupling, and Δ = 50 meV for strong coupling. The first interesting feature that emerges is that the current drive *increases* as Δ increases. Higher Δ induced broadening

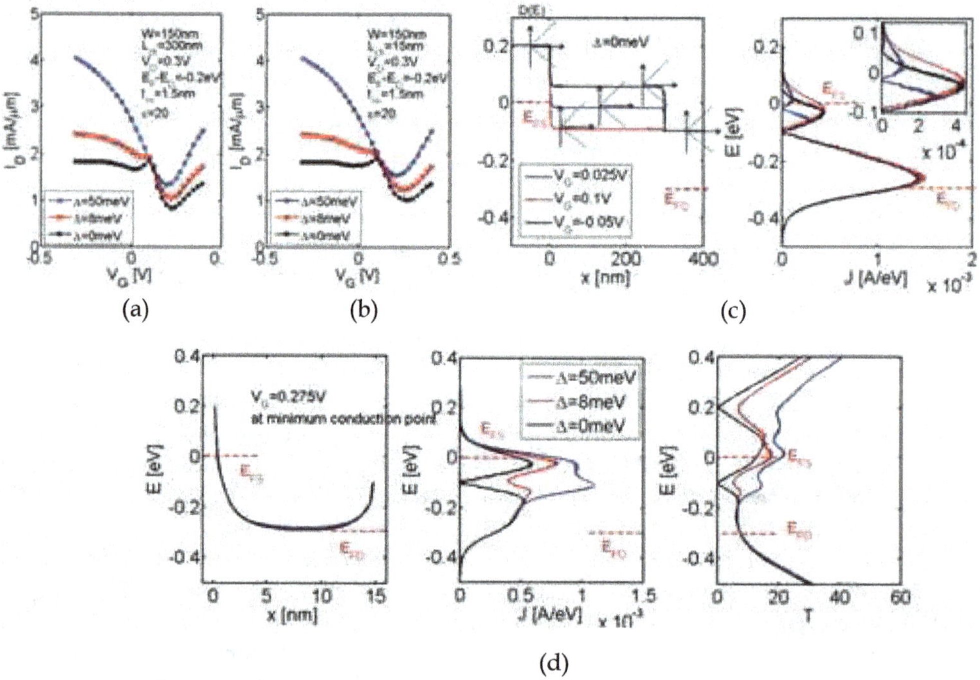

Fig. 6. (a) and (b) Effect of different coupling strength Δ at V_{DS} = 0.3 V. At about 0.2 V away from the minimum conduction point a distortion appears when metal/graphene coupling is weak (black and red lines). (c) shows S/D and channel DOS (cartoon) in series decide how total carriers transport through the channel. As Δ increases, DOS of graphene contact are broadened and distortion disappears. (d) For L_{ch} = 15 nm at V_G = 0.275 V, the drain Dirac point is between E_{FS} and E_{FD}, large Δ broaden DOS at drain contact, which increase the $T(E)$and total current.

of the DOS of graphene contacts increases the transmission coefficient $T(E)$. Larger $T(E)$ leads to higher on current. When $\Delta = 0$ eV (black curve in Fig. 6 (a)), beside the primary minimum conduction point, a distortion appears. The distortion corresponds with the minimum density-of-states (DOS) at the Dirac point of graphene. As shown in Fig. 6 (c), source, drain and channel contribute three cones in series, which decide states allowed for transport through the channel. At $V_G = 0.025$ V, both electron and hole cones of the channel graphene are involved, which gives a local minimum. When $V_G = -0.05$ V, carriers pass the channel through the channel hole half cone. In this case, the DOS minimum point of the channel is not in the transport energy windows between E_{FS} and E_{FD}, thus current increases compared with the local minimum case. When $V_G = 0.1$ V, carriers move through the channel electron half cone and current also increases. When the metal/graphene coupling strength increases, the DOS of the graphene contact regions are broadened. The current monotonically changes with V_G and the distortion disappears. A recent experimental work reports the presence of this type of distortion before annealing, and the disappearance of the distortion after annealing (Nouchi & Tanigaki, 2010). I_{min} also shows a dependence on Δ here. Fig. 6 (d) shows the minimum conduction point for $L_{ch} = 15$ nm. The drain Dirac point is between E_{FS} and E_{FD}. The broadened DOS at the drain contact increases the total current density. $T(E)$ along the channel Dirac point at about -0.3 eV does not show a dependence on different coupling strengths Δ. Thus the increase of I_{min} corresponds to DOS broadening in the contacts, and not the MIS in the channel. We point out that when $V_{DS} > -(E_F-E_D)/q$ the minimum current will increase with higher metal/graphene coupling.
The performance of the FET upon scaling of the channel length is a crucial point, which is discussed in Fig. 7. As mentioned earlier, our model assumed a high-κ insulator to avoid short channel effect. I_{on} remains almost the same at different channel lengths, but I_{min} increases with short L_{ch}. As discussed in Fig. 6, at short channel lengths, I_{min} is mainly decided by direct source/drain tunnelling and the effect of MIS in the channel is weak. Here, when L_{ch} is reduced from 150 nm to 40 nm, the slight increase of I_{min} corresponds to the increase in the probability of source/drain tunnelling. When $L_{ch} = 15$ nm, source/drain tunnelling becomes more severe, leading to higher I_{min}.

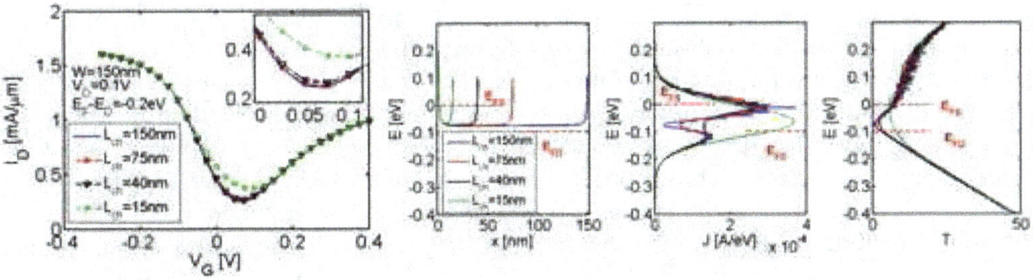

Fig. 7. Effect of channel length scaling at $V_{DS} = 0.1$ V. I_{on} stays the same when L_{ch} scaling down. As L_{ch} reduces I_{min} shows a slight increasing due to the direct source/drain tunnelling and at $L_{ch} = 15$ nm it becomes more severe thus I_{min} increase about 2 times.

2.5 High field transport model of 2D graphene

In this section, we will answer two questions: what is the value of the saturation velocity in graphene and what are the dominant scattering mechanisms under high field. We will discuss the role of all scattering mechanisms in high field transport. Simple models are used to estimate the saturation currents in graphene and also compared with the Ensemble Monte Carlo simulation (EMC) results.

The scattering rates of both elastic and in-elastic mechanisms at 300 K are shown in Fig. 8 (a). Two dominant modes of SiO_2 SO phonons, 60 meV and 148 meV, are considered in the plot. In order to capture high field effects, the scattering rates for 0 to 1 eV electron energies are calculated for all the scattering mechanisms. At high field, electrons are accelerated by the electric field and populate high energy states. Thus, the scattering rates at high energy states determine the current saturation. The intrinsic optical phonon scattering rate increases linearly with energy, whereas SO phonon scattering rate 'saturates' for high energy states. The impurity scattering rate decreases with the energy and is negligible at high energy. Although the acoustic phonon scattering rate increases with energy, it is smaller than the optical phonon scattering at high energies. In Fig. 8 (a), the bar plot shows the relative magnitudes of scattering rates at E = 0.6 eV. From these observations, optical phonon scattering is dominant for graphene under high field. In exfoliated graphene the substrate temperature can be heated to as high as 1000 K so that the 60 meV mode SO phonon scattering is more than one magnitude higher than at room temperature. As a result, this low energy mode plays a dominant role in energy relaxation of electrons at high temperature. Optical phonon emission is not the dominant energy relaxation mechanism in those situations. On the other hand, in epitaxial graphene grown on SiC substrate, the lowest SO phonon mode energy is 116 meV (see Section 2.2), which is higher than that in SiO_2 and close to the intrinsic optical phonon energy of graphene. The scattering rate is also lower than that in SiO_2 due to the weak polarization of Si-C bonds. Furthermore, the thermal conductivity of SiC is much higher than SiO_2. Thus, the high field transport property in epitaxial graphene is limited by its intrinsic optical phonon scattering. From Fig. 8 (a), the phonon scattering at high energy is more than 10 ps^{-1}. Optical phonons generated by electrons have to relax to other phonon modes before the heat can propagate into the substrate or contacts. The characteristic lifetime of optical phonon decay into acoustic phonons is around 1 to 5 ps in carbon-based sp^2 crystals. Since optical phonon lifetime is much longer than phonon generation time (<1 ps), the optical phonon population in graphene is out of equilibrium. The fast generation rate and slow decay rate of optical phonons create the non-equilibrium phonon effect in graphene.

To begin, we assume the phonon scattering is instantaneous and ignore all other elastic scatterings. Under this assumption, two possibilities of electron distribution functions are shown in Fig. 8 (b). The first model assumes that the Fermi circle keeps the shape but drifts rigidly along the electric field direction. Optical phonon emission empties the high energy states and fills the low energy states. As such, the instantaneous phonon emission stops further movement of the Fermi circle and a steady state distribution is achieved. The difference of highest energy state (dotted circle) and lowest energy state (solid circle) is exactly the optical phonon energy $\hbar\omega_0$. Fig. 8 (b) shows another possibility of the steady state distribution function. The energy difference between these dotted and solid circles is the optical phonon energy. Optical phonon emission by electrons in states between the two circles is prohibited by the Pauli exclusive principle. Optical phonon emission is allowed only for electrons approaching the dotted energy circle. The optical phonon emission continues to scatter the electrons from high energy states back to the low energy states. As a result, the Fermi circle is squeezed due to the Pauli principle. Both of the models capture

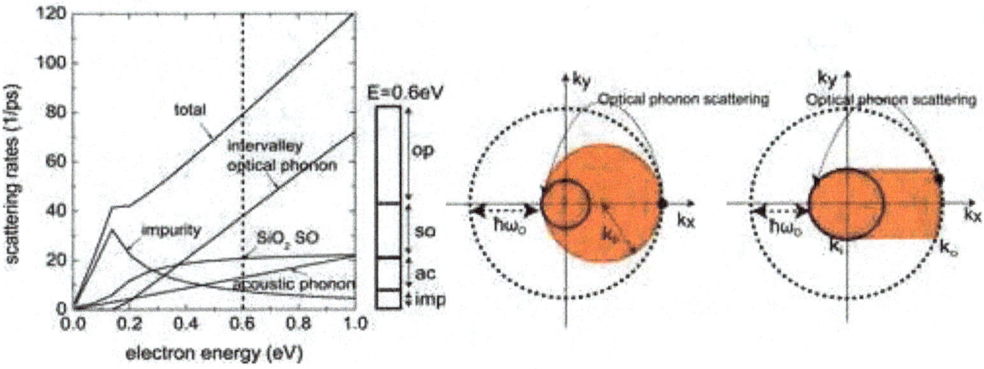

(a) (b)

Fig. 8. (a) Scattering rates vs. energy in 2D graphene: impurity density $n_{imp} = 5{\cdot}10^{11}\,cm^{-2}$, T=300 K. The bar shows the relative scattering rates at an energy of 0.6 eV. (b) Current saturation model in 2D graphene: (left) The drift Fermi circle model. k_F is the Fermi wave vector. (right) Squeezed distribution function model. The outer circle radius is k_o and the inner circle radius is k_i.

the effect of carrier degeneracy, which has not been discussed in non-degenerate conventional semiconductors.

The calculated saturation currents versus carrier density for the two models are shown in Fig. 9 (a); the result of the two models are very similar. Both of the models give a roughly square root relation of saturation current versus carrier density. From $J_{sat} = en\,v_{sat}$, the ensemble saturation velocity v_{sat} is proportional to $n^{1/2}$. Saturation velocity in conventional semiconductors is typically a constant, e.g. $v_{sat} = 10^7$ cm/s in Si. The primary reason for the difference is the degeneracy of carriers and the Pauli principle. Non-degeneracy of carriers leads to constant saturation velocity and fully degenerate carrier leads to a square root relation for 2D systems. This theoretical model is in good agreement with the measurement for graphene on SiO_2 substrate (Dorgan et al., 2010). The agreement of the model and measurement results indicates that the carriers are degenerate in such graphene devices under high field.

The analytical models capture the carrier degeneracy effect but ignore the rise of carrier temperature due to the applied electric field. In reality, the optical phonon emission rate is not instantaneous, so that carriers have opportunities to populate high energy states giving rise to a tail of the distribution function at high energies. At the same time, other scattering mechanisms spread the carrier distribution over a wider range in k space. Moreover, electron electron (e-e) scattering is strong in graphene at high carrier densities, causing carriers to thermalize to a Fermi-Dirac distribution with a finite temperature. In order to capture all of these effects, we applied an electron temperature approximation to high field transport. Since the carriers are degenerate in graphene, a shifted Fermi-Dirac distribution function is assumed, rather than the Boltzmann distribution used for conventional semiconductors (Lundstrom, 2000).

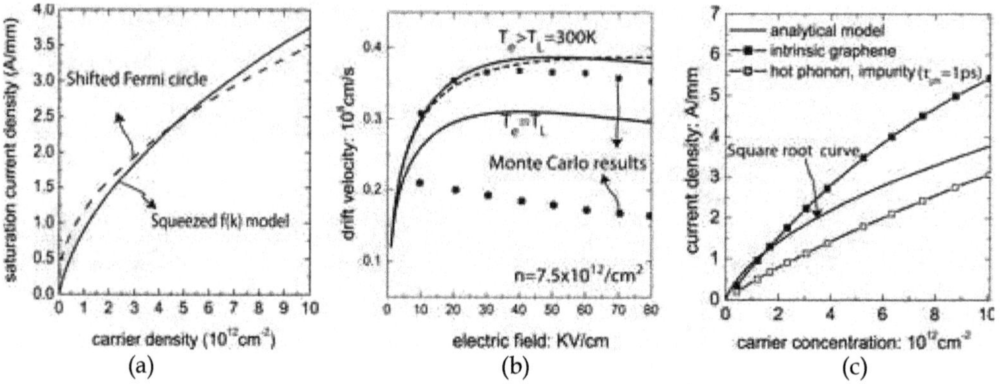

Fig. 9. (a) Current vs. carrier density of the two models. (b) Velocity vs. electric field based on electron temperature approximation. Solid lines: (upper) $T_e > T_L = T_{op} = 300$ K and (lower) $T_L = T_e$, $T_{op} = 300$ K; Dashed line: $T_{op} = T_e$; $T_L = 300$ K; Squares: MC simulation, $T_{op} = T_L = 300$ K; Circles: MC simulation, with hot phonon $\tau_{ph} = 5$ ps, $T_L = 300$ K. (c) Current vs. carrier density. Solid line: analytical current saturation model; Solid square: no hot phonon, $T_L = T_{op} = 300$ K, $F = 2{\cdot}10^4$ V/cm; Open square: with hot phonon, $T_L = 300$ K, $F = 8{\cdot}10^4$ V/cm, $\tau_{ph} = 1$ ps;

The 2D electron gas, acoustic phonon and optical phonon are not fully in thermal equilibrium with each other at high field. The acoustic phonon temperature T_L is the lattice temperature. The optical phonon temperature T_{op} describes the optical phonon vibrations. In Fig. 9 (b), the upper solid curve shows the currents by assuming lattice temperature is in equilibrium with environmental temperature, $T_e > T_L = 300$ K. The current decreases if assuming the graphene lattice temperature increases with electron's temperature, $T_L = T_e$, which is shown in the lower solid curve. In these two curves, the optical phonon temperature is still assumed to be in equilibrium with room temperature, $T_{op} = 300$ K. From the comparison of the two solid curves, the current drops substantially due to lattice heating. This indicates that acoustic phonon scattering could limit the current if graphene is heated to high lattice temperature. The Monte Carlo simulation results (square and circle symbols) are also shown on Fig. 9 (b). The black square dots are in good agreement with the model prediction. This indicates that the electron temperature approximation is a reasonable model for high field transport in graphene. The optical phonon temperature T_{op} rises due to the emission of phonon by electrons. The effect of optical phonon temperature increase is considered in the dashed curve in Fig. 9 (b), in which the optical phonon temperature is in equilibrium with electron's temperature, $T_{op} = T_e$. However, the current does not show obvious degradation. This indicate that temperature rise of optical phonons has less impact on the current compared to lattice temperature. On the other hand, the squares considering non-equilibrium (hot) optical phonon show more degradation of the current. This finding indicates that the non-equilibrium phonon is dominant in high field transport in graphene.

In Fig. 9 (c), the saturation current versus carrier density curves are shown. The solid black curve shows the analytical result from the squeezed distribution function model for comparison. The square-line is the intrinsic saturation currents of grapheme without hot phonon effect. In the absence of hot phonons and impurity scattering, the current-field curves show strong negative differential resistance (NDR). The currents of the squares are at a field of

$F = 2{\cdot}10^4$ V/cm. The saturation current is lower than the analytical result (solid line) when hot phonon exists in grapheme (open square line). In MC simulation, we also find that the saturation currents vary with the phonon lifetime. The longer is lifetime of phonon the lower is the saturation current. At high field, the impurity scattering has an effect on saturation current, but it is minor compared to the hot phonon effect. Saturation current density around 3 A/mm has been measured in epitaxial graphene on SiC substrate (n~10^{13} cm^{-2}) (Moon et al., 2009), which is consistent with our simulation result considering hot phonon effect (open square line). At low carrier densities, J_{sat} as a function of n curves are approximately linear; the slopes correspond to the ensemble saturation velocities. The saturation velocity thus extracted for intrinsic graphene (neglecting charged impurity scattering and hot phonon effect) is $5.2{\cdot}10^7$ cm/s and $2.7{\cdot}10^7$ cm/s when hot phonons are considered. From the simulation results, the hot phonon effect determines the saturation currents in graphene on substrates without strong SO phonon coupling. The (semi-) analytical models best serve as a rough guideline, since elastic scattering processes, and especially the hot phonon effect are not captured in the model, which is done in the EMC approach.

2.6 2D graphene FETs based on CVD grown graphene on Cu

Availability of high-quality large-area graphene on suitable handling substrates will ultimately determine the success of graphene as a device platform. Recently, the formation of single layer graphene via catalyzed-chemical vapour deposition (CVD) on copper foils has generated a rapidly growing body of research (Li et al., 2009). A few groups have reported the CVD growth of graphene on copper, and transport properties including quantum Hall effect (Levendorf et al., 2009; Bae et al., 2010) in layers subsequently transferred to insulating substrates. So far, mobility in Cu-CVD based graphene transferred to SiO_2/Si wafers has been measured around 4 000 cm^2/Vs while mobility of exfoliated graphene on SiO_2/Si has been demonstrated to exceed 15 000 cm^2/Vs. The reduced mobility for Cu-CVD graphene could come from the quality of the graphene itself (eg. lower uniformity, defects, small domain size) or from contamination introduced during the wet transfer process. There have been no careful studies of FETs fabricated from CVD grown graphene. In this work, we report the characteristics of single-layer graphene FETs whose channels were grown by CVD on copper.

We grow Cu-catalyzed CVD graphene in a quartz tube furnace under flow of CH_4 and H_2 with an Ar background on copper foils and on evaporated copper films. Pre-growth chemical and thermal treatment of the copper catalyst material enables single layer graphene deposition at temperatures ranging from 800 - 1050 °C (Fig. 10). For our optimized growth step, the chamber conditions are: CH_4:H_2 gas flow of 100:20 sccm, pressure of 2 Torr, and a temperature of 1000 °C. The gas and pressure are kept at the growth conditions during cooldown until the system is below 500 °C when the growth gases are turned off and an Ar flow is introduced. Fig. 10 shows optical and SEM images of as grown graphene on a copper foil substrate. Following growth, graphene sheets are transferred to a range of host substrates (typically quartz or silicon) using wet etch methods similar to those described in other work (Li et al., 2009). Fig. 11 (a) shows an optical image of a graphene sheet following removal of the PMMA used to support the graphene during wet transfer. Large regions of clean graphene are apparent along with areas of visible contamination from processing. Improvement of the transfer process is underway to reduce contamination introduced during processing.

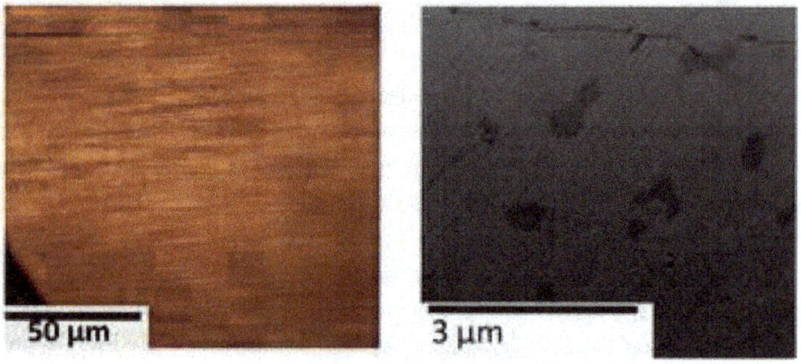

(a) (b)

Fig. 10. (a) Optical image and (b) SEM image of as grown graphene on Cu foil.

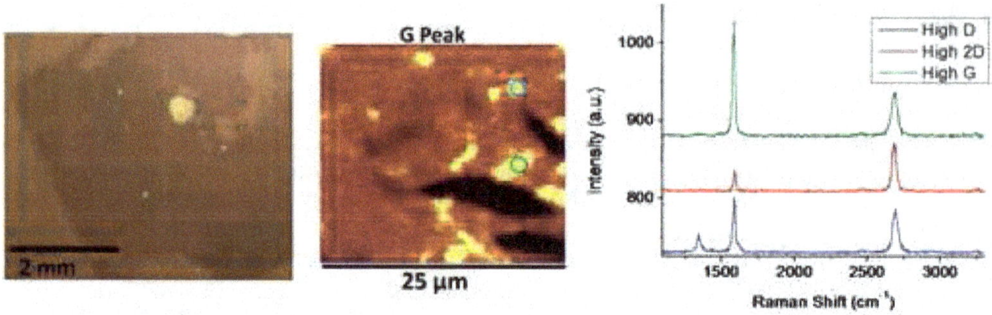

(a) (b) (c)

Fig. 11. (a) Optical image of as graphene following wet transfer to SiO_2/Si substrate. (b) Scanning Raman image of G peak of transferred graphene. (c) Raman spectra at indicated points from the G peak scan. The middle spectrum is representative of the main field of the graphene sample with characteristic 2D:G ratio of 2 of single layer graphene.

Following transfer, Raman mapping is used to determine the quality and uniformity of the CVD-graphene. Large area single-layer graphene regions are verified from the signature 2D:G peak ratio of ~2, that is uniform over large regions (Fig. 11 (a)). Complementary SEM and AFM characterization corroborate the Raman result. Fig. 11 (b) shows a map of the G peak intensity across a 25 μm x 25 μm field and the individual Raman spectra from the three circled locations are in Fig. 11 (c). The middle (red) spectrum is found in most of the graphene region consistent with the presence of high-quality single-layer graphene as indicated by the absence of a D peak at 1350 cm^{-1} and the characteristic 2D:G peak ratio > 2. The bright spots in the G peak scan have two types of associated spectra. The bottom (blue) spectrum is from a region with high defect peak intensity suggesting either contaminated or defective regions of graphene. The top (green) spectrum has a low defect peak intensity indicating high quality graphene, but has a low 2D:G peak ratio consistent with more than one layer of graphene.

Back-gated FETs were prepared by transferring the single-layer graphene to 300 nm SiO_2/conductive Si substrates, followed by conventional e-beam lithography, O_2 plasma etching of the graphene for device isolation, ohmic metal evaporation (Cr/Au), and lift-off.

The backside of the conductive Si wafer was metalized for back gating. The channel lengths of the FETs range from 1 - 8 μm and the channel widths vary between 1 - 4 μm. Based on the transfer characteristics (Fig. 12 (a)), the field-effect mobility is found to be ~4500 cm^2/Vs for electrons and slightly lower for holes. The Dirac point is at V_{BG} ~2 V, much closer to zero volt than in case of exfoliated graphene sheets, indicating close to equal amount of positive and negative charged contamination on the surfaces. The impurity concentration is estimated from the measurement to be ~$8{\cdot}10^{10}$ cm^{-2}. The devices exhibit field-effect current modulations of ~4 - 6 at 20 mV V_{DS} at gate sweeps of +/-30V (Fig. 12 (a)). Interestingly at very low applied source-drain bias (V_{DS} = 20 μV) the modulation increases significantly (Fig. 12 (b)). The current does not scale linearly; ~1000X lower bias causes less than ~10X smaller current. The apparent field-effect mobility is ~200 000 cm^2/Vs, but we stress that this is not a fundamental property of graphene, it just indicates the limitation of the traditional model to estimate the mobility of high purity graphene samples.

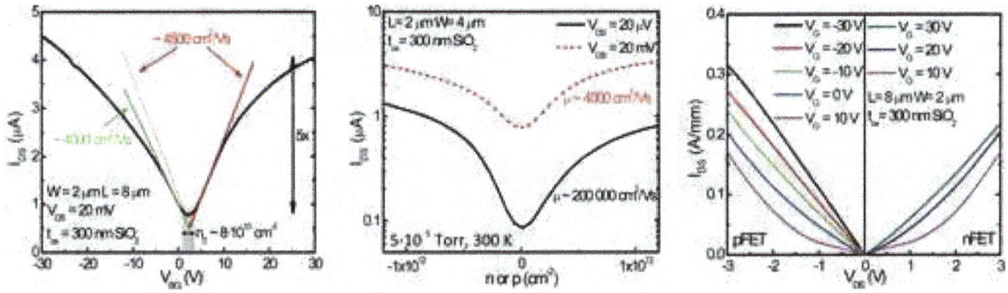

(a) (b) (c)

Fig. 12. (a) Transfer characteristic of a 2D graphene FET transferred to Si/SiO_2 substrate. The calculated impurity concentration is 8.10^{10} cm^{-2}. (b) Very low (V_{DS} = 20 μV) applied drain bias gives an extremely high apparent field-effect mobility. (c) High-field family I-V curves showing both sub-linear and super-linear behaviour.

At high drain bias voltages (Fig. 12 (c)), the DC output characteristics of the devices were linear up to ~1 V V_{DS}, and deviated significantly from linearity at higher drain biases. The super-linear behaviour is especially evident at gate biases close to the Dirac point, and is indicative of zero-gap single-layer graphene.

2.7 Epitaxial graphene RF FETs on SiC for analog application

While DC performance has been extensively studied just a few small signal characterizations have been done so far. The widely observed high carrier mobility naturally focuses our attention towards high frequency performance of graphene based devices. Gigahertz operation of exfoliated (Lin et al., 2009) and epitaxial (Moon et al., 2009) graphene FETs has been reported. Here we present the high frequency performance of graphene devices based on epitaxially grown graphene on SiC substrates. The measured $f_T{\cdot}L_G$ product of 8 GHz·μm and f_{max} of 16 GHz is in very good agreement with previously reported values (Moon et al., 2009). Surprisingly, in spite of the much lower electron mobility and transconductance, these FETs demonstrate remarkable small signal performance comparable to the small signal performance of higher mobility exfoliated graphene devices. Recently significant improvement was reported: device with $f_T{\cdot}L_G$ = 24 GHz·μm (Lin et al., 2010).

Epitaxial graphene FETs were fabricated on 2-3 layer graphene on Si-face 4H-SiC. According to AFM and LEEM characterization and Raman measurements the graphene thickness is 1.9 layers in average over the wafer (Luxmi et al., 2009). Graphene is patterned by optical lithography and etched it in O_2 plasma. Using the same resist-pattern the SiC is etched 100 nm deep in CF_4 plasma to facilitate the sticking of the metal contacts on the surface. Cr/Au source/drain contacts and e-beam evaporated Al_2O_3/Ti/Au top gate (t_{ox} = 15 nm) have been deposited to form field effect transistors (FETs). For increased RF performance short gates were fabricated using e-beam lithography. The channel lengths of the devices range from 1-4 μm and the gate length is 1 - 2 μm in optically defined and 40 - 500 nm in e-beam patterned devices. Standard ground-signal-ground probing pads are lithographically realized for the gate and drain. Open structures were used to de-embed the signals of the parasitic pad-capacitance.

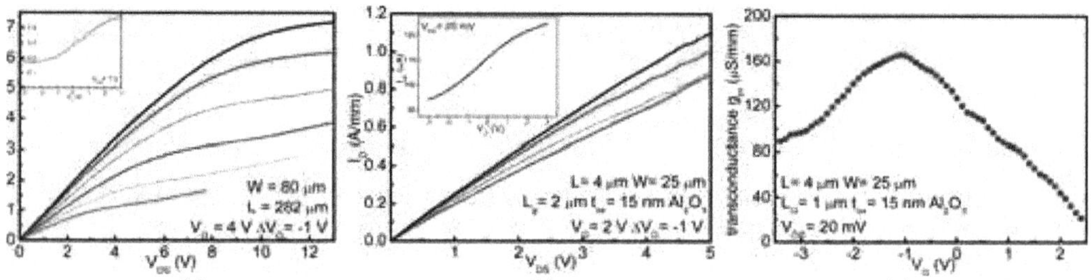

(a) (b) (c)

Fig. 13. Output characteristics of a saturating large area graphene transistor (a) and the graphene transistor (b), which RF characteristic is shown on Fig. 14. Insets: transfer characteristics at V_{DS} = 1 V (a) and at V_{DS} = 20 mV (b). (c) Transconductance as a function of gate voltage at V_{DS} = 20 mV.

Hall measurements reveal μ ~ 200 – 500 cm^2/Vs throughout the sample showing smooth variation and better performance in the middle of the wafer. The DC output characteristics of the devices were linear up to 5 V and slightly sublinear above 10 V. The drain current had a weak gate dependence as we swept the gate voltage between +/- 2.5 V. In the high source-drain bias range the gate modulation decreased indicating a non-reversible degradation mechanism by high electrical field. The transconductance of the devices varied between 120 – 400 μS/mm, which is significantly lower than in case of exfoliated devices having the same gate length. For very long gate lengths, reasonable current modulation and current saturation was observed as shown in Fig. 13 (a). However, with scaling of the gate length, the gate modulation and current saturation properties degraded substantially, indicating the need for improvement of material quality and gate material processing. In spite of the weak gate modulation, it is possible to envision usage of graphene FETs in low-noise amplifiers (LNAs) if RF performance improves. In addition, single layers of almost transparent graphene may be transferred to flexible insulating substrates, which can offer low-cost high frequency performance. Small signal performance of devices with gate-lengths between 2 μm and 0.5 μm on epitaxially grown graphene were measured. The used bias-conditions for the FETs are V_{DS} = 10 V and V_{GS} = 2 V.

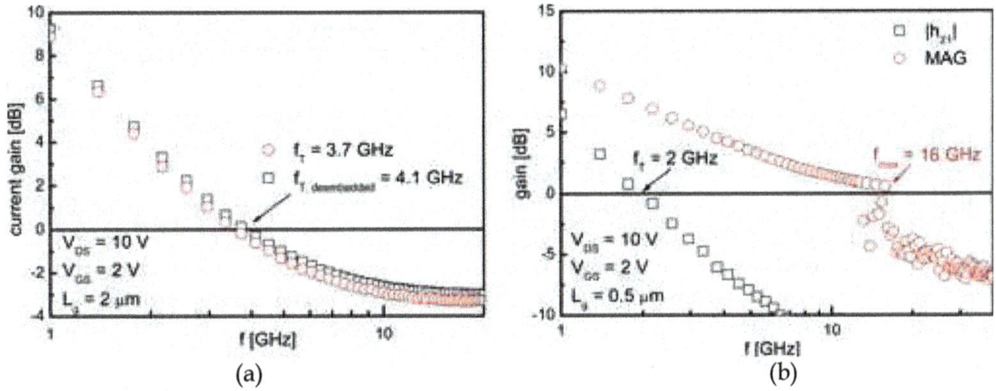

Fig. 14. (a) Current gain as function of frequency for a graphene FET with 2 μm long gate. (b) f_T and f_{max} of a graphene transistor with 0.5 μm long ebeam-defined gate.

A deembedded current gain cut-off frequency f_T of 4.1 GHz for devices with 2 μm long gates was achieved (Fig. 14 (a)). For a graphene FET with a 0.5 μm long gate an exceptional high power gain cut-off frequency f_{max} of 16 GHz is demonstrated. (Fig. 14(b)). These preliminary results are very encouraging and demonstrate the potential of the epitaxially grown graphene devices for high-frequency applications. As the quality of graphene on insulating substrates improved over time indeed cut-off frequency higher than 100 GHz was demonstrated (Lin et al., 2010). Moving from zero-bandgap single-layer 2D graphene to bilayer graphene or graphene nanostructures can enable controlled bandgaps. We present our efforts to date on the latter approach in the following section.

3. Graphene nanoribbon transistors

3.1 Band gap of graphene nanoribbons

Two-dimensional (2D) graphene sheets are nearly metallic, while ultrathin graphene nanoribbons (GNRs) can show semiconducting properties with the energy bandgap scaling inversely with the ribbon width. The achievable energy bandgap, superior transport properties, and the planar manufacturability establish GNRs as promising cornerstones beyond the Si CMOS technology (Zhang et al., 2008; Murali et al., 2009). Although sub−10 nm GNRs have been demonstrated by chemical approaches (Li et al., 2008; Jiao et al., 2009), the ability to form GNRs lithographically will facilitate the fabrication compatibility with the conventional planar integrated circuit (IC) manufacturing technology. GNRs were fabricated on exfoliated graphene flakes on SiO_2/Si. The oxide thickness is 300 nm and the Si substrate is heavily doped n-type. Al metal masks were patterned by EBL and e-beam deposition. The exposed graphene was removed by O_2 plasma etching and the metal masks were removed by Al etchant. GNRs connected to two 2D graphene sheets were thus achieved. Fig. 15 (a) shows the SEM images of a 29 nm wide Al strip mask (top) and a 21 nm wide GNR (bottom). The fact that the obtained GNR is thinner than the Al mask indicates the effect of lateral etching in O_2 plasma. Cr/Au was e-beam deposited as source (drain) contacts on the 2D graphene areas to reduce the contact resistance. Al/Au was used as the back-gate contact. The length of GNRs in this work is 2 μm. Current-voltage (I-V) measurements were performed in vacuum (~10^{-6} Torr) with the temperature ranging from 4.2 K to 300 K.

The temperature dependent transfer characteristics of GNRs at V_{DS} = 20 mV are shown in Fig. 15 (b). It is clearly seen that the minimum current decreases with decreasing temperature while the maximum current (due to the hole conduction) remains almost unchanged. As a result, the back-gate modulation increases from ~ 12 at room temperature to > 10^6 at 4.2 K, indicating the formation of energy bandgap. The positive Dirac point indicates the presence of holes, which can be induced by negatively charged impurities, at zero gate bias. It is worth noting that the transfer curve does not show the symmetric V shape, with the hole conduction much stronger than the electron conduction, a phenomenon often seen in GNRs (Chen et al., 2007). This is attributed to surface impurities based on the fact that surface passivation by Al_2O_3 grown by atomic layer deposition (ALD) was found to be able to improve the electron conduction. But the exact origin remains unclear currently and is under investigation. Fig. 15 (c) shows the high field (corresponding to 30 kV/cm at V_{DS} = 6 V) family I-Vs measured at 77 K. The drain current exhibits tendency to saturate at high drain bias and the maximum current density is ~1.4 A/mm. The inset shows I-Vs measured at 4.2 K. The maximum current density reaches 2 A/mm at V_{DS} = 15 V (equivalent electric field: 75 kV/cm), indicating high current drive capability of GNRs. The source and drain electrodes were found to be exchangeable as can be seen from the symmetric I-Vs in the first and third quadrant. This is not unexpected since the back gate controls the whole GNR channel universally.

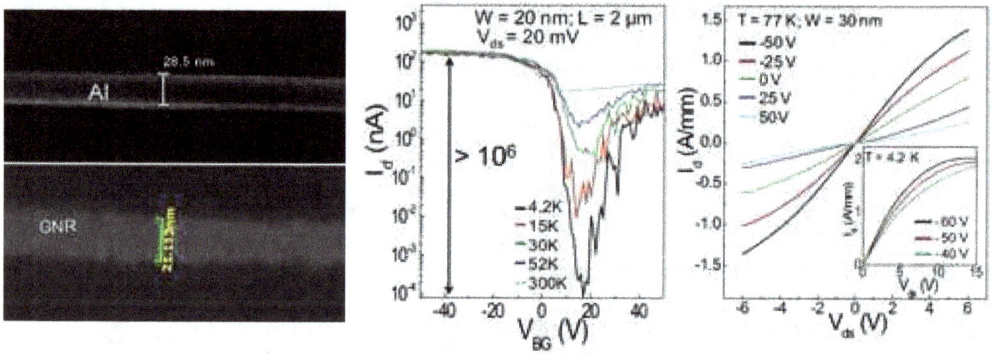

(a) (b) (c)

Fig. 15. (a) SEM image of an Al strip (top) and a GNR (bottom) formed using the Al strip as etching mask. (b) Temperature-dependent transfer characteristics at 20 mV drain bias and (c) high field family I-Vs measured at 77 K and 4.2 K (inset).

The quantum confinement in the quasi-1D GNR channel creates discrete subband energy levels, which are filled by carriers in sequence as the gate voltage increases. As a result, staircase-like features are expected in the transfer curve. Such quantum conductance was observed in the patterned GNRs. The conductance quantization was seen in different GNR devices and Fig. 16 (a) shows a typical transfer curve measured at 30 K plotted in the linear scale exhibiting conductance plateaus with roughly equal spacing. The conductance modulation by the back gate is about 200 at this temperature. GNRs with armchair edges have band structure expressed as:

$$E(k_n, k_y) = \hbar v_f \sqrt{k_n^2 + k_y^2} \tag{1}$$

where v_f (10^8 cm/s) is the Fermi velocity, $\hbar$ is the reduced Plank's constant, and k_y is the wave vector in the GNR length direction. The transverse wave vector k_n can only take allowed discrete values defined at $n\pi/3W$ (n is non-zero integer)(Fang et al., 2008). Therefore, perfect armchair GNRs should have quantized energy levels separated by:

$$\Delta E = \hbar v_f \pi / 3W \tag{2}$$

For GNRs with W = 20 nm, ΔE is calculated to be ~ 34 meV. At zero temperature, only the energy levels below the Fermi level E_f populated with electrons in the conduction band, or those above E_f populated with holes in the valence band, contribute to the conductance. Each 1D energy level (or mode) has a finite transmission probability. The overall device conductance at a finite temperature is the summation of all the available conducting modes, and is described by the well-known Landauer formula:

$$G = \left(\frac{2e^2}{h}\right) \sum_i t_i(E) \left(-\frac{\partial f}{\partial E}\right) dE \tag{3}$$

where t_i is the transmission probability of the i-th mode and f is the Fermi-Dirac statistics. The nearly equal spacing of the plateaus shown in Fig. 16 (a) reveals that each conduction mode should have similar transmission probability. To further analyze the measured conductance quantization using Landauer formula, the back-gate voltage needs to be correlated with the Fermi energy. The link between them is the carrier density in the channel:

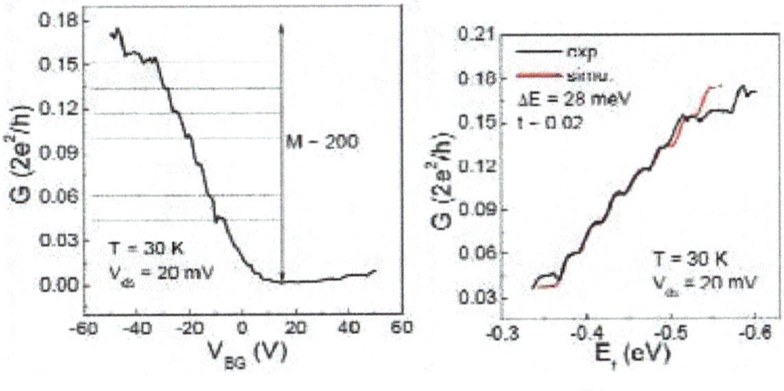

(a) (b)

Fig. 16. (a) a typical transfer curve measured at 30 K showing conductance plateaus and (b) fitting of the experimental data using Landauer formula.

$$C_g(V_{BG} - V_{Dirac}) = E_f^2 / \pi\hbar v_f^2 \tag{4}$$

where C_g is effective gate capacitance per area (F/cm^2). Both sides of the above equation give the carrier sheet density (cm^{-2}) in the GNR channel (Fang et al., 2007). The experimental data were fitted using Eq. (3), with ΔE, t, and C_g as adjustable parameters. Excellent agreement was achieved as shown in Fig. 16 (b). The obtained subband energy separation from the fitting is 28 meV, close to the calculated value (34 meV) for W = 20 nm. The discrepancy could be due to the likely existence of a mixture of both armchair and zigzag orientations in GNRs. The average transmission probability is $t \sim 0.02$, similar to that reported in (Lin et al., 2008). The very small t can be explained by the scattering due to GNR edge/bulk disorder which can severely suppress the conductance causing G much smaller than $2e^2/h$ (Mucciolo et al., 2009).

3.2 Low field transport in GNR

GNR with width down to 20 nm by lithography (see Section 3.1) or 2-5 nm by chemical preparation (Li et al., 2008) has been achieved. The measured carrier mobility (~200 cm^2/Vs) in the ultrathin GNRs are reported to be much lower than in corresponding 2D graphene sheets. GNR opens a bandgap by the edge confinement, but the edges induce line edge roughness (LER) scattering besides the normal scattering mechanisms present in 2D graphene. The LER scattering has W^{-4} dependence on the width of GNR, which is due to the linearity of graphene band structure (Fang et al., 2009). The effect of LER scattering depends on the width of GNR and the height of the roughness, which related to the fabrication processes. The mobility due to various scattering mechanisms was calculated in the Boltzmann transport scheme (Fang et al., 2009). Fig. 17 shows the mobility in a 3 nm wide GNR. The roughness height 0.5nm is used in the calculation. Scattering from surface impurity located at interface of GNR and substrate and line edge roughness (LER) are dominant at low carrier density shown in Fig. 17 (a). However, LER scattering becomes more important at higher carrier density (Fig. 17 (c)). This is due to the increasing of screening of the impurity charge in GNR at high carrier density. At the same time, acoustic phonon scattering also plays important role in limiting the mobility at 300K or higher temperature. The mobility from the theoretical calculation qualitatively explains the mobility measurement results in ultra-thin GNRs (Li et al., 2008).

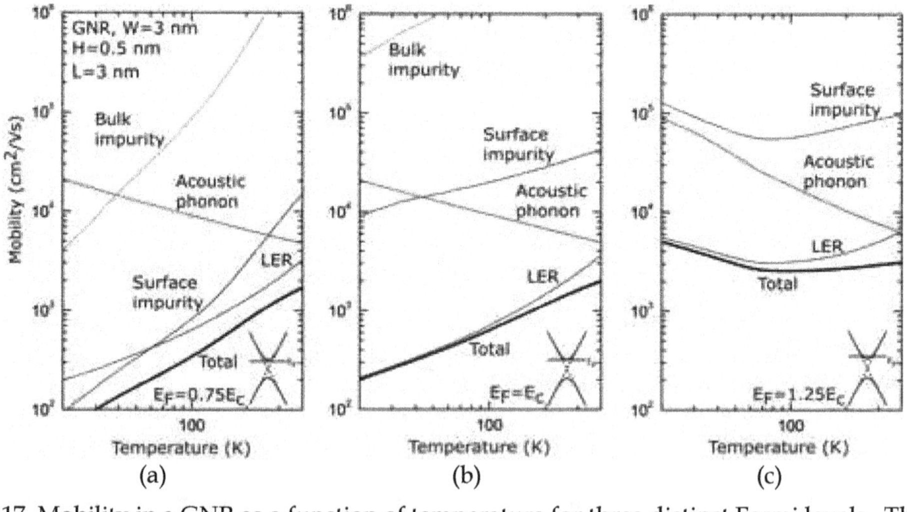

Fig. 17. Mobility in a GNR as a function of temperature for three distinct Fermi levels. The relative importance of the various scattering rates can be ascertained from the plots.

More recently, the LER scattering was investigated elaborately by measuring the carrier mobility in different width GNRs (Yand & Murali, 2010). The mobility is found to decrease at rate W^{-4} when the width is narrower than 60 nm, which indicates the height of LER in GNR fabricated by lithography is larger than 0.5 nm. In SEM the height of LER is estimated to be 2~3 nm, so that the theoretical work and experimental measurements are in agreement. The LER also induces states in the band gap, which enhance the Zener tunnelling probability. This effect has been studied in the tunnelling field effect transistors structure by numerical simulations (Luisier & Klimeck, 2009). The LER increases the off state current, which degrades the performance of FETs by increasing the sub-threshold voltage. The experimental works to achieve thinner GNR with better edges are future tasks to realize devices based on GNRs.

3.3 Graphene nanoribbon p-n junction transistors

For the reason outlined above GNR FET devices with both top- and back-gates were fabricated. The fabrication started by back gated GNR FETs as described in Section 3.1. Then using 1 nm e-beam evaporated Al as a seed layer, 30 nm ALD Al_2O_3 was deposited. The channel length of the fabricated devices is 2 μm and the length of the top gate electrode is about 1 μm, covering half of the channel. The SEM image on Fig. 18 shows a typical FET.

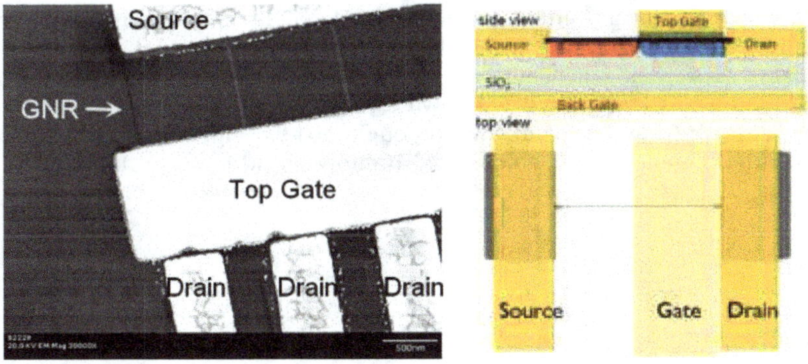

(a) (b)

Fig. 18. (a) SEM image and (b) schematics of the GNR FET. Half of the channel is top-gated, while the whole device is back-gated. Device parameters: L = 2 μm, L_G = 1 μm, W = 30 nm.

In this section the measurements of a representative device at room temperature will be presented. It is found that at high applied source drain current the temperature has very little effect on the device performance, namely at cryogenic temperature the high-field modulation is only ~10% more. Many similar devices were fabricated and similar characteristics were observed in all cases. Fig. 19 (a) and (b) shows the source-drain conductance at V_{DS} = 20 mV and at V_{DS} = 1 V respectively as a function of the top-gate and the back-gate voltage.

The back-gate capacitance is ~11 nF/cm^2 assuming κ = 3.9 and the top gate capacitance is 209 nF/cm^2 calculated based on the measurement of the minimum conductance (Dirac) point shown in Fig. 19 (b). The Dirac point is defined by the ratio of the applied voltages on both gates, the slope defining the value of C_{top}/C_{back} ~19. From either a constant back-gate or constant top-gate slice we can determine V_{TG}^0 = -6 V and V_{BG}^0 = -70 V, indicating strong n-type doping. While the devices before the top gate deposition were slightly p-type doped one can conclude that the strong n-doping is caused by trapped charges in the top gate oxide and other impurities on the graphene oxide interface.

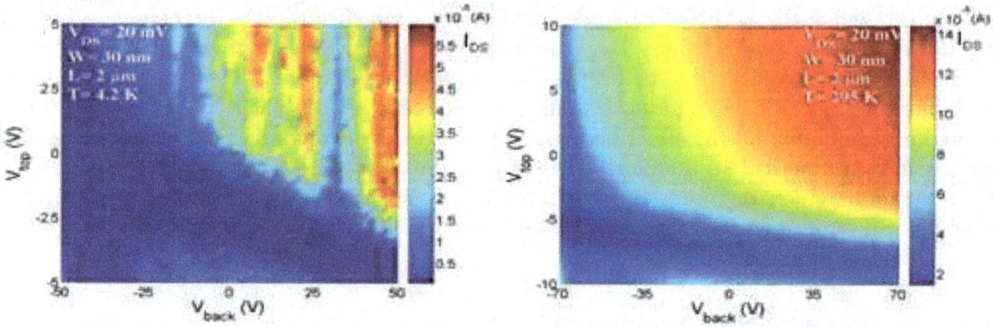

(a) (b)

Fig. 19. (a) Conductance steps at 4.2 K at low source-drain voltage (linear scale); (b) At room temperature the modulation is smaller even at low bias (V_{DS}= 20 mV).

Temperature dependent measurements have confirmed band-gap opening of >26meV depending on GNR widths, and result in 20x and 10^3x modulation at room temperature and 4 K, respectively, by varying the top gate potential between +/- 5V.

The source and drain contacts are in contact with a large 2D graphene region to ensure low contact resistance. The 2D graphene has no bandgap so the Cr/Au metal can form good ohmic contacts easily. The back-gate may vary the carrier concentration of the graphene at the contacts, but due to its large size it will always have much lower resistance than the GNR channel, always supplying the channel with enough carriers. Fig. 20 (a), (b) and (c) shows the measured I_{DS} for different top gate voltages at V_{back} = -70, 0 and +70 V respectively.

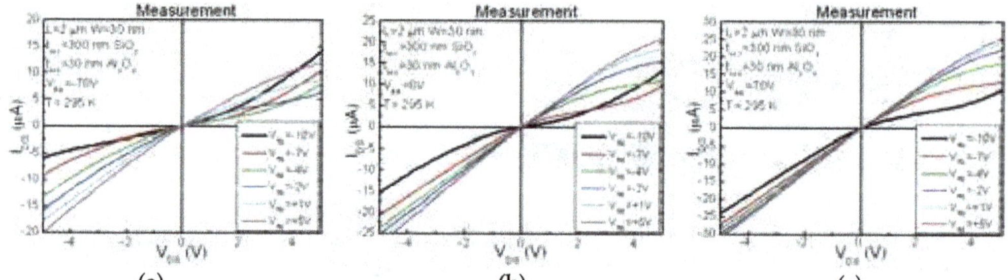

(a) (b) (c)

Fig. 20. I_{DS} versus V_{DS} of the GNR p-n junction FET at -70 V (a), 0 V (b) and +70 V (c) back gate bias. Plateauing of the IV characteristics can be observed in all three cases, but at different bias conditions.

To understand the device operation, examine the device model in detail. The channel of the devices can be separated into two distinct regions: one with a back-gate only, and the other with both top- and back-gates. A model have been developed to explain the transistor characteristics of the p-n channel GNR FET by extending the work of Meric et al., 2008. A field effect model was applied to both side of the junction, defining the only back-gated region similarly as the top-gate having V_{TG} = 0 V:

$$n(x) = \sqrt{n_0^2 + \left(\frac{C_{tg}}{q}(V_{tg} - V_{tg0} - V(x)) + \frac{C_{bg}}{q}(V_{bg} - V_{bg0} - V(x)) \right)^2}, \quad (5)$$

where is x is the distance along the channel and $V(x)$ is the potential in the channel due to the applied source-drain voltage. The current in the channel is expressed by:

$$J(x) = qn(x)\mu F(x), \tag{6}$$

where μ is the mobility and $F(x)$ is the electric field along the channel. The I-V characteristics are obtained by forcing current continuity in a self-consistent electrostatic and transport model:

$$I(V_{DS}) = q\frac{W}{L}\mu \int_0^{V_{DS}} n(x)dV(x), \tag{7}$$

where L is the channel length and W is the GNR width. Carrier drift velocity and mobility saturation is considered depending on the carrier concentration based on the work of Dorgan et al., 2010:

$$v_{sat} = \frac{\omega_{op}}{\sqrt{\pi n(x)}}, \tag{8}$$

where ω_{op} is the optical phonon wavelength of the dominant scattering phonons, which are surface optical phonons of the high-κ oxide. And

$$\mu = \frac{\mu_0}{\sqrt{1 + \left(\mu_0 V_{DS} / L v_{sat}\right)^2}}. \tag{9}$$

For simplicity, we assumed equal mobility and saturation velocity to describe both the electrons and holes. The solution not only provides the I-V characteristics of the device but detailed information about the carrier concentration under various bias conditions. Figs. 20 & 21 show the experimental and modelling data, displaying close agreement. As input parameters in the model we used mobility of 300 cm^2/Vs, a minimum carrier density of $5{\cdot}10^{11}$ cm^{-2} and source-drain series resistance of 0.5 Ω.mm. The saturation of current at high positive bias is caused by depletion of the carriers even in those cases when the gate biases set a high initial carrier concentration. At high negative drain bias on the other hand lack of saturation is observed. This is because the drain bias in this case further increases the carrier concentration instead of depleting. Depending on the gate bias conditions the flattening of the current occurs at different applied source-drain bias. Superlinear current increase is caused by the sharp increase of carrier concentration close to the Dirac point in conjunction with the increase of the accelerating field (V_{DS}).

In conclusion, p-n junction in GNR FETs has been experimentally demonstrated. The analysis of the device operation shows that sublinear and superlinear features observed in the I-V characteristics is caused by the device electrostatics. We point out here that the gate induced barrier at the source end of the GNR FET can be no more than the bandgap. Since the bandgap is small, interband tunnelling prevents the modulation of drain current at high drain biases. Though this is detrimental for a traditional FET, this problem can be turned around and actually used in a GNR based tunnelling FET. Such FETs have been modelled and have proposed GNR based TFETs, which is where current experimental effort is headed. (Zhang et al., 2008).

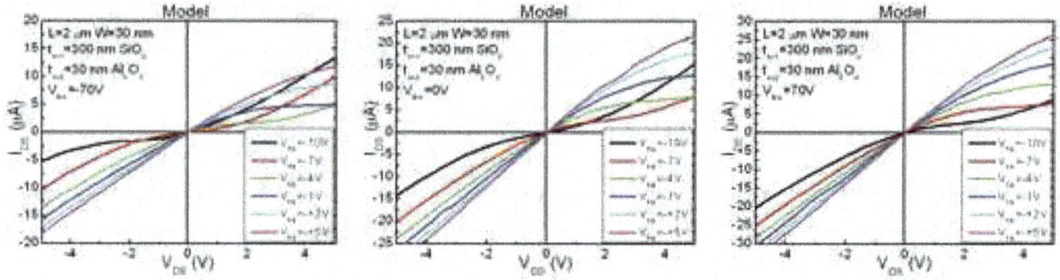

(a) (b) (c)

Fig. 21. Simulated device characteristic at the same bias conditions as the device shown in Fig. 20. I_{DS} versus V_{DS} at -70 V (a), 0 V (b) and +70 V (c) back gate bias.

3.4 Tunnelling in graphene and GNR p-n junction

Tunnelling in 2D graphene attracted a lot of attention due to the new phenomenon, Klein tunnelling of the massless Dirac particle (Stander et al., 2009). From an engineering point of view, graphene/GNR could provide higher tunnelling current if used as a channel in devices, e.g. tunnelling field-effect transistor. The current-voltage characteristics of GNR have been studied in (Low et al., 2009). The tunnelling probability in GNRs or in carbon nanotube (CNT) was found to be similar in magnitude with conventional parabolic semiconductors. However, two facts tilt the tunnelling currents decisively in favour of CNTs and GNRs. First is that the transverse kinetic energy of carriers can be large and leads to an exponential decrease in carrier tunnelling probability in bulk 3D p-n junctions, which is avoided in 1D structures. Second, if normal parabolic-bandgap semiconductors are shrunk to length scales comparable to those of CNTs and GNRs, their bandgaps and effective masses increase further due to quantum confinement. In 2D graphene, p-n junction has been realized in FETs with back and top gates. The conductance across such p-n junction has been studied both theoretically (Low et al., 2009) and experimentally (Huard et al., 2007). Here, we present the full I-V characteristics of p-n junction in 2D graphene. The tunnelling probability in 2D graphene is calculated by WKB approximation (Cheianov & Falko, 2006). At low temperature (T~ 0 K), the tunnelling current density in a biased p-n junction is calculated as

$$J_t = -\frac{e^2 V_a}{\pi^2 \hbar}\sqrt{\frac{E_{fn} + E_{fp} - qV_a}{\hbar \upsilon_F \lambda_{pn}}}, \tag{10}$$

where e is the electron charge, V_a is the applied voltage, E_{fn}/E_{fp} is the doping level at n/p region, λ_{pn} is the screening length of p-n junction (Low et al., 2009). Fig. 22 shows the full I-V characteristics of both abrupt and gradual p-n junctions. The screening length of gradual p-n junction is assumed to be λ_{pn} = 80 nm, which is achievable in real devices (Huard et al., 2007). The current includes both tunnelling and thermal emission parts. The analytical result shown in Eq. (10) gives a good estimation of the junction I-V curves. The tunnelling current is dominant at reverse bias. At high forward bias (>0.3 V) the current rises exponentially due to thermal emission. The full I-V curves show the current capability through graphene p-n junction has the magnitude around 1 A/mm, due to the zero band gap and high Fermi velocity (cm/s) in this material.

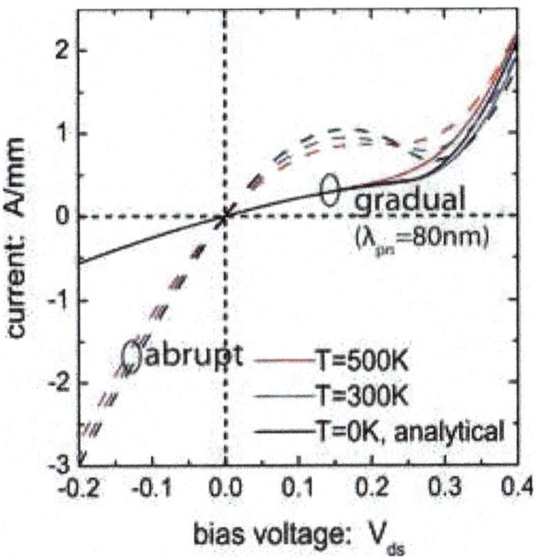

Fig. 22. I-V characteristics of graphene p-n junctions (n_1=n_2= cm^{-2}).

4. Summary

In summary, a wide variety of graphene devices and electronic device problems have been discussed. First 2D graphene transistors were made, which exhibited poor on/off ratio but promising mobility. Short channel effect has been observed, which invoked the detail exploration of the doping effect of the metal-graphene contacts. The poor saturation characteristics of the 2D graphene FETs led the attention toward the detailed study of the high field transport of graphene. The exfoliated graphene FETs were compared to CVD graphene based devices and were found to have similar performance but offering the advantage of wafer scale size. SiC based epitaxial graphene has the same size advantage along with the exceptional RF characteristics. For further improvement it became clear that band gap has to be opened by 1D confinement, which was experimentally demonstrated. The successful fabrication of p-n junction in GNR FET is a major achievement and an important step toward the realization of the GNR TFET.

5. References

Bae, S.; Kim, H.; Lee, Y.; Xu, X.; Park, J-S.; Zheng, Y.; Balakrishnan, J.; Lei, T.; Kim, H. R.; Song, Y. I. et al. (2010). Roll-to-roll production of 30-inch graphene films for transparent electrodes. *Nature Nanotechnology*, Vol. 5, pp. 574–578

Cheianov, V. V.; Falko, V. I. (2006). Selective transmission of Dirac electrons and ballistic magnetoresistance of np junctions in graphene. *Physical Review B,* Vol. 74, p. 041403

Chen, Y-F. & Fuhrer, M. (2005). Electric-Field-Dependent Charge Carrier Velocity in Carbon Nanotubes. *Physical Review Letters*, Vol. 95, pp. 236803

Chen, Z.; Lin, Y.; Rooks, M. J. & Avouris, P. (2007). Graphene nano-ribbon electronics. *Physica E*, Vol. 40, p. 228

Datta, S. (2005). *Quantum transport: atom to transistor*, Cambridge University Press, USA

Dorgan, V.; Bae, M-H. & Pop, E. (2010). Mobility and Velocity-Field Relationship in Graphene. *IEEE DRC Conf. Digest.*, Vol. 68, pp. 73-74

Fang, T.; Konar, A.; Xing, H. & Jena, D. (2007). Carrier statistics and quantum capacitance of graphene sheets and ribbons. *Applied Physics Letters*, Vol. 91, 092109

Fang, T.; Konar, A.; Xing, H. & Jena, D. (2008). Mobility in semiconducting graphene nanoribbons: Phonon, impurity, and edge roughness scattering. *Physical Review B*, Vol. 78, p. 205403

Geim, A. K. & Novoselov, K. S. (2007). The rise of graphene. *Nature Materials*, Vol. 6, p. 183

Giovannetti, G.; Khomyakov, P. A.; Brocks, G.; Karpan, V.; Brink, J. v d; & Kelly, P. J. (2008). Doping Graphene with Metal Contacts. *Physical Review Letters*, Vol. 101, p. 026803

Golizadeh-Mojarad, R. & Datta, S. (2009). Effect of contact induced states on minimum conductivity in graphene. *Physical Review B*, Vol. 79, p. 085410

Guo, J.; Datta, S.; Lundstrom, M. & Anantram, M. P. (2004). Towards Multi-Scale Modeling of carbon Nanotube Transistors. *The International Journal on Multiscale Computational Engineering*, Vol. 2, p. 257

Huard, B.; Sulpizio, J.A.; Stander, N.; Todd, K.; Yang, B.; Goldhaber-Gordon, D. (2007). Transport Measurements Across a Tunable Potential Barrier in Graphene. *Physics Review Letters*, Vol. 98, p. 236803

Huard, B.; Stander, N.; Sulpizio, J. A. & Goldhaber-Gordon, D. (2008). Evidence of the role of contacts on the observed electron-hole asymmetry in graphene. *Physical Review B*, Vol. 78, p. 121402

Hwang, E.; Adam, S.; & Das Sarma, S. (2007). Carrier Transport in Two-Dimensional Graphene Layers. *Physical Review Letters*, Vol. 98, p. 186806

Jena, D.; Fang, T.; Zhang, Q. & Xing, H. (2008). Zener tunneling in semiconducting nanotube and graphene nanoribbon p−n junctions. *Applied Physics Letters*, Vol. 93, p. 112106

Jiao, L.; Zhang, L.; Wang, X.; Diankov, G.; & Dai, H. (2009). Narrow graphene nanoribbons from carbon nanotubes. *Nature*, Vol. 458, p. 877

Levendorf, M. P.; Ruiz-Vargas, C. S.; Garg, S.; & Park, J. (2009). Transfer-Free Batch Fabrication of Single Layer Graphene Transistors. *Nano Letters*, Vol. 9, p. 4479

Lee, E. J. H.; Balasubramanian, K.; Weitz, R. T.; Burghard, M. & Kern, K. (2008). Contact and edge effects in graphene devices. *Nature Nanotechnology*, Vol. 3, pp. 486 – 490

Li, X.; Cai, W.; An, J.; Kim, S.; Nah, J.; Yang, D.; Piner, R.; Velamakanni, A.; Jung, I.; Tutuc, E. et al. (2009). Large-Area Synthesis of High-Quality and Uniform Graphene Films on Copper Foils. *Science*, Vol. 324, p. 1312

Li, X.; Wang, X.; Zhang, L.; Lee, and S.; & Dai, H. (2008). Chemically Derived, Ultrasmooth Graphene Nanoribbon Semiconductors. *Science*, Vol. 319, p. 1229

Lin, Y.; Perebeinos, V.; Chen, Z. & Avouris, P. (2008). Electrical observation of subband formation in graphene nanoribbons. *Physical Review B*, Vol. 78, p. 161409

Lin, Y.-M.; Jenkins, K. A.; Valdes-Garcia, A.; Small, J. P.; Farmer D. B.; & Avouris P. (2009). Operation of Graphene Transistors at Gigahertz Frequencies. *Nano Letters*, Vol. 9, pp. 422–426

Lin, Y.-M.; Dimitrakopoulos, C.; Jenkins, K. A.; Farmer, D. B.; Chiu, H.-Y.; Grill, A.; & Avouris, Ph. (2010). 100-GHz Transistors from Wafer-Scale Epitaxial Graphene. *Science*, Vol. 327, p. 662

Low, T.; Hong, S.; Appenzeller, J.; Datta, S. & Lundstrom, M. (2009). Conductance asymmetry of graphene pn junction. *IEEE Trans. on Electronic Devices*, Vol. 56, p. 129

Luisier, M. & Klimeck, G. (2009). Performance analysis of statistical samples of graphene nanoribbon tunneling transistors with line edge roughness. *Applied Physics Letters*, Vol. 94, p. 223505

Luo, X.; Lee, Y.; Konar, A.; Fang, T.; Snider, G.; Xing, H.; & Jena, D. (2008). Current-carrying capacity of long and short-channel 2D graphene transistors. *IEEE DRC Tech. Digest*, Vol. 67, p. 29

Lundstrom, M. (2000). *Fundamentals of Carrier Transport*. Cambridge University Press.

Luxmi; Nie, S.; Fisher, P. J.; Feenstra, R. M.; Gu, G.; & Sun, Y. J. (2009). Temperature Dependence of Epitaxial Graphene Formation on SiC(0001). *Journal of Electronic Materials*, Vol. 38, p. 718

Meric, I.; Han, M. Y.; Young, A. F.; Oezyilmaz, B.; Kim P.; & Shepard, K. L. (2008). Current saturation in zero-bandgap, top-gated graphene field-effect transistors. *Nature Nanotecnology*, Vol. 3, pp. 654-659

Moon, J.S.; Curtis, D.; Hu, M.; Wong, D.; McGuire, C.; Campbell, P.M.; Jernigan, G.; Tedesco, J.L.; VanMil, B.; Myers-Ward, et al. (2009). Epitaxial-Graphene RF Field-Effect Transistors on Si-Face 6H-SiC Substrates. *IEEE Electron Device Letters*, Vol. 30, p. 650

Mucciolo, E. R.; Castro Neto, A. H. & Lewenkopf, C. H. (2009). Conductance quantization and transport gaps in disordered graphene nanoribbons. *Physical Review B*, Vol. 79, p. 075407

Mueller, T.; Xia, F.; Freitag, M.; Tsang, J. & Avouris, Ph. (2009). Role of contacts in graphene transistors: A scanning photocurrent study. *Physical Review B*, 79, 245430

Murali, R.; Brenner, K.; Yinxiao Yang; Beck, T.; & Meindl, J.D. (2009). Resistivity of Graphene Nanoribbon Interconnects. *IEEE Electron Device Letters*, Vol. 30, p. 611

Nagashio, K.; Nishimura, T.; Kita, K. & Toriumi, A. (2009). Metal Graphene Contact as a Performance Killer of Ultra-high Mobility Graphene a Analysis of Intrinsic Mobility and Contact Resistance. *IEEE IEDM Tech. Dig., Baltimore, MD, Dec. 2009*, pp. 1-4

Nemec, N.; Tománek, D. & Cuniberti, G. (2008). Modeling extended contacts for nanotube and graphene devices. *Physical Review B*, Vol. 77, p. 125420

Nouchi, R. & Tanigaki, K. (2010). Charge-density depinning at metal contacts of graphene field-effect transistors. *Applied Physics Letters*, Vol. 96, p. 253503

Novoselov, K. S.; Geim, A. K.; Morozov, S. V.; Jiang, D.; Zhang, Y.; Dubonos, S. V.; Grigorieva, I. V. & Firsov, A. A. (2004). Electric field effect in atomically thin carbon films. *Science*, Vol. 306, pp. 666-669

Stander, N.; Huard, B. & Goldhaber-Gordon, D. (2009). Evidence for Klein Tunneling in Graphene p-n Junctions. *Physical Review Letters*, Vol. 102, p. 026807

Venugopal, R.; Ren, Z.; Datta, S.; Lundstrom, M. & Jovanovic, D. (2002). Simulating quantum transport in nanoscale transistors: Real versus mode-space approaches. *Journal of Applied Physics*, Vol. 92, p. 3730

Yand, Y. & Murali, R. (2010). Impact of Size Effect on Graphene Nanoribbon Transport. *IEEE Electron Device Letters*, Vol. 31, p.237

Zhang, Y.; Tan, Y.-W.; Stormer, H. L.; & Kim, P. (2005). Experimental observation of the quantum Hall effect and Berry's phase in graphene. *Nature*, Vol. 438, pp. 201-204

Zhang, Q.; Fang, T.; Seabaugh, A.; Xing, H.; & Jena, D. (2008). Graphene Nanoribbon Tunnel Transistors. *IEEE Electron Device Letters*, Vol. 29, pp. 1344-1346

Zhang, Q.; Lu, Y.; Xing, H.; Koester, S. & Koswatta, S. (2010). Scalability of Atomic-Thin-Body (ATB) Transistors Based on Graphene Nanoribbons. *IEEE Electron Device Letters*, Vol. 31, p. 531 - 533

5

Measuring Disorder in Graphene with Raman Spectroscopy

Ado Jorio[1], Erlon H. Martins Ferreira[2], Luiz G. Cançado[3], Carlos A. Achete[4] and Rodrigo B. Capaz[5]

[1,3]*Departamento de Física, Universidade Federal de Minas Gerais*
[2,4]*Instituto Nacional de Metrologia, Normalização e Qualidade Industrial*
[5]*Instituto de Física, Universidade Federal do Rio de Janeiro*
Brazil

1. Introduction

Since the experimental discovery of an easy method to produce graphene (Novoselov et al., 2004; 2005), there is a continuous rising of interest on this fascinating material. Due to its own peculiar electronic band structure, graphene is a promising material for the future electronic nanodevice industry. In any production process is of fundamental importance to control the quality of materials in a fast and non-destructive manner. Quantifying defects in graphene related systems is a major problem for the application of such materials (Neto et al., 2009).
In the field of carbon-based nanostructures, Raman spectroscopy has shown to be the most suitable technique since it accomplishes those desirable goals and also because it can be used to differentiate the many types of sp^2 carbon structures and to identify the presence of defects (Ferrari & Robertson, 2000; Pimenta et al., 2007). Defects are of great importance since they modify the electronic and optical properties of the system, sometimes in a desirable way. Raman spectroscopy is also one of the most sensitive techniques to probe disorder in carbon structures. The development of an accurate methodology to study disorder with Raman spectroscopy will certainly facilitate the control of the amount of induced disorder in a target system. By combining experiment and theory, a phenomelogical model has been established to explain the evolution of the Raman spectrum of graphene with increasing amount of disorder caused by low energy ion bombardment. Using this model one can extract information about the density of defects, the average defect size and also the Raman process relaxation length and its dependence with the laser excitation energy and temperature. How the ion bombardment procedure affects graphene samples with two or more layers and graphene read edges has also been studied.

2. Review of Raman spectroscopy in graphene-related systems

Graphene has two atoms in the unit cell and, therefore, six phonon branches. Three are acoustic branches and three are optical branches. From the three optical branches, one gives rise to an infra-red active mode at the Γ point, while the two other branches are degenerate at the Γ point and Raman active. Therefore, zone center ($q = 0$) phonons would generate a one peak Raman spectra. However, the electronic structure of graphene generates special electron-phonon induced resonance conditions with non-zone center modes ($q \neq 0$), known in the literature as the double-resonance Raman scattering process. This double-resonance process is responsible for the graphene related systems to have a Raman spectra with many features (Saito et al., 2002; Thomsen & Reich, 2000).

Although the double resonance process can activate phonons from all the six branches, the main features in the Raman spectra of graphene come from the phonon branch related to the zone-center Raman-active mode, i.e. to the optical phonon branch related to in-plane stretching of the C-C bondings. The strongest Raman peaks in crystalline graphene are the so-called G (~1584 cm^{-1}) and G' (~2400-2800 cm^{-1}) bands. The first is the first-order Raman-allowed mode at the Γ point, and the second is a second-order Raman-allowed mode near the K point, activated by the double-resonance process. Furthermore, the presence of disorder in the crystalline lattice causes changes in the graphene Raman spectra, the most evident being the appearance of two new peaks, the so-called D (~1200-1400 cm^{-1}) and D' (1600-1630 cm^{-1}) bands. Both bands come from the in-plane optical branches (iTO and LO, respectively) and both are related to the double-resonance process. The D band comes from the iTO phonon near the K point, while the D' band comes from the LO phonon near the Γ point.

Since 1970, large efforts have been devoted to develop disorder-quantification methods using Raman spectroscopy for nano-graphite (Cançado et al., 2006; Ferrari & Robertson, 2000; Pimenta et al., 2007; Tuinstra & Koenig, 1970), carbon nanotubes (Chou et al., 2007; Hulman et al., 2005) and graphene (Jorio et al., 2010; Lucchese et al., 2010; Martins Ferreira et al., 2010; Teweldebrhan & Baladin, 2009). The most advanced protocols have been developed for the edge-defects in nano-graphite, where the in-plane crystallite dimensions ranging from $L_a = 20\,\text{nm}$ up to values larger than a microscope spot size can be obtained by measuring the intensity ratio between the disorder induced D band and the first-order allowed tangential G band, including the dependence of I_D/I_G with the excitation laser wavelength, as given by (Cançado et al., 2006)

$$L_a(\text{nm}) = (2.4 \times 10^{-10})\lambda_{laser}^4 (I_D/I_G)^{-1}. \quad (1)$$

Such behavior is shown in Figure 1. Equation 1 gives the state-of-the-art for using Raman spectroscopy to quantify nano-graphite crystallite sizes, which is related to a certain type of disorder that are the graphene edges. Of course the result given above is an averaged measure of all possible edge structures present in a bulk nanographite sample. It has been shown that the D band intensity depends on the atomic structure at the edge, and it is actually absent in graphite edges when the atomic structure exhibit the zigzag arrangement (Cançado et al., 2004). Figure 2 shows an example of a graphite edge. The atomic orientation of the lattice was identified with scanning tunnelling microscopy, thus showing that the armchair and zigzag orientations are parallel to edges 1 and 2 in Figure 2, respectively. This is consistent with the observed change in the D band intensity, but the observation of a small but non-null D band at edge 2 shows that the structure at the edge is not perfectly zigzag. This result has been used to identify zigzag vs. armchair edges in graphene as well (Neubeck et al., 2010), although up to date there is no evidence of perfect zigzag graphene edged. Such a perfect zigzag edge structure would be evidenced by the observation of a D' band in the Raman spectra with a complete absence of the D band (Cançado et al., 2004).

Finally, the picture discussed above does not hold for ion-bombarded induced defects in graphene samples, as shown by (Lucchese et al., 2010), because of the simple and fundamental geometric difference between defects related to the size of a nano-crystallite and point defects in the lattice, which changes the I_D/I_G dependence on the amount of disorder. Basically, the

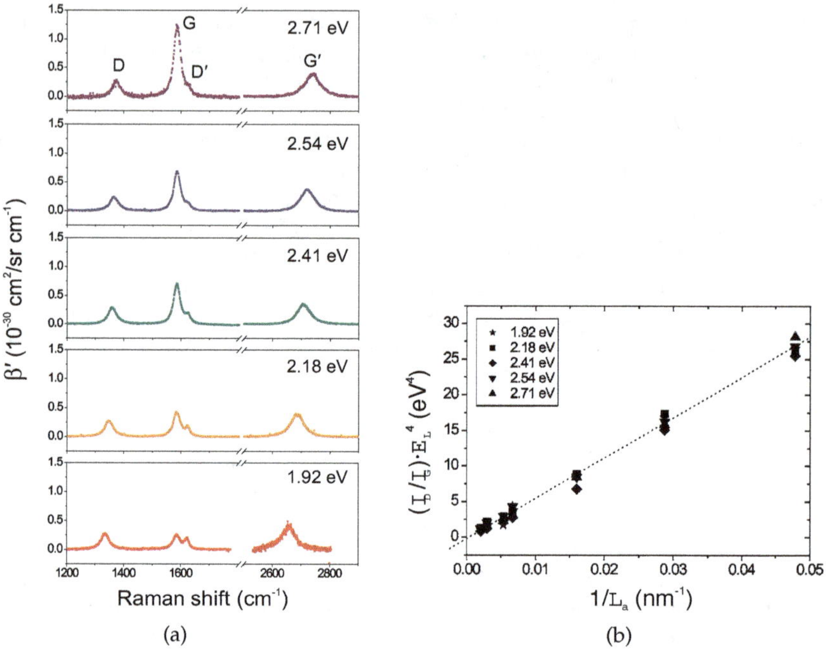

Fig. 1. (a) The spectral differential cross section β' of the D, G, D', and G' bands for the nanographite sample with $L_a = 35$ nm, using five different values of E_L, each indicated at the top of the respective spectrum. The same vertical scale was used for the five spectra for comparison. (Cançado et al., 2007) (b) The intensity ratio I_D/I_G for nanographite samples normalized by E_L^4, plotted versus $1/L_a$ using five different laser excitation energies (Cançado et al., 2006).

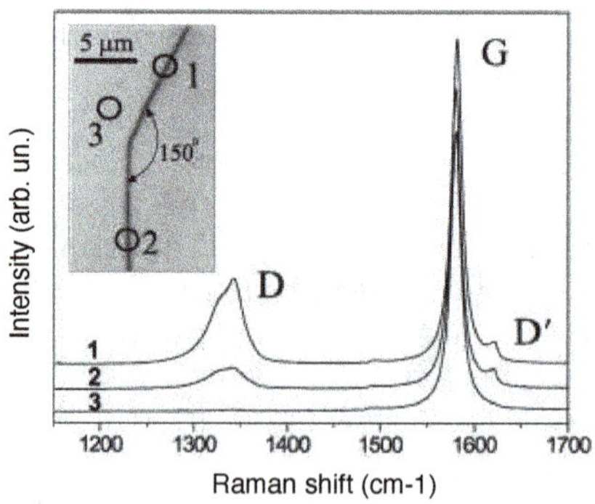

Fig. 2. Raman spectra obtained in three different regions of a highly oriented pyrolytic graphite (HOPG) sample. The inset shows an optical image of the step and the regions where spectra 1, 2, and 3 were taken (open circles). The atomic structure of the sample was identified using scanning tunneling microscopy, and the edges 1 and 2 were identified as parallel to the armchair and zigzag orientations, respectively. (Cançado et al., 2004)

disorder in a nano-crystallite can be quantified by the amount of border (one-dimensional defects) with respect to the total crystallite area, and this is a measure of the average inverse nano-crystallite size $1/L_a$. In ion bombarded graphene, the disorder is better quantified by the defect concentration, defined by $1/L_D^2$, where L_D is a typical interdefect distance, with the defect being a point-like (zero-dimensional) structure. In the following sections we will discuss these concepts in depth to develop the methods to quantify disorder in graphene related systems.

3. Creating defects on graphene by ion bombardment

There are many different techniques that could be used to investigate the amount of disorder in graphene, but Raman spectroscopy is certainly a fast and easy way to do it. However, to extract quantitative information from a Raman spectrum we need to calibrate the procedure. Such a calibration can be achieved with the help of a more direct technique such as scanning tunneling microscopy (STM), which can show the defects on the surface of a material. Doing STM on an exfoliated graphene is still a challenging task, since we need a graphene sample deposited on a conductive surface or make an electrical contact. The first method may complicate (if not prevent) to find the graphene on the surface and the second method can cause non-desirable defects on the sample (Ishigami et al., 2007; Neubeck et al., 2010). On the other hand, performing STM on graphite is a much simpler procedure, since the bulk graphite is conductive. One can then irradiate the graphite with low energy ions, do the STM image to count the number of defects per unit area and then perform the Raman measurement to correlate the relative intensity between the *D* and *G* bands with the measured ion dose. In a second step one can perform the same kind of irradiation on a graphene sample and use a HOPG flake as a reference for determining the actual ion dose (Lucchese et al., 2010).

Figure 3 shows STM images of a bulk HOPG surfaces, both pristine (a) and bombarded with 90 eV argon ions (b-e). Using the same parameters of current and ion energy, the increasing ion dose was obtained by increasing the bombardment time. From (b) to (e) the ion dose has been increased from 10^{11} Ar^+/cm^2 up to 10^{14} Ar^+/cm^2, which corresponds to an average distance between defects from $L_D = 32$ nm down to $L_D = 1$ nm. The right panel shows the Raman spectra measured for pristine HOPG and for the same HOPG right after each bombardment procedure shown in panels (b-e) (increasing bombardment dose from bottom to top). Although for the highest dose the HOPG surface seems already completely damaged, the Raman spectrum shows a well defined *G* band, indicating that the ions affect mostly the upper layers. To fully calibrate the bombardment procedure that is further going to be applied to the graphene samples, ion bombardment and consecutive Raman measurements on HOPG were performed three times, for up to eight different ion doses between 10^{11} and 10^{15} Ar^+/cm^2. The Raman response was measured at up to 7 locations within the HOPG specimen for each ion dose (Jorio et al., 2009).

Using this procedure, the same bombardment can be applied to a graphene sample. When obtaining a graphene sample by the micro-mechanical cleavage of graphite and further deposition on a SiO_2 substrate ("scotch tape" method), one can always find some bulk graphite flake next to a graphene sample. This small flake can then be used as a "standard" for the ion dose calibration by comparing its Raman spectrum with the previously HOPG measured spectrum discussed in Fig. 3. Thus one can skip the STM procedure on the graphene sample to actually count the number of defects on the sample. Figure 4(a) shows the Raman spectra of such a ion bombarded monolayer graphene subject to the same ion doses measured in the HOPG case. It is clear from those spectra that the graphene sample is much more

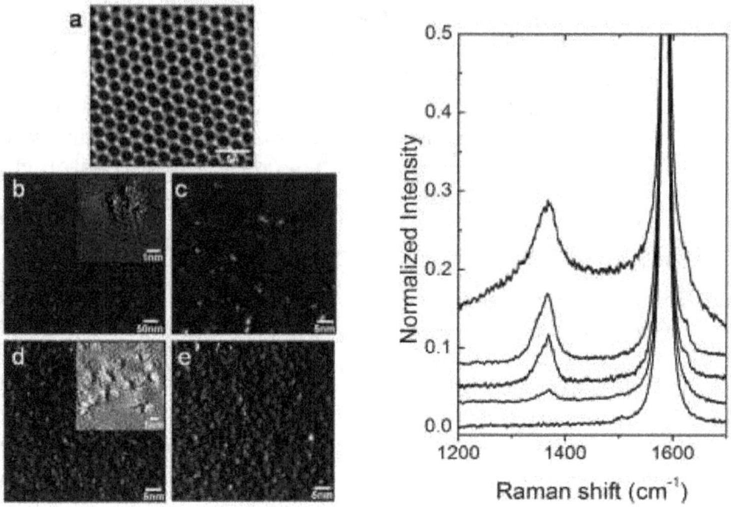

Fig. 3. Left panel are the STM images of (a) pristine HOPG surface and (b)–(e) after increasing ion bombardment showing how the defects are spread on the surface. (a) to (e) corresponds to zero, 10^{11}, 10^{12}, 10^{13}, 10^{14} Ar^+/cm^2 dose. Insets to (b) and (d) show some defects in detail. The right panel shows the Raman spectra of pristine HOPG and the same HOPG right after each bombardment in (a-e), increasing ion dose from bottom to top, showing the increasing D band.

sensitive to the ion bombardment procedure than the bulk graphite. Because it has just one layer, there is no extra contribution to the G band intensity as in the case of the inner layers of the HOPG sample. Then the relative intensity I_D/I_G increases up to $\sim$ 3.5 in the monolayer case while it never reaches 1 in the HOPG sample, with an excitation laser energy of 2.41 eV (λ = 514 nm). A more complete behavior of those relative intensities can be seen in Figure 4(b). This graph shows the evolution of the relative intensity I_D/I_G with increasing ion dose (decreasing distance between defects) from 1.5×10^{11} to 2.5×10^{15} Ar^+/cm^2 (26 nm down to 0.63 nm) and compares with both the original Tuinstra-Koenig relation (Equation 1) for the graphite and a modified version introduced by (Lucchese et al., 2010). It is clear from the plot that Equation 1 does not hold for graphene because the topology of defects created by the ion bombardment is different from those boundaries defects. The relation introduced for point-like defects is (Lucchese et al., 2010).

$$\frac{I_D}{I_G} = \frac{C(\lambda)}{L_D^2}, \tag{2}$$

where, in this case, $C(514\ \text{nm}) \sim 107\ \text{nm}^2$. Still, the limit of validity of this relation is for the regime of low defect density (or $L_D > 6\,\text{nm}$). In the graph we can see two well defined disorder regimes, one in the right side of the curve, which is the low disorder limit, and the one in the left side, which is typical of amorphous carbon structures. The complete model to describe the transition between the two regimes was described also in (Lucchese et al., 2010) and will be presented in Section 5.

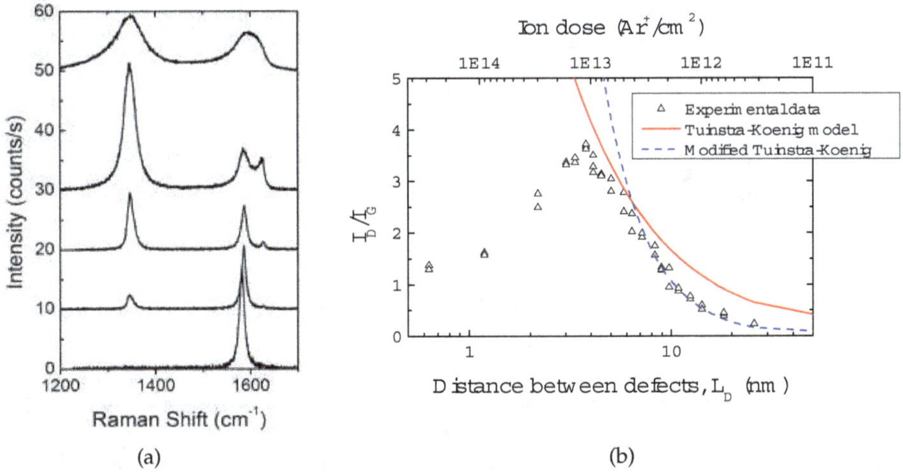

Fig. 4. (a) Raman spectra of a monolayer graphene bombarded with low energy (90 eV) argon ions measured with a 514 nm laser. From bottom to top, curves correspond to zero, 10^{11}, 10^{12}, 10^{13} and 10^{14} Ar^{+}/cm^2 dose. (b) Comparison between experimental I_D/I_G data and both original and modified Tuinstra-Koenig relations for disorder in carbon structures (see text).

When studying disorder by means of Raman spectroscopy, the relative intensity between the D and G bands is the main feature that is taken into consideration. However, disorder is also responsible for the changes in the intensities of other bands, such as D' and G', and also affects the position and shape of the Raman peaks. Figure 5 plots the relative integrated areas of the main Raman bands, namely D, D' and G' with respect to G, and Figures 6(a) and 6(b) show the behaviour of the peak position and the full width at half maxium (FWHM) of those same bands for the same monolayer graphenes of Figure 4. A careful look at those plots shows that in the limit of low disorder (large L_D) there is no significant change in the peak position and just a slightly increase of the FWHM of all bands. However, for increasing disorder ($L_D \lesssim 4$ nm), there is an abrupt increase of the FWHM and also a small downshift in the peak positions. This suggests that even at moderately large disorder regimes, the C–C bonds are still sp^2 kind, with no significant strain in the bonds. When looking at the disorder in graphene, not only the ratio I_D/I_G must be considered, as it is clear that it is possible to have the same value of I_D/I_G for two different values of L_D. Therefore, it is important to look also at the values of the FWHM which can tell if the system is in lower or higher disorder regime.

4. Laser energy dependency

The experimental data presented in the previous Sections show that disorder introduced by a random distribution of defects causes significant changes in the relative intensities of the resonance Raman bands and it leads to broadening and shifting of those bands. Moreover, these effects appear to be strongly dependent on the laser excitation energy. It was a work by (Mernagh et al., 1984) that showed for the first time a strong dependency of the ratio I_D/I_G on the excitation laser energy E_L used in the Raman scattering experiment (Mernagh et al., 1984). Later on, (Cançado et al., 2006) measured the I_D/I_G ratio dependence on E_L for nanographites, and generated Equation 1.

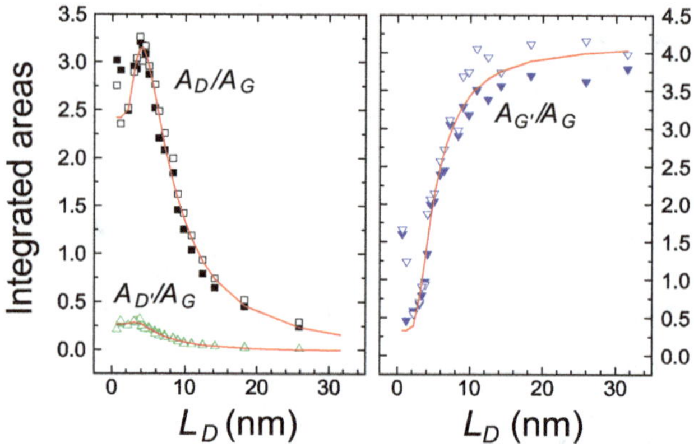

Fig. 5. Evolution of the integrated areas of the D, D' and G' bands normalized by the G band for a monolayer graphene bombarded with low energy argon ions. Symbols are the experimental data and the curves are the fitting with Equations 3 and 4. Open and filled symbols stand for two different graphene samples. (Martins Ferreira et al., 2010)

In this section we describe recent efforts to fully accomplish the protocol for quantifying the amount of point-like defects in graphene (or equivalently, L_D) (Cançado et al., 2011). For that, different excitation laser lines have been used for measuring the Raman spectra of ion bombarded samples with different L_D values and their respective I_D/I_G ratios.
Figures 7(a-c) show the Raman spectra of five distinct ion-bombarded graphene samples which were exposed to different ion bombardment doses in the range of 10^{11} Ar^+/cm^2 (one defect per 4×10^4 C atoms) to 10^{15} Ar^+/cm^2. The Raman spectra shown in panels (a), (b), and (c) were taken using the excitation laser energies (wavelengths) $E_L = 2.41$ eV ($\lambda_L = 514.5$ nm), $E_L = 1.96$ eV ($\lambda_L = 632.8$ nm), and $E_L = 1.58$ eV ($\lambda_L = 785$ nm), respectively. The samples have $L_D = 24$ nm, 14 nm, 7 nm, 5 nm and 2 nm. Figure 7(d) shows the Raman spectra of the ion-bombarded sample with $L_D = 7$ nm obtained using the three different laser energies. Notice that the ratio I_D/I_G decreases as the laser energy increases, showing that a complete analysis for the I_D/I_G ratio in graphene must take into account the excitation laser energy. However, before going into details, we have to introduce the model to analyse the evolution of the D and G peaks.

5. Theoretical model

5.1 The local activation model

A unified theoretical description of the disorder induced effects from a more fundamental point-of-view is still an open problem, although some advances have been made on the related problem of the D band arising from graphene edges (Basko, 2009; Casiraghi et al., 2009). Nevertheless, some understanding of the problem at hand has been achieved through phenomenological models, as we describe in this Section.
Both experiments (Cançado et al., 2006; Tuinstra & Koenig, 1970) and theory (Basko, 2009; Casiraghi et al., 2009) describe the D band as being "locally activated", i.e., it only becomes active in the near vicinity of an imperfection (such as an edge or a point defect). Therefore, for the specific case of disorder induced by impact of individual Ar^+ ions, we can associate a characteristic length scale r_A, measured from the impact point, as the typical decay distance

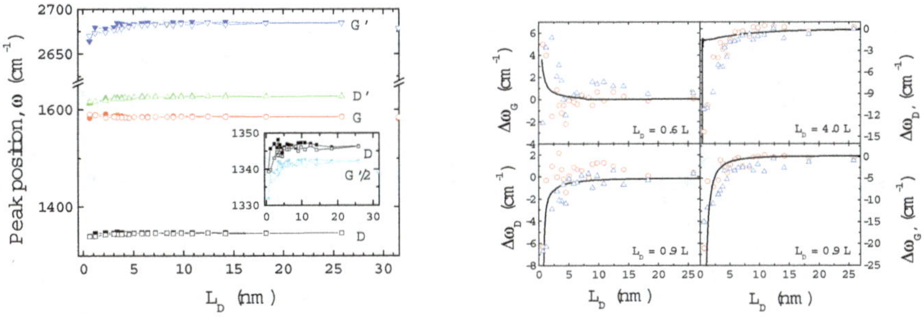

(a)

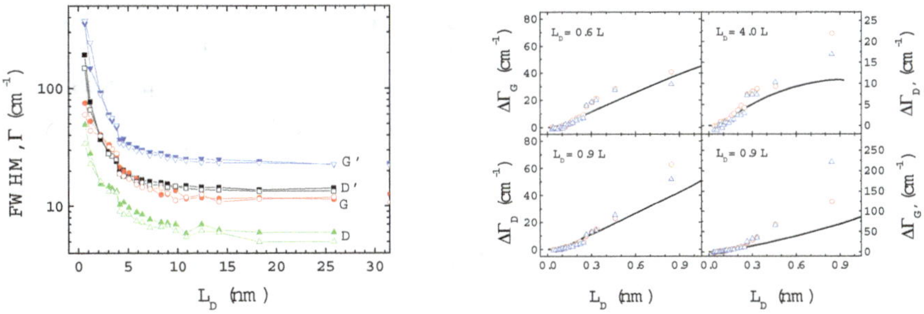

(b)

Fig. 6. Peak position (a) and full width at half maximum (b) of the four main Raman bands D, G, D' and G' of a monolayer graphene bombarded with low energy ions as a function of the average distance between the defects (L_D). Left panels show the absolute experimental data, and right panels the shifts in peak position and peak width with respect to the initial values (from the first bombardment). Symbols are experimental data and curves are the fitting from the theoretical model presented in Section 5. (Martins Ferreira et al., 2010)

of the D band intensity (green region in Figure 8). Another length scale is given by r_S, also measured from the impact points, which describes the region where the graphene sp^2 network disorganizes due to the ion impact (red region in Figure 8).

The activated region leads to an increase of the A_D/A_G ratio (equivalently on I_D/I_G), whereas the disorganized region reduces this ratio. By solving rate equations describing the evolution of the green and red regions with the ion dose, an analytical expression is obtained for A_D/A_G as a function of L_D (Lucchese et al., 2010; Martins Ferreira et al., 2010):

$$\frac{A_D}{A_G}(L_D) = C_A \frac{r_A^2 - r_S^2}{r_A^2 - 2r_S^2}\left[e^{-\frac{\pi r_S^2}{L_D^2}} - e^{-\frac{\pi(r_A^2 - r_S^2)}{L_D^2}}\right] + C_S\left[1 - e^{-\frac{\pi r_S^2}{L_D^2}}\right], \tag{3}$$

In this expression, the coefficients C_A and C_S correspond to the A_D/A_G ratio in two idealized limits. C_A refers to the situation in which the D band is activated in the whole of the graphene layer. Therefore, it embodies important information on the relative strengths

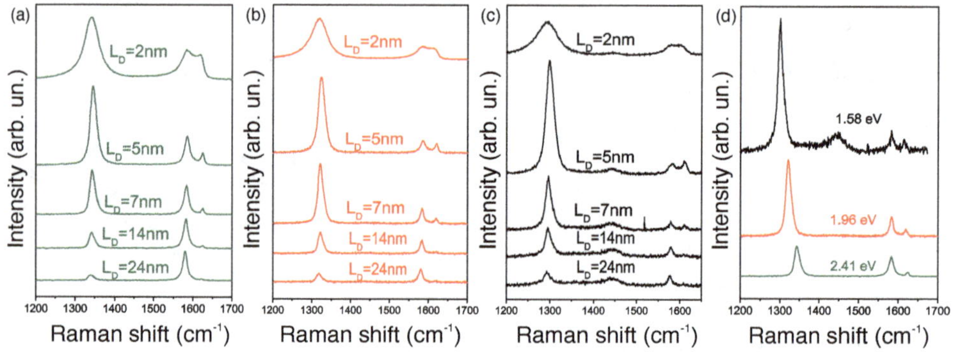

Fig. 7. (a-c) Raman spectra of five distinct ion-bombarded graphene samples using the excitation laser energies (wavelengths) $E_L = 2.41\,\text{eV}$ ($\lambda_L = 514.5\,\text{nm}$), $E_L = 1.96\,\text{eV}$ ($\lambda_L = 632.8\,\text{nm}$), and $E_L = 1.58\,\text{eV}$ ($\lambda_L = 785\,\text{nm}$), respectively. (b) Raman spectra of an ion-bombarded sample with $L_D = 7\,\text{nm}$ obtained using the three different excitation laser energies. (Cançado et al., 2011)

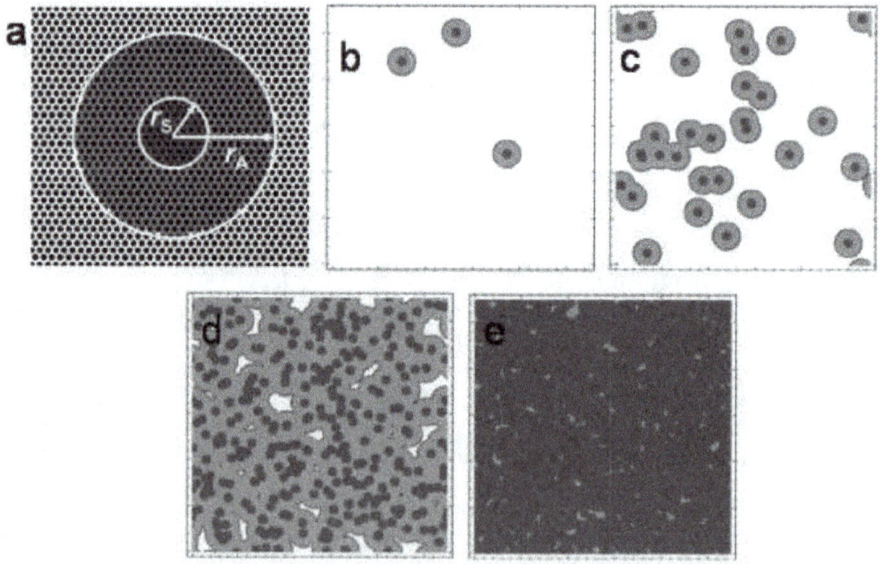

Fig. 8. (a) The two length scales (r_A and r_S) describing the evolution of the D band intensity with disorder induced by the impact of Ar^+ ions in graphene. Panels (b) to (e) correspond to the same ion doses in panels (b) to (e) in Figure 3. (Lucchese et al. (2010))

of the electron-phonon scattering near the Γ and K points in the Brillouin zone and on the electron-defect scattering cross-section for the D band. In particular, the strong energy dependence of C_A is revealing of the physics of Kohn anomaly near the K-point phonons in graphene (Piscanec et al., 2004) and of electron-defect scattering on short-range imperfections of graphene (Basko, 2008). The coefficient C_S corresponds to the A_D/A_G ratio in the highly disordered limit.

Notice from Figure 5 that this analytical function describes the experimental evolution of A_D/A_G with the amount of disorder, including its energy dependence. Moreover we can also extract physical meaning from the two length scales obtained from the fits. For instance r_S appears to be energy-independent, which is consistent with its interpretation as geometrical, structure-related length. The fitted values of r_S are in accordance with the typical defect-size estimates found from STM analysis (Martins Ferreira et al., 2010), which shows a sparse size distribution with an average value of 1.8 nm. The length $r_A - r_S$ is a rough measure of the length traveled over the lifetime of the electron-hole pair, v_F/ω_D, where v_F is the graphene Fermi velocity (Casiraghi et al., 2009). This length is energy-dependent, as imposed by the double-resonance condition and the strong phonon dispersion near the K point.

The analytical function in Equation 3 also describes the evolution of D' band integrated area with disorder (Martins Ferreira et al., 2010). Similar ideas can be applied for the $A_{G'}/A_G$ ratio. However, the G' band is already active for pristine graphene, so the integrated area ratio is only affected by the progressive disorganization of the graphene network, leading to:

$$\frac{A_{G'}}{A_G}(L_D) = \frac{A_{G'}}{A_G}(\infty) - B\left[1 - e^{-\frac{\pi r_S^2}{L_D^2}}\right], \tag{4}$$

where $A_{G'}/A_G(\infty)$ is the area ratio for pristine graphene. The fitting of the experimental data also shows an excellent agreement in this case.

5.2 The excitation energy dependence

Now we can consider the excitation energy dependence in more depth. Figure 9(a) shows the I_D/I_G data (bullets) for all graphene samples and laser energies used in the experiment. The lines are the fitting curves following Equation 3. The C_A values obtained from the fitting shown in Figure 9(a) are plotted in the inset to Figure 9(a) (down triangles). Notice that C_A decrease as the laser energy increases. This behavior is ruled by a strong dependency of the strength of the D band on the wavevector of phonons involved in the double-resonance scattering process (Cançado et al., 2011). The solid line in the inset to Figure 9(a) is the fitting of the experimental data using an inverse fourth power dependence on the excitation laser energy, which gives $C_A = 140\,E_L^{-4}$, following the results on nanocrystallites (Cançado et al., 2006). This dependence is not yet understood and may be restricted to this energy range.

Although C_S could present some dependency with the excitation laser energy, our experimental data set does not allow us a clear determination of this dependency. In this case, we have considered $C_S = 0.9$ for all three excitation laser energies used in the experiment, in agreement with the more detailed results displayed in Figure 4(b). The fitting also gives $r_A = 3.1$ nm, and $r_S = 1$ nm, which is in excellent agreement with the values obtained by (Lucchese et al., 2010) and (Beams et al., 2010).

Figure 9(b) shows the plot of the product $E_L^4(I_D/I_G)$ versus L_D for the experimental data shown in Figure 9(a). It is clear from the plot depicted in Figure 9(b) that the data with $L_D > 10$ nm obtained with different laser energies collapse in the same curve. The same is not valid for

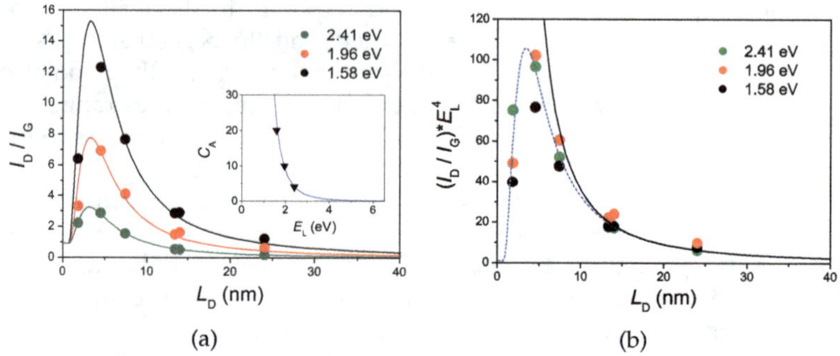

Fig. 9. (a) I_D/I_G data (bullets) for all graphene samples and laser energies used in the experiment. The solid lines are the fitting curves according to Equation 3. The inset shows the plot of C_A versus E_L. (b) Plot of the product $(I_D/I_G)E_L^4$ versus L_D for the experimental data shown in Figure 9(a). (Cançado et al., 2011)

$L_D < 10\,\text{nm}$, indicating that C_S might also play a role in the E_L dependence. The dashed blue line is the plot obtained from the substitution of the relation $C_A = 140/E_L^4$ in Equation 3.
We now turn our attention to the low defect density regime ($L_D \geq 10$ nm). In this regime, for which $L_D > 2r_A$, the total area contributing to the D band scattering is simply proportional to the number of point defects, giving rise to $I_D/I_G \propto 1/L_D^2$ (see Equation 2). By considering large values of L_D, and also taking in account that $C_A(r_A^2 - r_S^2) \gg C_S r_S^2$, Equation 3 takes the form

$$\left(\frac{I_D}{I_G}\right) \simeq C_A \frac{\pi(r_A^2 - r_S^2)}{L_D^2}. \tag{5}$$

By taking the values $r_A = 3.1$ nm, $r_S = 1$ nm, and also the relation $C_A = 140/E_L^4$ obtained from the fit of the experimental data shown in Figure 9(a), Equation 5 can be rewritten as

$$L_D^2(\text{nm}^2) = \frac{3600}{E_L^4}\left(\frac{I_D}{I_G}\right)^{-1}. \tag{6}$$

The above relation is valid for Raman data obtained from graphene samples that present point defects with $L_D \geq 10$ nm using excitation laser lines in the optical range. In terms of the defect density $n_D(\text{cm}^{-2}) = 10^{-14}/L_D^2$, Equation 6 becomes

$$n_D(\text{cm}^{-2}) = (2.8 \times 10^{-18})E_L^4\left(\frac{I_D}{I_G}\right). \tag{7}$$

The solid dark line in Figure 9(b) is the plot of the product $E_L^4(I_D/I_G)$ versus L_D according to Equation 6. The plot shown in Figure 9(b) clearly validates this relation for samples with $L_D > 10\,\text{nm}$.

5.3 Evolution of peak frequency and width

The evolution of the Raman FWHM and frequency shifts with increasing disorder in graphene can also be described by a phenomenological model, the so-called "spatial-correlation model". This model was originally developed to describe the evolution of the Raman spectra of disordered semiconductors (Richter et al., 1981; Tiong et al., 1984). In a crystalline system, the Raman intensity $I_0(\omega)$ associated to a vibrational mode of wavevector $\mathbf{q}_0$ and frequency $\omega(\mathbf{q}_0)$ is described by a Lorentzian:

$$I_0(\omega) \propto \frac{1}{[\omega - \omega(\mathbf{q}_0)]^2 + [\Gamma_0/2]^2}, \tag{8}$$

where the FWHM Γ_0 is the inverse phonon lifetime. In a perfect system, Γ_0 has contributions from both electron-phonon and phonon-phonon (anharmonic effects) interactions. A disordered distribution of point defects will add a contribution to the FWHM by coupling phonons of different momenta, so the phonon wave packet in k-space can be described by a Gaussian function $\exp[-(\mathbf{q} - \mathbf{q}_0)^2 L^2/4]$ centered in $\mathbf{q}_0$ and having a width proportional to $1/L$. In real space, L is a measure of the phonon coherence length, which should be a good measure of the average distance between point defects. Then, the Raman intensity for the disordered graphene $I(\omega)$ is written as (Tiong et al., 1984):

$$I(\omega) \propto \int_{BZ} d^2q \frac{W(\mathbf{q}) \exp\left[\frac{-(\mathbf{q}-\mathbf{q}_0)^2 L^2}{4}\right]}{[\omega - \omega(\mathbf{q})]^2 + [\Gamma_0/2]^2}, \tag{9}$$

where, in our case, the integral is taken over the two-dimensional graphene Brillouin Zone and $W(\mathbf{q})$ is a weighting function that describes wavevector dependence of the electron-phonon coupling for the Raman process.
Within this phenomenological model, and using experimentally available phonon dispersion relations the full lineshape of $I(\omega)$ can be calculated, and from that the disorder-induced peak shifts $\Delta\omega_{\mathbf{q}_0}$ and the increase in FWHM $\Delta\Gamma_{\mathbf{q}_0}$ can be extracted. Details of the application of this model to specific Raman bands can be found in (Martins Ferreira et al., 2010). As seen in Figures 6(a) and 6(b), this simple model describes the main qualitative features of experimental data on the evolution of the FWHM and peak-frequency shifts of the D, D', G and G' bands with disorder. Notice that the agreement is better for large values of L_D, as expected. In the figures, we also show the best relationships between L and L_D in each case (obtained by the fits). Notice that L and L_D are similar to each other and this behavior arises naturally from the fitting procedure. That means the disordered-induced phonon coherence length is of the same order of the typical inter-defect distance, as we argued. Also, it seems that the D' modes are the most affected by disorder, showing a smaller coherence length than the other modes for the same amount of disorder. Finally, the model allows us to explain the greater increase in FWHM of the modes near K with respect to the modes near Γ as simply a consequence of the larger magnitude of phonon dispersions near K.

6. Effect of ion bombardment on few-layer graphenes

Differently from the monolayer, where the G' and D bands are described by a single Lorentzian function, in a graphene with 2 or more layers, those bands become more complex due to the interaction between the layers, and, therefore, a quantitative description of the evolution of the Raman bands with increasing disorder is a tougher task (Martins Ferreira et al., 2010). Figure 10 shows the Raman spectra of a bilayer and a trilayer graphene sample subjected to the same ion bombardment of the previously discussed monolayer sample. In comparison, the effect of ion bombardment in this sample is less pronounced than in monolayer. The I_D/I_G ratio grows faster and reaches a higher value for the monolayer than for the bilayer, and the same applies to the bilayer as compared to the trilayer. This behavior is a confirmation that the defects are mainly on the surface (Jorio et al., 2010). Since the Raman signal gets the contribution of all the layers, the G band intensity must be stronger for the bilayer and trilayer, while the D band intensity, which depends mostly on the number of defects, must be approximately the same. This can explain the behavior of the intensity ratios I_D/I_G. As for the case of the monolayer, we see no significant change in the frequency of the G band, and we observe a broadening of the peaks for the higher ion doses (Martins Ferreira et al., 2010).

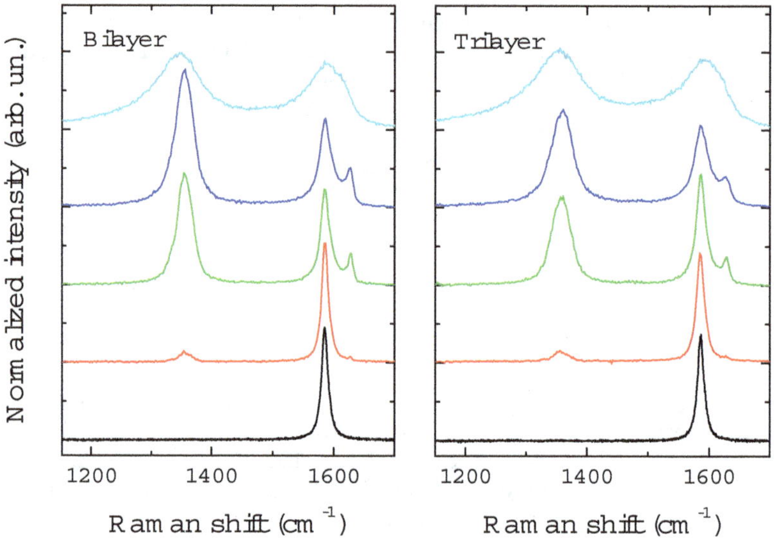

Fig. 10. Raman spectra of bilayer and trilayer graphenes bombarded with low energy (90 eV) argon ions. From bottom to top, curves correspond to zero, 10^{11}, 10^{12}, 10^{13} and 10^{14} Ar^+/cm^2 doses. (Jorio et al., 2010)

7. Raman at edges

The borders of a graphene sheet act like defects in the crystallographic structure and may have great impact in transport properties of graphene. As discussed in Section 2, previously studies on graphite edges have shown that the Raman D band intensity depends on the crystallographic orientation of the edge and the polarization of the laser beam (Cançado et al., 2004). Figure 11 shows the Raman mapping of the D band of a graphene sample showing the presence of defects at the borders (Carozo et al., 2011). For graphene, it has also been shown that the intensity of the D band is stronger for an incident light with polarization parallel to the border and is proportional to $\cos^2\theta$, where θ is the angle between the polarization of incident light and the edge (Casiraghi et al., 2009; Gupta et al., 2009). Theoretically it is expected that the D band intensity would vanish for a perfectly oriented edge, such as armchair or zigzag, when the incident light polarization is perpendicular to the edge. However, both (Casiraghi et al., 2009) and (Gupta et al., 2009) failed to found a null D band intensity with a perpendicular polarization and they could not tell the difference between the a zigzag and armchair borders by the evaluation of the relative intensity I_D/I_G, contrary to what have been found by (Cançado et al., 2004). The conclusion was that the edges were not a perfect zigzag or armchair, but a mix of them. A more recent work by (Neubeck et al., 2010) determined the crystallographic orientation of a graphene sheet by STM measurements and found that the intensity of the D band is higher for the armchair edge than for the zigzag edge when the polarization of the laser beam was parallel to the edges, in accordance with have been published by (Cançado et al., 2004). Yet, it is expected that a perfect zigzag border would not present a D band, due to phonon momentum conservation (Cançado et al., 2004). It is to conclude then, that although at a microscopy level the border may be mainly zigzag or armchair, at the atomic level it is not perfect.

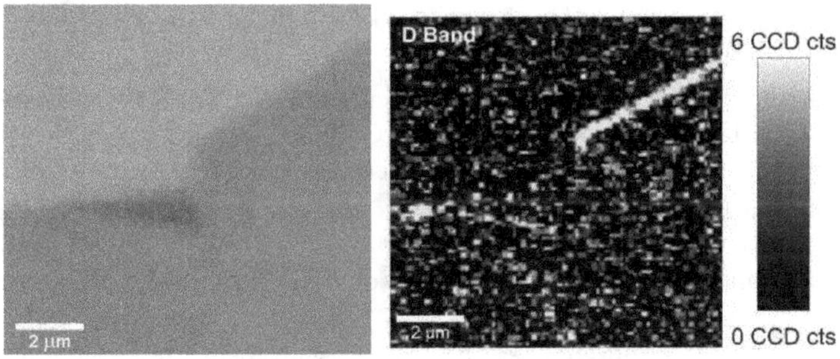

(a) (b)

Fig. 11. (a) Optical image of a monolayer graphene and (b) the Raman mapping of the D band of the same graphene sample showing the presence of the defects at the border (Carozo et al., 2011).

Finally, a theoretical work of (Sasaki et al., 2010) proposed that the intensity of the G band at the edges may be different depending on the orientation of the border, however no experimental data has corroborated it so far. The major difficulty in these studies to the moment is to obtain a perfect zigzag or armchair edge in sample produced by micro-mechanical cleavage. It is expected that graphene samples produced via chemical exfoliation may produce a much smoother edge and then those theories could be tested.

8. Conclusions and perspectives

The possibility of performing experiments on an isolated graphene sample improved substantially our understanding of disorder-induced effects in the Raman spectra of graphene related systems, even though such effects have been studied since 1970. Performing joint spectroscopy and microscopy experiments on a single layer of atoms allowed us to have a clear picture of the disorder effect in both real and reciprocal spaces, setting us free from volume-related aspects that were not easily addressable by microscopy and depend on light penetration depth. The parameters ruling the intensity of disorder-induced Raman bands, named r_A, r_S, C_A, C_S and defect topology, are now clearly established, so that accurate models can be developed for fully understanding the importance of electron-phonon and electron-defect interactions in this matter. The evolution of peak frequencies and linewidths are also clearly understood now for the low disorder limit, as discussed in this chapter. However, for understanding the peak behavior at the large disorder limit, more experimental and theoretical developments are needed. Finally, all this development has shown that, up to date, no perfect zigzag graphene edge has been produced. Edge perfection will be important for the electronic properties of graphene nanoribbons, and Raman spectroscopy might play an important role on testing the development of such edges.

9. References

Basko, D. M. (2008). Resonant low-energy electron scattering on short-range impurities in graphene, *Phys. Rev. B* 78: 115432.

Basko, D. M. (2009). Boundary problems for dirac electrons and edge-assisted Raman scattering in graphene, *Phys. Rev. B* 79: 205428.

Beams, R., Cançado, L. G. & Novotny, L. (2010). Optical measurement of the phase-breaking length in graphene. arXiv:1008.1563v.

Cançado, L. G., Jorio, A., Martins Ferreira, E. H., Stavale, F., Achete, C. A., Moutinho, M. V. O., Capaz, R. B., Lombardo, A. & Ferrari, A. C. (2011). Laser energy dependency of the disorder-induced band in the Raman spectrum of graphene. Umpublished.

L. G., Takai, K., Enoki, T., Endo, M., Kim, Y. A., Mizusaki, H., Jorio, A., Coelho, L. N., Magalhães-Paniago, R. & Pimenta, M. A. (2006). General equation for the determination of the crystallite size L_a of nanographite by Raman spectroscopy, *Applied Physics Letters* 88(16): 163106.

Cançado, L. G., Jorio, A. & Pimenta, M. A. (2007). Measuring the absolute Raman cross section of nanographites as a function of laser energy and crystallite size, *Phys. Rev. B* 76(6): 064304.

Cançado, L. G., Pimenta, M. A., Neves, B. R. A., Dantas, M. S. S. & Jorio, A. (2004). Influence of the atomic structure on the Raman spectra of graphite edges, *Phys. Rev. Lett.* 93(24): 247401.

Carozo, V., Almeida, C. M., Martins Ferreira, E. H., Achete, C. A., Jorio, A. (2011). Raman Spectroscopy of Folded Graphene. Umpublished.

Casiraghi, C., Hartschuh, A., Qian, H., Piscanec, S., Georgi, C., Fasoli, A., Novoselov, K. S., Basko, D. M. & Ferrari, A. C. (2009). Raman spectroscopy of graphenes edges, *Nano Lett.* 9(4): 1433–1441.

Chou, S. G., Son, H., Kong, J., Jorio, A., Saito, R., Zheng, M., Dresselhaus, G., & Dresselhaus, M. S. (2007). Length characterization of DNA-wrapped carbon nanotubes using Raman spectroscopy, *Appl. Phys. Lett.* 90: 131109.

Ferrari, A. C. & Robertson, J. (2000). Interpretation of Raman spectra of disordered and amorphous carbon, *Phys. Rev. B* 61(20): 14095–14107.

Gupta, A. K., Russin, T. J., Gutiérrez, H. R. & Eklund, P. C. (2009). Probing graphene edges *via* Raman scattering, *ACS Nano* 3(1): 45–52.

Hulman, M., Skakalova, V., Roth, S. & Kuzmany, H. (2005). Raman spectroscopy of single-wall carbon nanotubes and graphite irradiated by γ rays, *J. of Appl. Phys.* 98(2).

Ishigami, M., Chen, J. H., Cullen, W. G., Fuhrer, M. S. & Williams, E. D. (2007). Atomic structure of graphene on SiO_2, *Nano Letters* 7(6): 1643–1648.

Jorio, A., Lucchese, M. M., Stavale, F. & Achete, C. A. (2009). Raman spectroscopy study of Ar^+ bombardment in highly oriented pyrolytic graphite, *Phys. Status Solidi B* 246(11–12): 2689–2692.

Jorio, A., Lucchese, M. M., Stavale, F., Martins Ferreira, E. H., Moutinho, M. V. O., Capaz, R. B. & Achete, C. A. (2010). Raman study of ion-induced defects in N-layer graphene, *J. Phys.: Condens. Matter* 22: 334204.

Lucchese, M. M., Stavale, F., Martins Ferreira, E. H., Vilani, C., Moutinho, M. V. O., Capaz, R. B., Achete, C. A. & Jorio, A. (2010). Quantifying ion-induced defects and Raman relaxation length in graphene, *Carbon* 48(5): 1592–1597.

Martins Ferreira, E. H., Moutinho, M. V. O., Stavale, F., Lucchese, M. M., Capaz, R. B., Achete, C. A. & Jorio, A. (2010). Evolution of the Raman spectra from single-, few-, and many-layer graphene with increasing disorder, *Phys. Rev. B* 82(12): 125429.

Mernagh, T. P., Cooney, R. P. & Johnson, R. A. (1984). Raman spectra of graphon carbon black, *Carbon* 22(1): 39–42.

Neto, A. C., Guinea, F., Peres, N. M. R. & Geim, K. S. N. A. K. (2009). The electronic properties of graphene, *Rev. Mod. Phys.* 81: 109–162.

Neubeck, S., You, Y. M., Ni, Z. H., Blake, P., Shen, Z. X., Geim, A. K. & Novoselov, K. S. (2010). Direct determination of the crystallographic orientation of graphene edges by atomic resolution imaging, *Appl. Phys. Lett.* 97(5): 053110.

Novoselov, K. S., Geim, A. K., Morozov, S. V., Jiang, D., Zhang, Y., Dubonos, S. V., Grigorieva, I. V. & Firsov, A. A. (2004). Electric field effect in atomically thin carbon films, *Science* 306(666–669).

Novoselov, K. S., Jiang, D., Schedin, F., Booth, T. J., Khotkevich, V. V., Morozov, S. V., & Geim, A. K. (2005). Two-dimensional atomic crystals, *Proc. Natl. Acad. Sci. USA* 102(30): 10451–10453.

Pimenta, M. A., Dresselhaus, G., Dresselhaus, M. S., Cançado, L. G., Jorio, A. & Saito, R. (2007). Studying disorder in graphite-based systems by Raman spectroscopy, *Physical Chemistry Chemical Physics* 9: 1276–1291.

Piscanec, S., Lazzeri, M., Mauri, F., Ferrari, A. C. & Robertson, J. (2004). Kohn anomalies and electron-phonon interaction in graphite, *Phys. Rev. Lett.* 93(18): 185503.

Richter, H., Wang, Z. P. & Ley, L. (1981). The one phonon Raman spectrum in microcrystalline silicon, *Solid State Commun.* 39: 625.

Saito, R., Jorio, A., Filho, A. G. S., Dresselhaus, G., Dresselhaus, M. S. & Pimenta, M. A. (2002). Probing phonon dispersion relations of graphite by double resonance Raman scattering, *Phys. Rev. Lett.* 88(02): 027401.

Sasaki, K., Saito, R., Wakabayashi, K. & Enoki, T. (2010). Indentifying the orientation of edge of graphene using G band Raman spectra, *J. of the Phys. Soc. of Japan* 79(4): 044603.

Teweldebrhan, D. & Baladin, A. A. (2009). Modification of graphene properties due to electron-beam irradiation, *Appl. Phys. Lett.* 94: 013101.

Thomsen, C. & Reich, S. (2000). Double resonant Raman scattering in graphite, *Phys. Rev. Lett.* 85(24): 5214.

Tiong, K. K., Amirtharaj, O. M., Pollak, F. H. & Aspnes, D. E. (1984). Effects of as^{+} ion implantation on the Raman spectra of GaAs: spatial correlation interpretation, *Appl. Phys. Lett.* 44: 122.

Tuinstra, F. & Koenig, J. L. (1970). Raman spectrum of graphite, *J. Chem. Phys.* 53(3): 1126–1130.

6

Graphene-Supported Platinum and Platinum-Ruthenium Nanoparticles for Fuel Cell Applications

Lifeng Dong[1,2], Qianqian Liu[1], Li Wang[2] and Kezheng Chen[2] [1]*Department of Physics, Astronomy, and Materials Science, Missouri State University,* [2]*College of Materials Science and Engineering,*
Qingdao University of Science and Technology,
[1]*U.S.A.*
[2]*P. R. China*

1. Introduction

As one of the most promising alternative energies, fuel cells have been receiving increased attention recently due to the depletion of fossil fuels and the increase in environmental pollution. Among different types of fuel cells, direct ethanol fuel cells (DEFCs) and direct methanol fuel cells (DMFCs) are excellent power sources due to their high energy density, low pollutant emission, low operating temperature, and ease of handling liquid ethanol/methanol fuel (Lamy et al., 2002; Long et al., 2000; Maillard et al., 2003). However, critical obstacles remain that inhibit broad applications of DEFCs and DMFCs, including low electrocatalytic activity of anode/cathode electrodes and the high cost of noble metal, platinum (Pt)-based catalysts. In order to enhance catalytic activity and to reduce the usage of Pt-based catalysts, one strategy is to explore novel carbon materials as catalyst supports and to effectively disperse Pt-based particles on these supports (Chen et al., 2007; Sanganna Gari et al., 2010; Girishkumar et al., 2006; Halder et al., 2009; Maiyalagan, 2008; Selvaraj & Alagar, 2007; Rajesh et al., 2003; Tang et al., 2004; Tsai et al., 2006; Zhao et al., 2004). Vulcan XC-72R carbon black has been the most widely-used catalyst support for the preparation of fuel cell catalysts because of its exceptional electronic conductivity and surface area (Guo et al., 2008; Lin et al., 2005; Maiyalagan et al., 2005). During the past several years, a few research groups have investigated graphene, one type of novel carbon nanostructures, as a catalyst support and have demonstrated that graphene can improve the electrocatalytic activity of Pt nanoparticles more effectively for methanol and ethanol oxidation than Vulcan XC-72R carbon black (Dong et al., 2010; Seger et al., 2009; Yoo et al., 2009).

In this chapter, a series of atomic force microscopy (AFM) and electron microscopy techniques, including scanning electron microscopy (SEM) and scanning transmission electron microscopy (STEM), were employed to characterize the morphology of Pt nanoparticles, platinum-ruthenium (Pt-Ru) nanoparticles, and graphene sheets as well as the distribution of Pt and Pt-Ru nanoparticles on the graphene supports. Effects of graphene supports on electrocatalytic activity of Pt and Pt-Ru nanoparticles for methanol and ethanol oxidations were investigated using cyclic voltammetry. In comparison to Vulcan XC-72R carbon black, graphene-supported Pt and Pt-Ru catalysts demonstrate better electrocatalytic activities in terms of oxidation potential, forward oxidation peak current density, and the ratio of forward peak current density to reverse peak current density. This study indicates that graphene can serve as an effective catalyst support for both direct methanol and ethanol fuel cells. These systematic characterizations provide detailed information on the morphology and structures of Pt and Pt-Ru nanoparticles supported on graphene sheets, and this information will help to design and synthesize graphene-supported Pt/Pt-Ru nanoparticles for fuel cell applications.

2. Structural characterization and electrocatalytic measurements

2.1 Synthesis of Pt and Pt-Ru nanoparticles

For the synthesis of graphene-supported Pt and Pt-Ru catalysts, 1 g of graphene oxide powder was dispersed in 100 ml of ethylene glycol (EG) and sonicated for 30 min (Li et al., 2003). Graphene oxide was prepared from graphite powder in accordance with the Hummers-Offeman method (Hummers et al., 1958). Hexachloroplatinic EG solution (15 ml) (7.4 mg H_2PtCl_6 per ml of EG) was added into the graphene oxide solution and mechanically stirred for 2 h. Sodium hydroxide was added to adjust the pH of the solution to 11.0, and then the solution was fluxed under flowing argon at 130 °C for 3 h. The solid was filtered, washed with deionized water, and dried in an oven at 80 °C for 12 h. For the synthesis of graphene-supported Pt-Ru catalysts, the only difference was to add 15 ml of EG solution, including both H_2PtCl_6 and ruthenium chloride ($RuCl_3$) (7.4 mg H_2PtCl_6 and 7.4 mg $RuCl_3$ per ml of EG) into the graphene oxide solution. Similar procedures were used to prepare carbon black (Vulcan XC-72R) supported Pt and Pt-Ru catalysts.

2.2 Electron microscopy and microanalysis

The morphology of Pt and Pt-Ru nanoparticles and their distributions on carbon black and graphene were investigated using an FEI Quanta 200 field emission scanning electron microscope (FESEM). The FESEM system is equipped with an Everhart-Thornley secondary electron detector (SED), a low-vacuum SED, a gaseous SED, a backscattered electron detector (BSED), a scanning transmission electron microscopy detector (STEM), and an Oxford INCA 250 silicon drift X-ray energy dispersive spectrometer (EDS).

2.3 Atomic force microscopy

A Multimode Nanoscope IIIa atomic force microscopy (AFM) system was used to examine surface characteristics of graphene sheets and nanoparticles supported on the graphene sheets. The AFM system was equipped with an E-type scanner and a standard silicon cantilever with a length of 115-135 μm and a tip radius less than 10 nm. All images were obtained using tapping mode in air at room temperature with a 1.0 Hz scan rate, a 0° scan angle, and a 261.0 Hz tapping frequency. Both topographical and phase images were captured for each real-time scan. Nanoscope III 6.13rl software was used for off-line image flattening and analysis.

2.4 Electrocatalytic activity measurements

Electrocatalytic activities of Pt and Pt-Ru nanoparticles on different supports (Vulcan XC-72R carbon black and graphene) were measured in a conventional three-electrode cell using a CHI

600C electrochemical work station. Ag/AgCl served as a reference electrode, Pt wire as a counter electrode, and glassy carbon coated with carbon black- and graphene-supported nanoparticles as a working electrode. For the preparation of a working electrode, 3.8 mg of catalyst powder was dispersed in 1 ml of ethanol solution and sonicated for 30 min; 30 μl of the suspension was placed onto the surface of a glassy carbon electrode and dried for 15 min; and 10 μl of 5% Nafion 117 solution was sequentially applied and dried for 15 min.

Electrocatalytic oxidation of methanol on graphene and carbon black-supported Pt and Pt-Ru nanoparticles was measured in 1 M CH_3OH + 0.5 M H_2SO_4 electrolyte by cyclic voltammetry between -0.3 V and 1.0 V. The potentials were measured and reported with respect to the Ag/AgCl electrode. After methanol oxidation measurements, the electrodes were stored in distilled water for 30 min prior to ethanol oxidation measurements. Electrocatalytic oxidation of ethanol was measured in 1 M CH_3CH_2OH + 0.5 M H_2SO_4 electrolyte by cyclic voltammetry between -0.3 V and 1.0 V. Prior to electrochemical characterizations, all electrolyte solutions were de-aerated under flowing nitrogen gas for 20 min. Electrical current density was calculated by normalizing electrical current on the surface of 3 mm diameter glassy carbon electrodes.

In order to ensure the reproducibility of experimental results, all electrocatalyst powders, including graphene- and carbon black-supported Pt and Pt-Ru nanoparticles, were synthesized three times; the same amount of catalytic powder was cautiously loaded onto the electrodes.

3. Electron microscopy study of Pt nanoparticles

Pt-Ru nanoparticles were synthesized using the same procedures and the same reaction parameters as for Pt nanoparticles, and thereby had similar morphology and similar distributions on catalytic supports (graphene and carbon black) compared to Pt nanoparticles. In this study, we report only the study of the morphology of Pt nanoparticles and their distributions on catalytic supports using electron microscopy and atomic force microscopy.

3.1 Scanning electron microscopy characterization

The Everhart-Thornley secondary electron detector and the solid state backscattered electron detector are the two most commonly-used detectors for a scanning electron microscope (SEM). Low energy SEs (less than 50 eV, with a most probable energy of 3 - 5 eV) are emitted from the surface and immediate sub-surface of a specimen and provide information regarding the surface topography of the specimen; high energy BSEs reveal areas having different chemical compositions because heavy elements with high atomic number backscatter incident electrons more strongly than light elements with low atomic number (Goldstein et al., 2003). For a sample characterization using SE/BSE detectors, nanoparticle powder was directly affixed onto the surface of double-stick conductive copper tape, which in turn was attached to an aluminum SEM sample stub.

Fig. 1a and Fig. 1c are SE images of Pt nanoparticles supported on graphene and carbon black, respectively. Graphene sheets are rippled and crumpled, with a dimension of several hundred nm to several μm; Vulcan XC-72R carbon black supports are spherical particles with a dimension of 30 nm to 100 nm. Few small particles exist on the surface of both graphene and carbon black supports, yet it is difficult to identify the locations of these particles due to weak contrast between small particles and their supports. However, simultaneously obtained BSE images (Fig. 1b and Fig. 1d) clearly reveal the dimension and

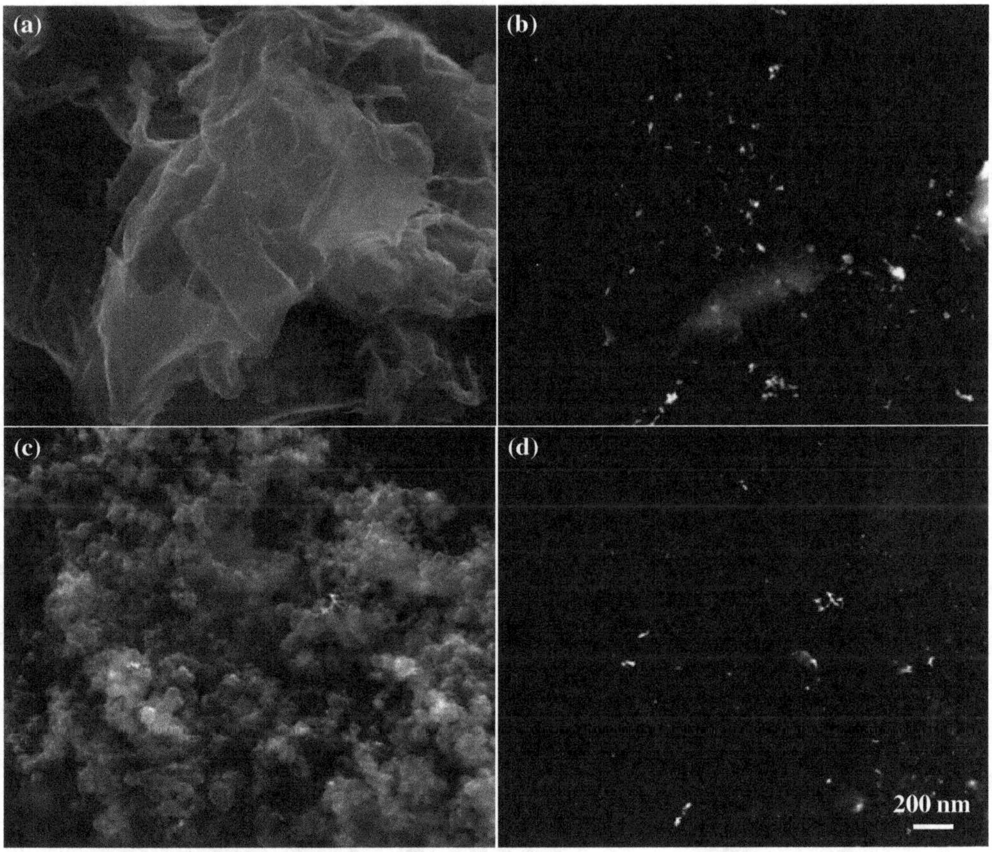

Fig. 1. SE (a, c) and BSE (b, d) images recorded simultaneously from Pt nanoparticles supported on graphene and carbon black, respectively. SE images demonstrate the surface morphology of graphene and carbon black, while the Z-contrast effect in the BSE imaging permits identification of Pt nanoparticles. All images were acquired at the same magnification for comparison.

distribution of those particles on the surface of graphene and carbon black, an observation which confirms that BSE imaging is a Z-contrast technique that yields compositional contrast information. Atomic numbers Z for Pt and C are 78 and 6, respectively. The signal from the BSE detector is proportional to the backscatter coefficient, which nearly monotonically increases with increasing Z. In Fig. 1b and Fig. 1d, the higher Z particles appear brighter than the lower Z background structures. The bright particles are Pt nanoparticles with a size of less than 10 nm. These nanoparticles are difficult to identify in SE images (Fig. 1a and Fig. 1c). SE images reveal the morphology and dimension of graphene and carbon black supports, while BSE images demonstrate the morphology and distribution of Pt nanoparticles associated with the surface of the graphene and carbon black support materials. SE and BSE imaging are complementary concurrent techniques for the characterization of Pt nanoparticles and supporting materials.

3.2 Scanning transmission electron microscopy characterization

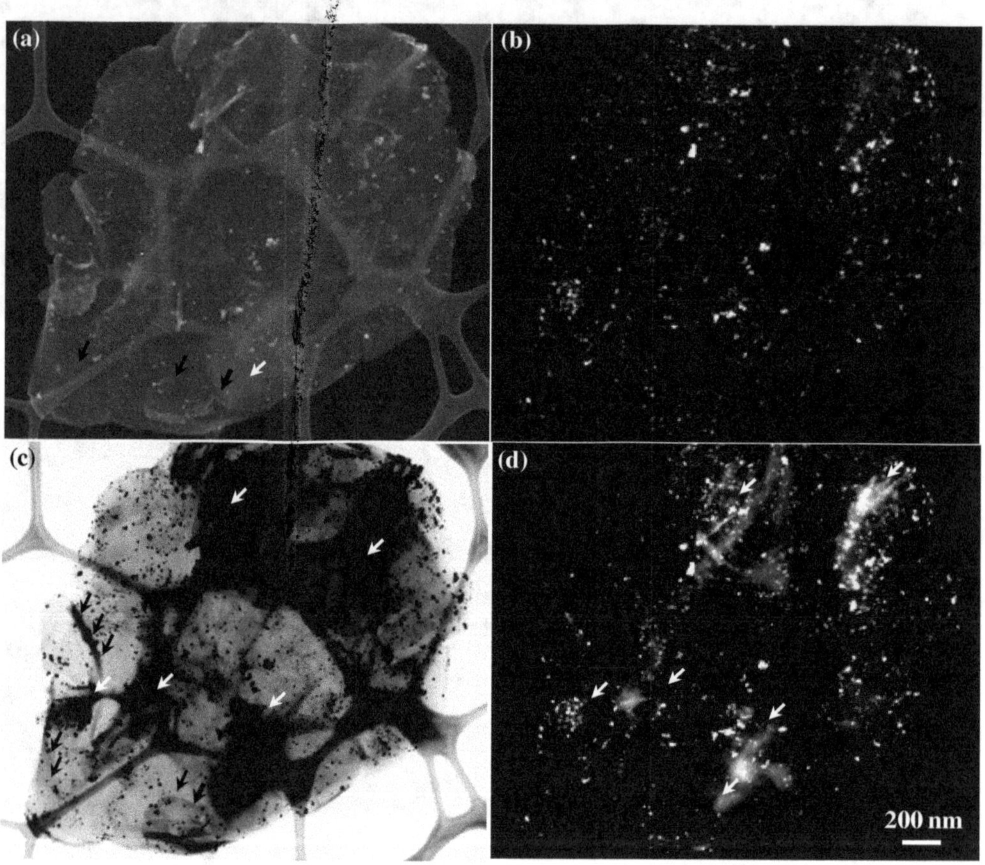

Fig. 2. Pt nanoparticles supported on graphene sheets were simultaneously imaged with SE (a), BSE (b), STEM-BF (c), and STEM-HAADF (d) detectors. Different detectors concurrently reveal different information regarding the morphology of graphene sheets and Pt nanoparticles as well as the distribution of Pt nanoparticles on the graphene sheets. Multiple graphene layers, white arrows; sheet folds and very few graphene layers, black arrows.

With the development of technology and improved manufacturing techniques, SEMs are more commonly equipped with a STEM detector for revealing internal structures of a specimen. A retractable STEM detector, attached to the FEI Quanta 200 FESEM system, consists of three different modes of detection: bright field (BF), dark field (DF), and high angle annular dark field (HAADF). The STEM-BF detector collects electrons transmitted through the specimen. Both mass-thickness and diffraction contrast contribute to the BF image formation; the STEM-DF detector gathers electrons diffracted in crystalline areas and incoherent Rutherford scattering electrons; and the STEM-HAADF detector collects electrons scattered to high angles (50 to 200 mrad) where the scattering is predominately incoherent Rutherford scattering and roughly proportional to the square of the atomic number (Z). SE and BSE in combination with STEM-BF/DF/HAADF provide information about the surface and throughout the specimen at the same time. One prerequisite for STEM imaging is that the specimen needs to be thin enough to enable the transmission of incident electrons. Usually, the thickness of the specimen is less than 100 nm. For a sample characterization using SE/BSE/STEM-BF/STEM-HAADF, sample powder was dispersed in an ethanol solution and sonicated for 5 min. Several drops of suspension were transferred subsequently onto a holey carbon film supported on a 3 mm copper grid.

As given in Fig. 2, SE, BSE, STEM-BF, and STEM-HAADF images of Pt nanoparticles supported on graphene sheets demonstrate different characteristics of various imaging techniques. A graphene sheet is a single atomic layer of carbon, which is thin enough to allow incident electrons to pass completely through it (Girit et al., 2009; Novoselov et al., 2004; Novoselov et al., 2005; Zhang et al., 2005). All four images in Fig. 2, which are different from the images in Fig.1, illustrate the distribution of Pt nanoparticles on the surfaces of graphene sheets. Different imaging techniques have their unique advantages for morphology and structural characterization. For instance, an SE image (Fig. 2a) shows several small pieces of graphene sheets (200 nm to 500 nm) and some small particles that exist on the surface of a large piece of graphene sheet (around 2.5 μm). In comparison to Fig. 2a, Z-contrast BSE image (Fig. 2b) clearly demonstrates the dimension and distribution of Pt nanoparticles, but no information on graphene sheets. However, STEM-BF imaging (Fig. 2c) provides information on both graphene sheets and Pt nanoparticles' dimension and distribution. In addition, STEM-BF imaging qualitatively provides information on the thickness of those graphene sheets: some areas with dark contrast resulted from their large thickness. These areas were too thick to allow incident electrons to pass through them and thereby, no information on Pt nanoparticles could be revealed in these regions (marked by white arrows in Fig. 2c); graphene sheets with a few layers or folded layers show gray contrast (marked by black arrows in Fig. 2c). In comparison to BSE imaging (Fig. 2b), Z-contrast STEM-HAADF imaging (Fig. 2d) provides similar information on Pt nanoparticles but some additional information about the thickness of graphene sheets: some thick regions appear brighter than surrounding backgrounds due to multi-scattering of electrons in multi-layers of graphene sheets (marked by white arrows in Fig. 2d).

Therefore, in comparison to conventional SE/BSE imaging, STEM-BF/HAADF imaging in combination with SE/BSE detectors provides much more information on the morphology of graphene sheets and nanoparticles as well as the distribution of nanoparticles on the surface of graphene sheets and their inside layers. A next step for electron microscopy characterization is to explore interactions between Pt nanoparticles and graphene supports through atomic imaging of their interfaces using transmission electron microscopy (TEM)/STEM with a high acceleration voltage (e.g., 200 kV) or aberration-corrected TEM/STEM (Dong, 2009; Pennycook et al., 2009; Zhu et al., 2009).

4. Atomic force microscopy characterization of Pt nanoparticles

Tapping mode atomic force microscopy (AFM) provides information about surface structure of a sample through topographical imaging and phase imaging. As illustrated in Fig. 3a, some individual particles and small clusters consisting of several particles exist on the surfaces of graphene sheets. A height line profile (Fig. 3c), recorded along two nanoparticles (marked as a dashed line in Fig. 3a), shows the dimensions of Pt nanoparticles were approximately 6 nm to 8 nm. The dimension and distribution of nanoparticles on graphene sheets are consistent with those observed using SEM/STEM. Beyond topographical

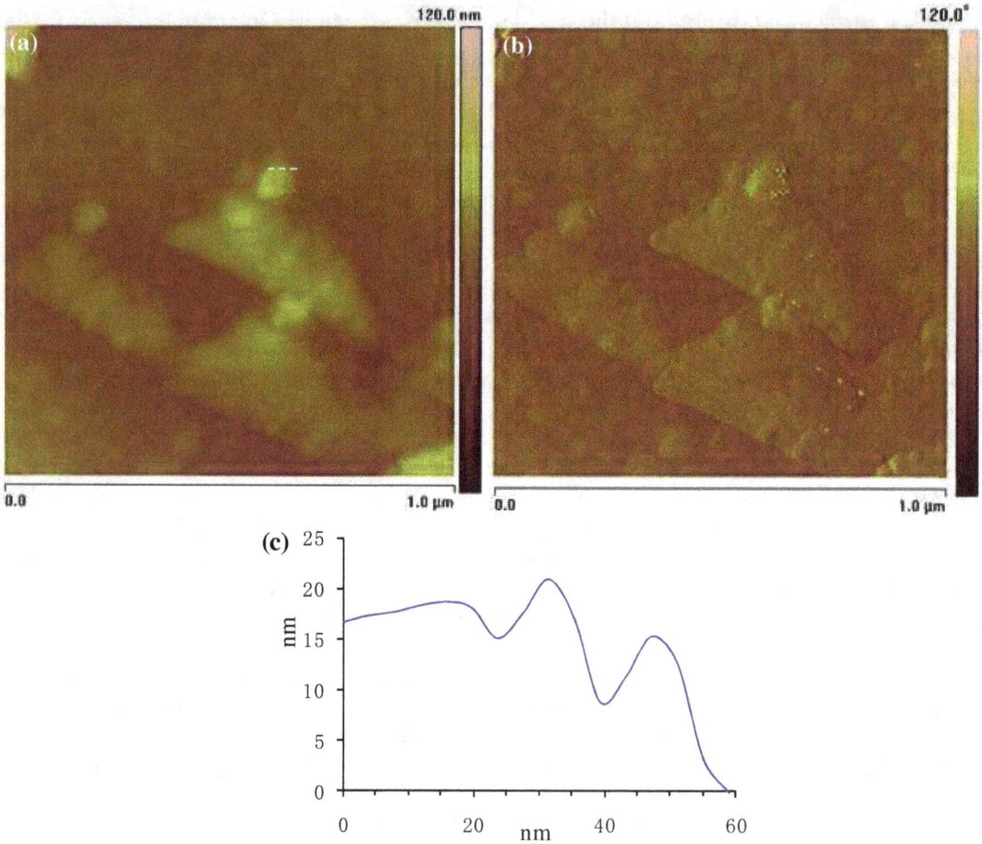

Fig. 3. AFM topography (a) and phase (b) images of Pt nanoparticles supported on graphene sheets. A line height profile (c), recorded along two nanoparticles (marked as a dashed line in Fig. 3a), shows that the dimensions of Pt nanoparticles were approximately 6 nm to 8 nm.

information, phasing imaging detects variations in composition, viscoelasticity, adhesion, and others, since these parameters affect the phase of the oscillation of a cantilever (Liu et al., 2009; Tamayo & Garcia, 1997). In comparison to topographical imaging (Fig. 3a), the phase imaging in Fig. 3b clearly provides more information on the morphology of Pt nanoparticles and their distribution on graphene sheets. Besides those nanoparticles observed in Fig. 3a, some individual particles with smaller dimensions were revealed through the phase imaging mode. However, it is challenging to obtain detailed information on large pieces of graphene sheets due to their intrinsic tendency to fold and crumple together. AFM also lacks depth of field, yet SEM can have a depth of field of several µm.

5. Electrocatalytic activity for methanol oxidation

The electrocatalytic activity of methanol oxidation on carbon black- and graphene-supported Pt and Pt-Ru nanoparticles was characterized by cyclic voltammetry in an electrolyte of 1 M CH_3OH and 0.5 M H_2SO_4 at 50 mV/s. The voltammograms of each sample became stable after the fifth cycle. The resulting voltammograms from the fifth cycle are shown in Fig. 4. The efficiencies of the Pt and Pt-Ru nanoparticles on methanol oxidation were compared with regard to oxidation potential, forward oxidation peak current density, and the ratio of the forward peak current density to the reverse peak current density; these data are summarized in Table 1.

Electrode	Forward Sweep		Reverse Sweep		I_F/I_R Ratio
	E (V)	I_F (mA/cm^2)	E (V)	I_R (mA/cm^2)	
Pt-graphene	0.65	19.1	0.46	2.93	6.52
Pt-CB	0.59	9.76	0.43	7.01	1.39
Pt-Ru-graphene	0.50	16.5	-	-	-
Pt-Ru-CB	0.58	5.23	0.36	1.51	3.46

Table 1. Comparison of electrocatalytic activity of methanol oxidation on graphene and carbon black-supported Pt and Pt-Ru nanoparticles.

Graphene-supported Pt nanoparticles have a higher activity for methanol oxidation and a better tolerance to CO in comparison to carbon black-supported Pt nanoparticles (Fig. 4a). On the forward potential sweep, the current increases slowly at lower potentials, then quickly increases at potentials higher than 0.5 V, and oxidation occurs at approximately 0.65 V. The magnitude of the peak current density is directly proportional to the amount of methanol oxidized at the electrode. The peak current density observed with graphene is approximately 19 mA/cm^2, nearly two times the peak current density for carbon black (9.8 mA/cm^2); this result suggests that graphene plays a critical role in promoting methanol oxidation of Pt nanoparticles. In the reverse potential sweep, an oxidation peak is observed around 0.46 V, putatively associated with the removal of CO and other residual carbon species formed on the electrode in the forward sweep. The reverse anodic peak current density (2.93 mA/cm^2) is much lower than those measurements for carbon black (7.01 mA/cm^2). The ratio of the forward anodic peak current density (I_F) to the reverse anodic peak current density (I_R) can be used to describe the tolerance of Pt catalysts to CO and other carbonaceous species. Graphene-supported Pt nanoparticles have a ratio of 6.52, much higher than carbon black-supported Pt nanoparticles (1.39), a result which suggests that graphene-supported Pt nanoparticles generate a more complete oxidation of methanol to carbon dioxide.

As shown in Fig. 4c, the addition of Ru results in a negative shift of oxidation potential of methanol at all two electrodes with the decreasing of I_R and the increase of the I_F to I_R ratio. The oxidation peak potential of methanol was 0.50 V compared with an oxidation of 0.65 V for graphene-Pt nanoparticles. This observation suggests that the presence of Ru can significantly decrease the barrier to methanol oxidation and that Pt-Ru nanoparticles perform in a better fashion than Pt nanoparticles. Conversely, the oxidation potential of methanol at carbon black-supported Pt-Ru nanoparticles minimally shifts 0.02 V (0.58 V to 0.56 V). These data indicate that graphene-supported Pt-Ru may be an ideal candidate for direct methanol fuel cell electrodes. The above data also clearly support the proposition that Ru-Pt nanoparticles can effectively advance the oxidation of methanol to CO_2 and inhibit the accumulation of CO and other carbonaceous species formed on the graphene and carbon black electrodes. For graphene electrodes especially, there is no obvious reverse peak in Fig. 4b. Similar effects of Ru on other catalysts for methanol oxidation have been reported by a number of research groups (Long et al., 2000; Maillard et al., 2003).

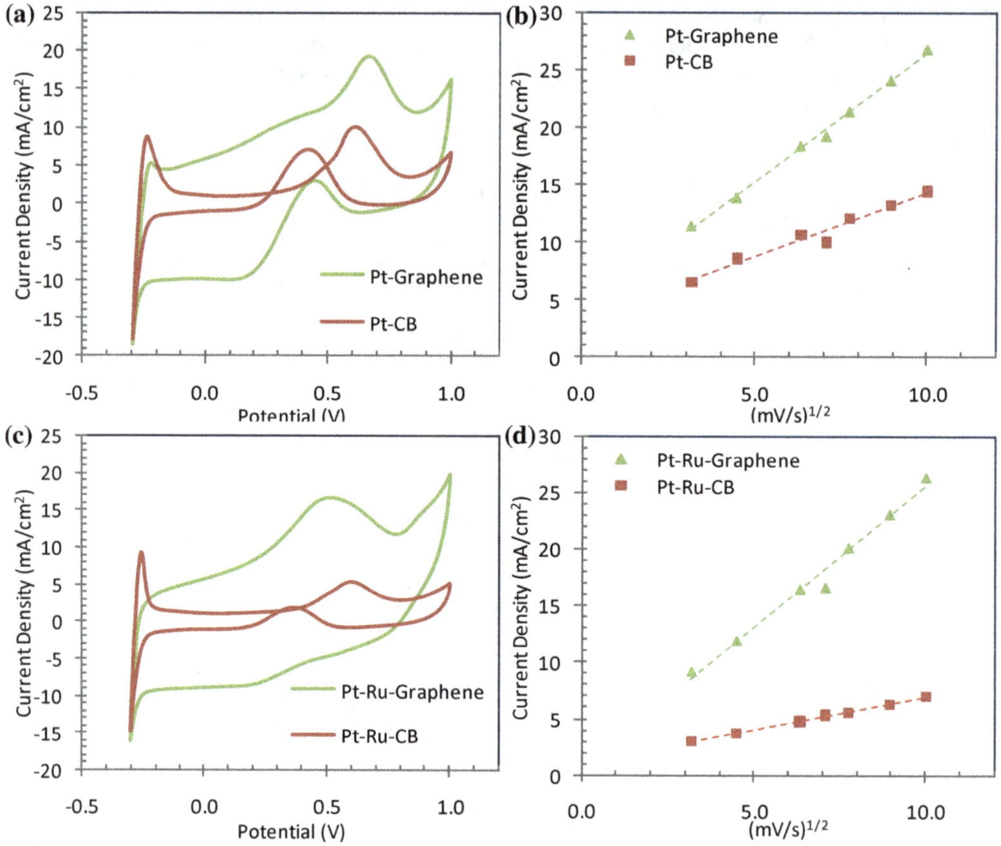

Fig. 4. Electrocatalytic activity of graphene and Vulcan XC-72R carbon black-supported Pt and Pt-Ru nanoparticles for methanol oxidation: cyclic voltammograms of Pt nanoparticles (a) and Pt-Ru nanoparticles (c) in 1 M CH_3OH/0.5 M H_2SO_4 at 50 mV/s between -0.3 V and +1.0 V vs. Ag/AgCl; and the relationship of peak current density vs. scan rate for Pt nanoparticles (b) and Pt-Ru nanoparticles (d).

As given in Fig. 4b and Fig. 4d, for all graphene- and carbon black-supported Pt and Pt-Ru nanoparticles, the forward oxidation current density (I_F) is proportional to the square root of the scan rate, suggesting that the oxidation behavior of methanol at all electrodes is controlled by diffusion processes. The slopes for graphene supports are larger than those for carbon black, indicative of a faster diffusion process of methanol on the surfaces of graphene sheets than that for methanol on the carbon black substrata.

Chronoamperometry was used to characterize the stability of the electrodes for methanol oxidation activity. The current density-time plots of graphene- and carbon black-supported Pt and Pt-Ru nanoparticles in 1 M CH_3OH/0.5 M H_2SO_4 at 0.4 V are shown in Fig. 5. During the 60-min test period, graphene-supported nanoparticles behave in a more stable fashion than those nanoparticles supported on carbon-black, and graphene-supported Pt-Ru nanoparticles were more stable than graphene-supported Pt nanoparticles. These results suggest that the addition of Ru onto Pt nanoparticles lowers not only oxidation potential (0.65 V to 0.5 V), but also increases the tolerance of Pt nanoparticles to CO and thereby improves the stability of Pt nanoparticles for methanol oxidation.

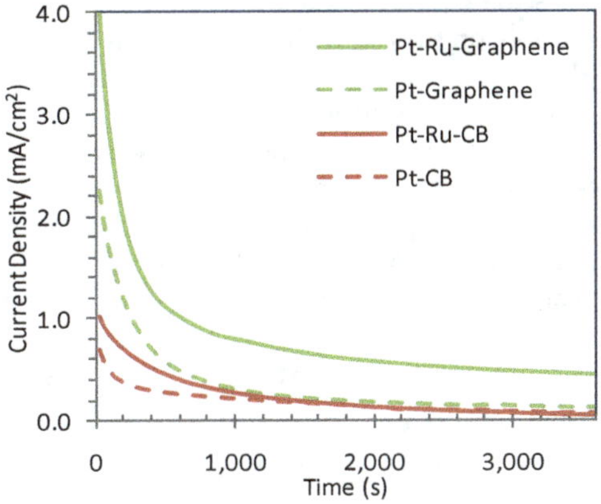

Fig. 5. Chronoamperogram curves for graphene- and carbon black-supported Pt and Pt-Ru nanoparticles in 1 M CH_3OH/0.5 M H_2SO_4. The potential was 0.4 V vs. Ag/AgCl.

6. Electrocatalytic activity for ethanol oxidation

In comparison to direct methanol fuel cells, fuel cells based on direct ethanol oxidation possess several advantages, such as demonstrating enhanced safety, having a higher energy density (8.01 vs. 6.09 kW/kg), and being more easily produced in large quantities from agricultural products and from the fermentation of biomass (Lamy et al., 2002). A complete ethanol electro-oxidation involves the release of 12 electrons per molecule and the cleavage of the C-C bond. Experimental studies have revealed that ethanol is largely and partially oxidized to acetaldehyde and acetic acid (two- and four-electron oxidation, respectively) instead of being fully oxidized (12-electron oxidation) (Wang & Liu, 2007). The ethanol electro-oxidation was considered more complicated than that for methanol, and consequently, needs more active and selective catalysts. In this study, we investigated effects of graphene as a catalyst support for ethanol electro-oxidation of Pt and Pt-Ru nanoparticles in comparison to the extensively-used Vulcan XC-72R carbon black.

Electrode	Forward Sweep		Reverse Sweep		I_F/I_R Ratio
	E (V)	I_F (mA/cm^2)	E (V)	I_R (mA/cm^2)	
Pt-graphene	0.72	16.2	0.48	4.43	3.66
Pt-CB	0.65	13.8	0.46	15.3	0.90
Pt-Ru-graphene	0.66	18.1	0.46	5.23	3.46
Pt-Ru-CB	0.64	7.50	0.36	4.93	1.52

Table 2. Comparison of electrocatalytic activity of ethanol oxidation on graphene and carbon black-supported Pt and Pt-Ru nanoparticles.

Compared with carbon black, graphene enhances considerably the electrocatalytic activity of Pt nanoparticles for ethanol oxidation by increasing the forward oxidation current density and decreasing the reverse peak current density (Fig. 6 and Table 2). As discussed above, the

I_F to I_R ratio describes the catalyst tolerance to CO-like intermediates formed on electrodes during the forward potential sweep. The high forward peak current density (16.2 mA/cm^2) and the high I_F to I_R ratio (3.66) for graphene-supported Pt nanoparticles indicate that graphene effectively enhances the complete oxidation of ethanol to CO_2, and fewer CO-like carbonaceous species accumulate on the electrode surface. The forward peak current density and the I_F to I_R ratio for carbon black-supported Pt nanoparticles were 16.2 mA/cm^2 and 0.9, respectively. As given in Fig. 6c, the addition of Ru effectively lowers the potentials for ethanol oxidations and increases the I_F to I_R ratios. The oxidation peak potential of ethanol was 0.66 V for graphene-supported Pt-Ru nanoparticles in comparison to an oxidation potential of 0.72 V for graphene-supported Pt nanoparticles, and the potential shift of ethanol oxidation is 0.06 V. However, no significant shifts of oxidation potential for ethanol were observed for carbon black-supported Pt nanoparticles (from 0.65 V to 0.64 V).

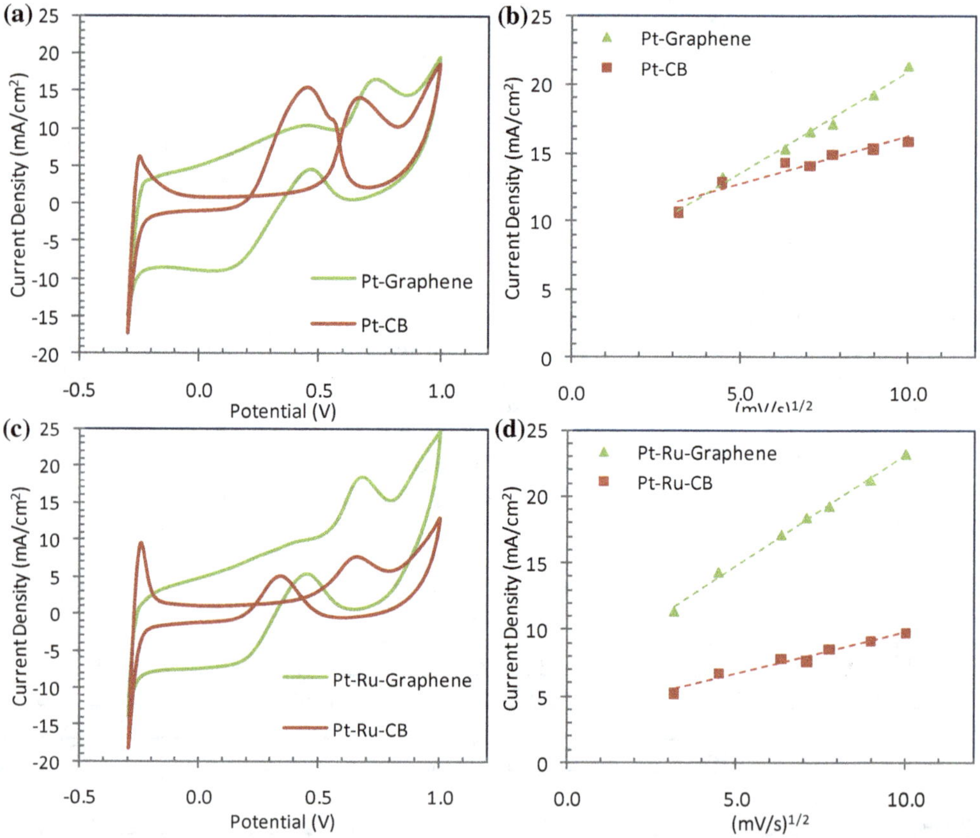

Fig. 6. Electrocatalytic activity of graphene- and Vulcan XC-72R carbon black-supported Pt and Pt-Ru nanoparticles for ethanol oxidation: cyclic voltammograms of Pt nanoparticles (a) and Pt-Ru nanoparticles (c) in 1M CH_3CH_2OH/0.5M H_2SO_4 at 50 mV/s between -0.3 V and +1.0 V vs. Ag/AgCl; the relationship of peak current density vs. scan rate for Pt nanoparticles (b) and Pt-Ru nanoparticles (d).

The forward peak potentials for ethanol oxidation are much higher in comparison to methanol oxidation. For example, the peak potentials for ethanol oxidations on graphene- and carbon black-supported Pt nanoparticles were 0.72 V and 0.65V, respectively; the peak potentials for methanol oxidations on graphene- and carbon black-supported Pt nanoparticles were 0.65 V and 0.59 V, respectively. These results suggest that ethanol oxidation needs more energy than does methanol oxidation. Similar phenomena occur on graphene- and carbon black-supported Pt-Ru nanoparticles (Table 1 and Table 2).

As shown in Fig. 6b and 6d, for both graphene and carbon black-supported Pt and Pt-Ru nanoparticles, the forward oxidation peak current density (I_F) is proportional to the square root of the scan rate, suggesting that the oxidation of ethanol at all electrodes is a diffusion-controlled phenomenon. Graphene has a larger slope than that for carbon black, suggesting that the diffusion efficiency of ethanol inside graphene sheets is higher than that inside carbon black and therefore, can enhance the efficiency and capacity of fuel cells.

The stability of the electrodes for ethanol oxidation was studied using chronoamperometry. The current density-time plots of graphene- and carbon black-supported Pt and Pt-Ru nanoparticles in 1 M CH_3CH_2OH/0.5 M H_2SO_4 at 0.4 V are shown in Fig. 7. Graphene-supported nanoparticles appear to be more stable than those nanoparticles supported on carbon-black, and graphene-supported Pt-Ru nanoparticles were more stable than graphene-supported Pt nanoparticles. Initial current density for carbon black-supported Pt-Ru nanoparticles was higher than that for graphene-supported Pt catalysts, but ethanol oxidation activity on carbon black supports deteriorated more rapidly than that on graphene. After 200 s, graphene-supported Pt nanoparticles demonstrated higher oxidation activity than those supported on carbon-black supports. These results indicate that the addition of Ru onto Pt nanoparticles lowers not only oxidation potential (0.72 V to 0.66 V), but also increases the tolerance of Pt nanoparticles to CO and thereby improves the stability of Pt nanoparticles for ethanol oxidation.

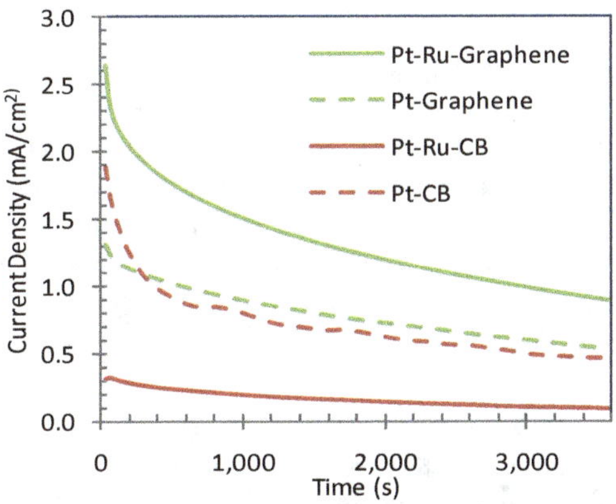

Fig. 7. Chronoamperogram curves for graphene- and carbon black-supported Pt and Pt-Ru nanoparticles in 1 M CH_3CH_2OH/0.5 M H_2SO_4. The potential was 0.4 V vs. Ag/AgCl.

Experimental results on electrocatalytic activity of both methanol and ethanol oxidations clearly demonstrates that graphene-supported Pt and Pt-Ru nanoparticles possess superior electrocatalytic activity compared to Vulcan XC-72R carbon black. Exceptional electrical conductivity and large surface area are two critical parameters for the development of novel catalyst supports for direct methanol and ethanol fuel cells. Graphene sheets possess both advantages when compared to Vulcan XC-72R carbon black.

7. Conclusion

Different imaging techniques have their various unique advantages for morphology and structural characterization. AFM tapping mode imaging can clearly reveal the morphology of nanoparticles and their distributions on graphene supports; however, it is challenging to image folded or crumpled graphene sheets. In comparison to both AFM imaging and conventional SE/BSE SEM imaging, concurrent STEM/SEM imaging can provide much more detailed information on the morphology of graphene sheets and nanoparticles as well as the distribution of nanoparticles on graphene supports. Future work on electron microscopy characterization will explore interactions between Pt nanoparticles and graphene supports through atomic imaging of their interfaces using high resolution TEM/STEM techniques.

In comparison to Vulcan XC-72R carbon black as a catalyst support, graphene-supported Pt and Pt-Ru nanoparticles have lower oxidation potentials, higher forward oxidation peak current density, higher ratios of forward peak current density to reverse peak current density, and better stability for both methanol and ethanol oxidation. These findings clearly demonstrate that as catalyst supports, graphene can more effectively enhance electrocatalytic activity of Pt and Pt-Ru nanoparticles for the oxidation of methanol and ethanol into CO_2. A next step will be the exploration of a facile and efficient method to synthesize graphene-supported Pt and Pt-Ru nanoparticles with controlled properties, which includes the formation of nanoparticles on defined locations, the density of nanoparticles located on graphene sheets, the size of nanoparticles, and the size and thickness of graphene sheets (Dong, 2009).

8. Acknowledgments

The authors thank Dr. Michael M. Craig for helpful discussions and Raghavendar Reddy Sanganna Gari and Zhou Li for their help with some experiments. This work was partially supported by a Summer Faculty Fellowship from Missouri State University, the American Chemical Society Petroleum Research Fund (47532-GB10), the Cottrell College Science Award from Research Corporation for Science Advancement, and the National Science Foundation (DMR-0821159). L. F. Dong also acknowledges financial support by the Taishan Scholar Overseas Distinguished Professorship program from the Shandong Province Government, P. R. China.

9. References

Chen, C. C.; Chen, C. F.; Chen, C. M. & Chuang, F. T. (2007). Modification of multi-walled carbon nanotubes by microwave digestion method as electrocatalyst supports for direct methanol fuel cell applications. *Electrochemistry Communications*, 9, 159-163.

Dong, L. F. (2009). DNA-templated synthesis of Pt nanoparticles on single-walled carbon nanotubes. *Nanotechnology*, 20, 465602.

Dong, L. F.; Sanganna Gari, R. R.; Li, Z.; Craig, M. M. & Hou, S. F. (2010). Graphene-supported platinum and platinum-ruthenium nanoparticles with high electrocatalytic activity for methanol and ethanol oxidation. *Carbon*, 48, 781-787.

Girishkumar, G.; Hall, T. D.; Vinodgopal, K. & Kamat, P. V. (2006). Single wall carbon nanotube supports for portable direct methanol fuel cells. *Journal of Physical Chemistry B*, 110, 107-114.

Girit, C.; Meyer, J. C.; Erni, R.; Rossell, M. D.; Kisielowski, C.; Yang, L.; Park, C. H.; Crommie, M. F.; Cohen, M. L.; Louie, S. G. & Zettl, A. (2009). Graphene at the edge: stability and dynamics. *Science*, 323, 1705-1708.

Goldstein, J.; Newbury, D. E.; Joy, D. C.; Lyman, C. E.; Echlin, P.; Lifshin, E.; Sawyer, L. & Michael, J. R. (2003). *Scanning Electron Microscopy and X-ray Microanalysis*, Springer, New York.

Guo, Y. L.; Zheng, Y. Z. & Huang, M. H. (2008). Enhanced activity of PtSn/C anodic electrocatalyst prepared by formic acid reduction for direct ethanol fuel cells. *Electrochimica Acta*, 53, 3102-3108.

Halder, A.; Sharma, S.; Hegde, M. S. & Ravishankar, N. (2009). Controlled attachment of ultrafine platinum nanoparticles on functionalized carbon nanotubes with high electrocatalytic activity for methanol oxidation. *Journal of Physical Chemistry C*, 113, 1466-1473.

Hummers, W. S. & Offeman, R. E. (1958). Preparation of graphitic oxide. *Journal of the American Chemical Society*, 80, 1339-1339.

Lamy, C.; Lima, A.; LeRhun, V.; Delime, F.; Cutanceau, C. & Leger, J. M. (2002). Recent advances in the development of direct alcohol fuel cells. *Journal of Power Sources*, 105, 283-296.

Li, W. Z.; Liang, C. H.; Zhou, W. J.; Qiu, J. S.; Zhou, Z. H.; Sun, G. Q. & Qin, X. (2003). Preparation and characterization of multiwalled carbon nanotube-supported platinum for cathode catalysts of direct methanol fuel cells. *Journal of Physical Chemistry B*, 107, 6292-6299.

Lin, Y. H.; Cui, X. L.; Yen, C. H. & Wai, C. M. (2005). PtRu/carbon nanotube nanocomposites synthesized in supercritical fluid: a novel electrocatalyst for direct methanol fuel cells. *Langmuir*, 21, 11474-11479.

Liu, H.; Fu, S. Y.; Zhu, J. Y.; Li, H. & Zhan, H. Y. (2009). Visualization of enzymatic hydrolysis of cellulose using AFM phase imaging. *Enzyme and Microbial Technology*, 4, 274-281.

Long, J. W.; Stroud, R. M.; Swider-Lyons, K. E. & Rolison, D. R. (2000). How to make electrocatalysts more active for direct methanol oxidation avoid PtRu bimetallic alloys! *Journal of Physical Chemistry B*, 104, 9772-9776.

Maillard, F.; Gloaguen, F. & Leger, J. M. (2003). Preparation of methanol oxidation electrocatalysts: ruthenium deposition on carbon supported platinum nanoparticles. *Journal of Applied Electrochemistry*, 33, 1-8.

Maiyalagan, T. (2008) Synthesis and electro-catalytic activity of methanol oxidation on nitrogen containing carbon nanotubes supported Pt electrodes. *Applied Catalysis B: Environmental*, 80, 286-295.

Maiyalagan, T.; Viswanathan, B. & Varadaraju, U. V. (2005). Nitrogen containing carbon nanotubes as supports for Pt-alternate anodes for fuel cell application. *Electrochemistry Communications*, 7, 905-912.

Novoselov, K. S.; Geim, A. K.; Morozov, S. V.; Jiang, D.; Zhang, Y.; Dubonos, S. V.; Grigorieva, I. V. & Firsov, A. A. (2004). Electric field effect in atomically thin carbon films. *Science*, 306, 666-669.

Novoselov, K. S.; Geim, A. K.; Morozov, S. V.; Jiang, D.; Katsnelson, M. I.; Grigorieva, I. V.; Dubonos, S. V. & Firsov, A. A. (2005). Two-dimensional gas of massless Dirac fermions in graphene. *Nature*, 438, 197-200.

Pennycook, S. J.; Varela, M.; Lupini, A. R., Oxley, M. P. & Chisholm, M. F. (2009). Atomic-resolution spectroscopic imaging: past, present, and future. *Journal of Electron Microscopy*, 58, 87-97.

Rajesh, B.; Thampi, K. R.; Bonard, J. M.; Xanthopoulos, N.; Mathieu, H. J. & Viswanathan, B. (2003). Carbon nanotubes generated from template carbonization of polyphenyl acetylene as the support for electrooxidation of methanol. *Journal of Physical Chemistry B*, 107, 2701-2708.

Sanganna Gari, R. R.; Li, Z. & Dong, L. F. (2010). Effects of different carbon nanotube supported catalysts on methanol and ethanol electro-oxidation, In: *Nanomaterials for Polymer Electrolyte Membrane Fuel Cells*, Swider-Lyons, K. & Park, B. (Ed.), T08-17.1.-17.6., Materials Research Society, Warrendale, PA.

Seger, B. & Kamat, P. V. (2009). Electrocatalytically active graphene-platinum nanocomposites: role of 2-D carbon support in PEM fuel cells. *Journal of Physical Chemistry C*, 113, 7990-7995.

Selvaraj, V. & Alagar, M. (2007). Pt and Pt-Ru nanoparticles decorated polypyrrole/multi-walled carbon nanotubes and their catalytic activity towards methanol oxidation. *Electrochemistry Communications*, 9, 1145-1153.

Tamayo, J. & Garcia, R. (1997). Effects of elastic and inelastic interactions on phase contrast images in tapping-mode scanning force microscopy. *Applied Physics Letters*, 71, 2394.

Tang, H.; Chen, J. H.; Huang, Z. P.; Wang, D. Z.; Ren, Z. F.; Nie, L. H.; Kuang, Y. F. & Yao, S. Z. (2004). High dispersion and electrocatalytic properties of platinum on well-aligned carbon nanotube arrays. *Carbon*, 42, 191-197.

Tsai, M. C.; Yeh, T. K. & Tsai, C. H. (2006). An improved electrodeposition technique for preparing platinum and platinum-ruthenium nanoparticles on carbon nanotubes directly grown on carbon cloth for methanol oxidation. *Electrochemistry Communications*, 8, 1445-1452.

Wang, H. F. & Liu, Z. P. (2007). Selectivity of direct ethanol fuel cell dictated by a unique partial oxidation channel. *Journal of Physical Chemistry C*, 111, 12157-12160.

Yoo, E. J.; Okata, T.; Akita, T.; Kohyama, M.; Nakamura, J. & Honma, I. (2009). Enhanced electrocatalytic activity of Pt subnanoclusters on graphene nanosheet surface. *Nano Letters*, 9, 2255-2259.

Zhang, Y. B.; Tan, Y. W.; Stormer, H. L. & Kim, P. (2005). Experimental observation of the quantum Hall effect and Berry's phase in graphene. *Nature*, 438, 201-204.

Zhao, X. S.; Li, W. Z.; Jiang, L. H.; Zhou, W. J.; Xin, Q.; Yi, B. L. & Sun, G. Q. (2004). Multi-wall carbon nanotube supported Pt-Sn nanoparticles as an anode catalyst for the direct ethanol fuel cell. *Carbon*, 42, 3263-3265.

7

Electronic Properties of Graphene Probed at the Nanoscale

Filippo Giannazzo, Sushant Sonde and Vito Raineri
Consiglio Nazionale delle Ricerche, Istituto per la Microelettronica e Microsistemi, Catania, Italy

1. Introduction

Graphene is a two-dimensional (2D) crystal made out of carbon atoms arranged on a hexagonal (honeycomb) structure, as shown schematically in Fig. 1(a). Although it has been experimentally isolated only recently (2004) [1], graphene has been the object of theoretical investigations for more than 60 years, because it represents the simplest structure to understand the electronic properties of all other allotropes of carbon with sp^2 hybridisation.

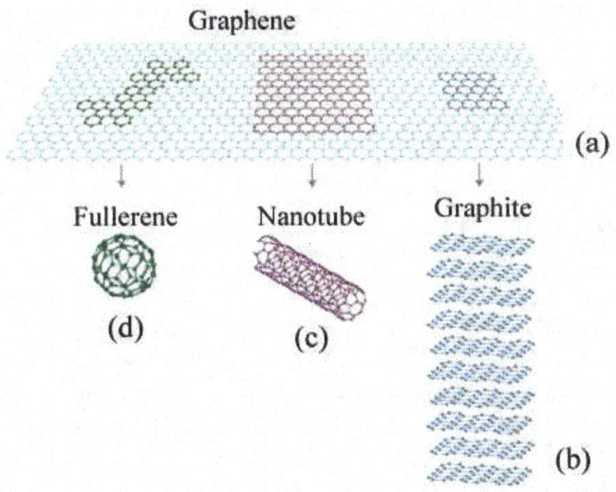

Fig. 1. Schematic structure of graphene (a), graphite (b), carbon nanotube (c) and fullerene (d).

In fact, fullerenes [2] are molecules where carbon atoms are arranged spherically (Fig.1(d)), and hence, from the physical point of view, are zero-dimensional (0D) objects with discrete energy states. Carbon nanotubes [3,4] are one-dimensional (1D) objects (Fig.1(c)) , and can be modelled as a rolled graphene sheet along a given direction with reconnected carbon bonds. Graphite, the three dimensional (3D) allotrope of sp^2 carbon, can be modelled as the stacking of graphene sheets (Fig.1(b)).
The graphene lattice structure is characterized by two types of C–C bonds (σ, π) constructed from the four valence orbitals (2s, $2p_x$, $2p_y$, $2p_z$), where the z-direction is perpendicular to

the sheet. Three σ-bonds join a C atom to its three neighbours. They are quite strong, leading to optical-phonon frequencies much higher than observed in diamond. In addition, the C-C bonding is enhanced by a fourth bond associated with the overlap of p_z (or π) orbitals. The electronic properties of graphene are determined by the bonding π- and antibonding π*-orbitals, that form electronic valence and conduction bands. Fig.2(a) shows a stereoscopic graph of the graphene valence (π) and conduction (π*) bands in the first Brillouin zone, while Fig.2(b) shows the energy of the π (π*) and σ (σ*) band along the high symmetry directions in the first Brillouin zone.

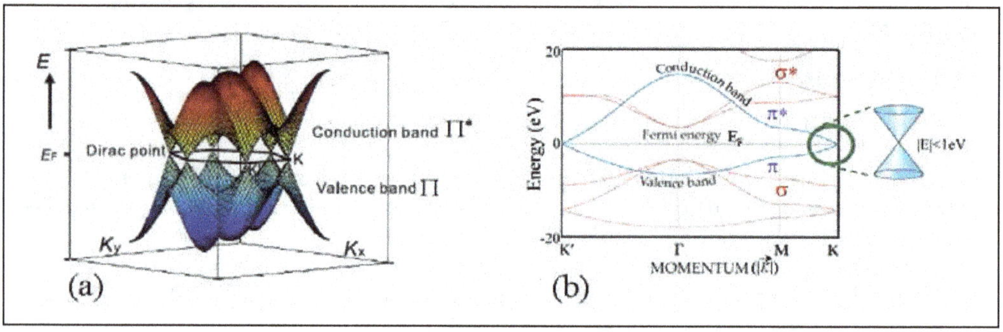

Fig. 2. Electronic bandstructure of graphene in the first Brillouin zone.

According to this bandstructure, graphene can be described as a zero-gap semiconductor. The π-band electronic dispersion for graphene at the six corners of the 2D hexagonal Brillouin zone is found to be linear,

$$E = \hbar v_F k. \tag{1}$$

where ħ is the reduced Planck's constant and v_F (≈10^6 m/s) is the electron Fermi velocity in graphene.

The linear electronic band dispersion leads to the description of carriers in graphene as "massless Dirac fermions". The six points where the cones touch are referred to as the "Dirac" points, and Fermi energy crosses these points in neutral graphene.

Graphene is a true 2D electron system. There are, broadly speaking, four qualitative differences between 2D graphene and 2D semiconductor systems found in Si inversion layers in MOSFETs, GaAs-AlGaAs heterostructures, quantum wells, etc.

i. First, 2D semiconductor systems typically have very large (> 1 eV) bandgaps so that 2D electrons and 2D holes must be studied using completely different electron-doped or hole-doped structures. By contrast, graphene is a gapless semiconductor with the nature of the carrier system changing at the Dirac point from electrons to holes (or vice versa) in a single structure. A direct corollary of this gapless (or small gap) nature of graphene, is of course, the "always metallic" nature of 2D graphene, where the Fermi level is always in the conduction or the valence band. By contrast, the 2D semiconductor becomes insulating below a threshold voltage, as the Fermi level enters the bandgap.

ii. Graphene systems are chiral, whereas 2D semiconductors are non-chiral. Chirality of graphene leads to some important consequences for transport behavior. For example, $2k_F$ -backscattering is suppressed in single layers of graphene at low temperature.

iii. Monolayer graphene dispersion is linear, whereas 2D semiconductors have quadratic energy dispersion. This leads to substantial quantitative differences in the transport properties of the two systems.
iv. Finally, the carrier confinement in graphene is ideally 2D, since the graphene layer is precisely one atomic monolayer thick. For 2D semiconductor structures, the quantum dynamics is two dimensional by virtue of confinement induced by an external electric field, and as such, 2D semiconductors are quasi-2D systems, and always have an average width or thickness $<t>$ ($\approx$ 5 nm to 50 nm) in the third direction with $<t>\leq \lambda_F$, where λ_F is the 2D Fermi wavelength (or equivalently the carrier de Broglie wavelength). The condition $<t> < \lambda_F$ defines a 2D electron system. In particular, for graphene, we have $\lambda_F \approx (2\times10^7\sqrt{\pi}/\sqrt{n})$ nm, where n is the carrier density in cm^{-2}, and since $t \approx 0.1$ nm to 0.2 nm (the monolayer atomic thickness), the condition $\lambda_F>t$ is always satisfied, even for an unphysically large $n=10^{14}$ cm^{-2}.

The classic technique to verify the 2D nature of a particular system is to show that the orbital electronic dynamics is sensitive only to a magnetic field perpendicular to the 2D plane of confinement. Therefore, the observation of magnetoresistance oscillations (Shubnikov-de Hass effect) proves the 2D nature. The most definitive evidence for 2D nature, however, is the observation of the quantum Hall effect, which is a quintessentially 2D phenomenon. Any system manifesting an unambiguous quantized Hall plateau is 2D in nature, and therefore the observation of the quantum Hall effect in graphene in 2005 by Novoselov et al. [5] and Zhang et al. [6] is an absolute evidence of its 2D nature. In fact, the quantum Hall effect in graphene persists to room temperature [7], indicating that graphene remains a strict 2D electronic material even at room temperature.

1.1 Graphene for electronic applications: opportunities and challenges

Graphene has been adopted as a material of choice by researchers involved in many fields. In particular, simultaneous observation of high mobility, sensitivity to field effect and ease of contacting made graphene an appealing candidate for field-effect transistor devices. A lot of progress has been made in that direction. Manufacturability of graphene transistor operating at 100 GHz has been demonstrated on wafer-scale [8]. Moreover, chemically stable, robust and foldable graphene sheets possess enough optical transparency to be considered as transparent conductors in applications as coatings for solar cells, liquid crystal displays etc. for the replacement for indium tin oxide [9].

However some open issues still remain and are currently the object of intense research activity.

i. Although the "intrinsic" mobility of graphene has been estimated to be $>2\times10^5$ $cm^2V^{-1}s^{-1}$, the mobility values commonly reported in the literature span a wide range, from $\sim10^2$ to $\sim10^4$ $cm^2V^{-1}s^{-1}$, depending on the synthesis method used to obtain graphene, on the kind of substrate where it is deposited, etc... It is evident that the intrinsically outstanding transport properties of this material are limited by extrinsic mechanisms. However, there is no consensus about the scattering mechanism that currently limits the mobility in graphene devices;
ii. Due to the zero bandgap of graphene electronic structure, it is not possible to reach high values of the ratio I_{on}/I_{off} between the on state and off state current in graphene field effect devices. To date several strategies have been developed to open a gap in graphene electronic bandstructure. Among them, the following must be cited.

a. The inclusion of sp^3 hydrocarbon defects in the sp^2 lattice [10],
b. The distortion of the graphene lattice under uniaxial strain [11],
c. The application of a transversal electric field to a bilayer graphene [12],
d. The influence of the SiC substrate on the epitaxial graphene on Si face of hexagonal SiC has also been shown to induce substantial band gap [13],
e. Moreover, a bandgap can also be induced in semimetallic graphene if it is shaped in nanoribbons by a direct consequence of the lateral confinement of the 2DEG [14]. However, this points to extremely challenging need to perform atomically precise tailoring of graphene over long distances for advanced graphene electronic architectures.

1.3 Need for a nanoscale control of graphene electronic properties

As ongoing efforts to incorporate graphene into electronic applications, attempts are already being made to gain consistent control over electronic properties in graphene down to nanometer scale so as to explore its possibility as a complementary material to Silicon for prospective nanoelectronics. Notable amongst these include intentional control of the density and character of its charge carriers by doping [15], and confining charge carriers to nanometer constrictions [14]. Typically, most experiments probe the global electrical response of graphene devices thus yielding properties averaged over the whole device. As a consequence, local properties and their distributions within the sample are inaccessible with standard electron transport measurements. In low dimensional electronic systems, like graphene, the local phenomenona/effects might have much stronger impact than previously believed. Hence, naturally, developing techniques to analyse and/or control the local effects in graphene has attracted wide interests. Consequently, such local probing techniques have helped gain significant insights into various aspects of functioning graphene devices. Notably, scanning probe microscopy (SPM) based measurements have provided the most direct probe of local electronic properties. For example, Scanning Photocurrent Microscopy [16] has been used to probe the 'electrostatic potential landscape' of graphene in view of analyzing the local changes in the electronic structure of the graphene sheets introduced by their interaction with deposited metal contacts and by local symmetry breaking/ disorder at the sheet edges. On the other hand Raman microscopy has been used to probe mesoscopic inhomogeneity in the doping [17], due, possibly, to the presence of charged impurities. Formation of sub-micron (resolution limited) electron and hole charge puddles near Dirac point in graphene, has been demonstrated by Martin et. al. [18] using Scanning Single Electron Transistor microscopy where they gauged the intrinsic size of the puddles to be ~30 nm. Atomic Force Microscope based scanning microscopy techniques have also been significantly used to get better insights in graphene electronic properties. For example, high-resolution scanning tunneling microscopy (STM) [19] experiments have directly imaged charge puddles of ~10 nm in size and suggested they originate from individual charged impurities underneath graphene.

In the wake of such proven local inhomogeneity in electrical properties of graphene the work presented in this *chapter* focuses on local evaluation of transport properties in substrate-supported graphene by the use of scanning probe microscopy based methods with nanometer scale lateral resolution.

In particular Section 2 deals with the most common graphene synthesis methods and, in particular, on those used in this study, i.e. mechanical exfoliation of HOPG and epitaxial growth by thermal treatments on hexagonal SiC. The characterization methods used for the identification and morphological and structural analyses of the produced graphene layers are also illustrated

Section 3 is focused on the evaluation of the local transport properties, electron mean free path and mobility, in graphene on insulating and semiconducting substrates. In particular the role of the interaction of 2DEG in graphene with the substrate (in particular, the role of dielectric environment) is discussed.

2. Graphene synthesis methods

Graphene has been adopted as a material of interest quite widely. This had led to many motivations for material synthesis. To date, the synthetic techniques fall into two categories. In the first approach the weak bonding between the graphene layers in HOPG is exploited and single or few layers of graphene are separated from the parent crystal either by mechanical [1] or chemical exfoliation [20]. Mechanical separation, the technique used to obtain and study graphene in the laboratory for the first time [1], is a simple, inexpensive method, yielding graphene flakes of very high crystalline quality. However, the yield of graphene production is extremely low and the flakes have small dimension (from 1 μm to 100 μm lateral size), so that this approach lacks the scalability required by mass device production. Exfoliation of graphene in solution from HOPG by chemical methods allows to improve the yield, but the size of the flakes remains too small and the crystalline quality is worse than in mechanically exfoliated graphene.

The second approach is based on the epitaxial growth of graphene on substrates. In particular, graphene has been obtained by chemical vapour deposition on catalytic metal films (Ni, Cu,..) from hydrocarbon precursors [21,22], or by controlled thermal decomposition of SiC [23]. This second approach has the potential of producing large-area lithography-compatible films and is rapidly advancing at the moment. The crystalline quality of epitaxial graphene is typically lower than in mechanically exfoliated graphene flakes, but it could be improved by improving the growth methods. In the case of graphene grown on metal substrates, the grown film needs to be transferred to an insulating substrate, to be useful in electronic applications. On the contrary, epitaxial graphene obtained by thermal treatments of SiC grows directly on a semiconducting or semiinsulating substrate, and is expecially suitable for electronic applications.

In the present study, the used graphene samples were obtained by mechanical exfoliation of HOPG and by epitaxial growth on hexagonal SiC substrate. The sample preparation and graphene identification for the two methods are discussed in the following.

2.1 Mechanical exfoliation of HOPG

Mechanical exfoliation, developed by Novoselov and Geim [1], involves a topdown approach to get graphene from highly oriented pyrolitic graphite (HOPG). Using a common adhesive tape a thin graphite layer is peeled off from HOPG. With an inter-layer van der Waals interaction energy of about 2 eV/nm^2, the force required to exfoliate graphite is extremely weak, about 300 nN/μm^2. After repeated thinning the flakes are transferred by gently pressing the tape onto suitable substrate. We used extensive characterization by Optical contrast microscopy imaging (OM), TappingMode AFM (t-AFM), and microRaman (μR) spectroscopy to analyze/identify manolayer flakes.

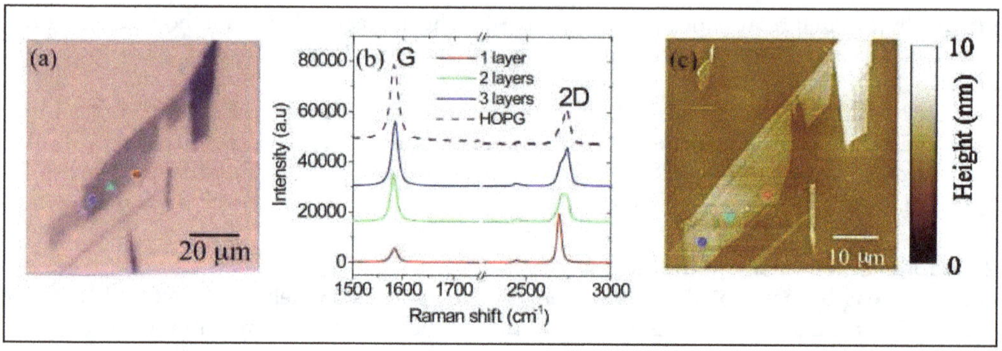

Fig. 3. Optical microscopy image (a), micro-Raman spectra (b) and atomic force microscopy (c) image of few layers of graphene obtained by mechanical exfoliation of HOPG (Data from Ref. [24].)

In Fig. 3(a), an OM image of a FLG sample is reported [24]. The variable contrast in the image is indicative of a variable number of layers in the different regions of the sample. In particular, a lower contrast is associated to thinner layers. After an appropriate calibration, optical contrast variations can be used for a quantitative determination of the number of layers. In Fig.3(b), μR spectra measured on some selected positions indicated by the red, green and blue dots in the OM of Fig. 3(a) are shown. As a reference, the spectrum obtained from HOPG is also reported. Raman spectroscopy is a very powerful method for the characterization of HOPG and FLG. In particular, it allows an unambiguous distinction between single layer, bilayer and multilayers in the case of graphene obtained by exfoliation of HOPG [25]. This is possible through a comparison of the relative intensity of the characteristics G peak (appearing at ~1580 cm^{-1}) and 2D peak (appearing at ~2680 cm^{-1}) and from the symmetry of the 2D peak in the Raman spectra. As it is evident from Fig.3(b), the spectrum measured in the flake region with the lowest optical contrast (position indicated by a red point in Fig.3(a)), exhibits a symmetric 2D peak with very high intensity compared with the G peak. These characteristic spectroscopic features have been associated uniquely to a graphene single layer. As it is evident, the relative intensity of the two peaks is very different in the case of a graphene bilayer, for which the 2D peak is not symmetric. For a number of layers larger than three the Raman spectrum becomes qualitatively similar to that of HOPG.

In Fig.4(a), an AFM morphological image on the same sample region as in Fig.3(a) is reported. In order to obtain quantitative information on the number of graphene layers from these morphological image, a height profile along the dashed line indicated in Fig.4(a) is reported in Fig.4(b). The zero height value is represented by the SiO_2 surface level. Hence, from the profile in Fig.4b it is possible to deduce the height associated to the single layer, bilayer and trilayer of graphene previously identified by Raman spectroscopy.

Interestingly, the measured height of a single layer with respect to the SiO_2 is ~0.7 nm, while the relative heights of the second layer with respect to the first one and that of the third layer with respect to the second are both ~0.35 nm. The latter thickness value is very close to the interlayer spacing between graphene planes within HOPG. Hence, it seems that the measured step-height for a single layer of graphene is affected by an "offset". This means that, in order to use AFM to determine the number of graphene layer in a sheet, it is

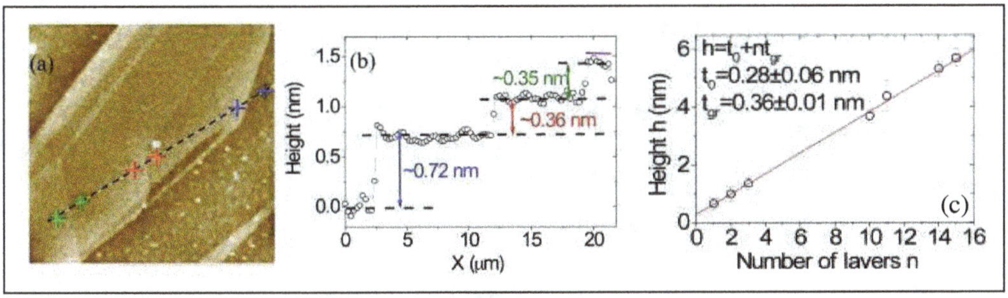

Fig. 4. AFM morphological image of a FLG (a) and height profile along the indicated dashed line (b). Average height h as a function of the estimated number of layers n (c). The data have been fitted with the linear relation, $h=t_{gr}\times n+t_0$, with the interlayer spacing t_{gr} and the offset t_0 fitting parameters. The best fit with the experimental data obtained for the values of $t_{gr}=0.36\pm0.01$ nm and $t_0=0.28\pm0.06$ nm (Data from Ref. [24]).

necessary to have an accurate estimate of this value. To do this, the step height with respect to SiO_2 was measured on a large number of flakes with variable number of layers, and with a "staircase" structure similar to that illustrated in Fig.4(b). For each flake, the single layer region was unambiguously determined by μR measurements. The number of layers for thicker regions could be obtained by counting the number of subsequent steps with ~0.35 nm height. In Fig.4c, the average height h is reported as a function of the estimated number of layers n. These data have been fitted with the linear relation, $h=t_{gr}\times n+t_0$, where the interlayer spacing t_{gr} and the offset t_0 are the fitting parameters. The best fit with the experimental data has been obtained for the values of $t_{gr}=0.36\pm0.01$ nm and $t_0=0.28\pm0.06$ nm. Based on such calibration curve (measured height versus number of layers), it is possible to determine in an accurate and straightforward way the number of layers in a multilayer sheet deposited on thermal SiO_2 by an AFM analysis. However, this calibration procedure has to be repeated if the substrate material is changed, since the offset t_0 depends on the specific substrate.

Quantitative information on number of graphene layers deposited on a substrate can also be obtained from optical images, if an accurate analysis of the contrast is carried out. The optical contrast C can be quantitatively defined as,

$$C = (R_{sub} - R_{gr}) / R_{sub} \tag{2}$$

where R_{gr} is the reflected light intensity by FLG and R_{sub} the reflected intensity by the substrate[26]. Typically a SiO_2/Si substrate with 100 or 300 nm thick SiO_2 is chosen, since for these thicknesses optical contrast is maximized for light wavelengths in the visible range (400-700 nm), due to a constructive interference effect.

Fig. 5(a) and (b) illustrate the procedure to extract the optical contrast from an OM on a single layer of graphene flake deposited on 100 nm thick SiO_2 and illuminated with light of 600 nm wavelength [24]. Based on this procedure, the optical contrast can be determined for flakes of different thickness. The number of layers in the flakes can be independently determined by AFM according to the calibration curve in Fig.4 (c). In Fig.5(c), the symbols represent the contrast as a function of the flake thickness (and the number of graphene

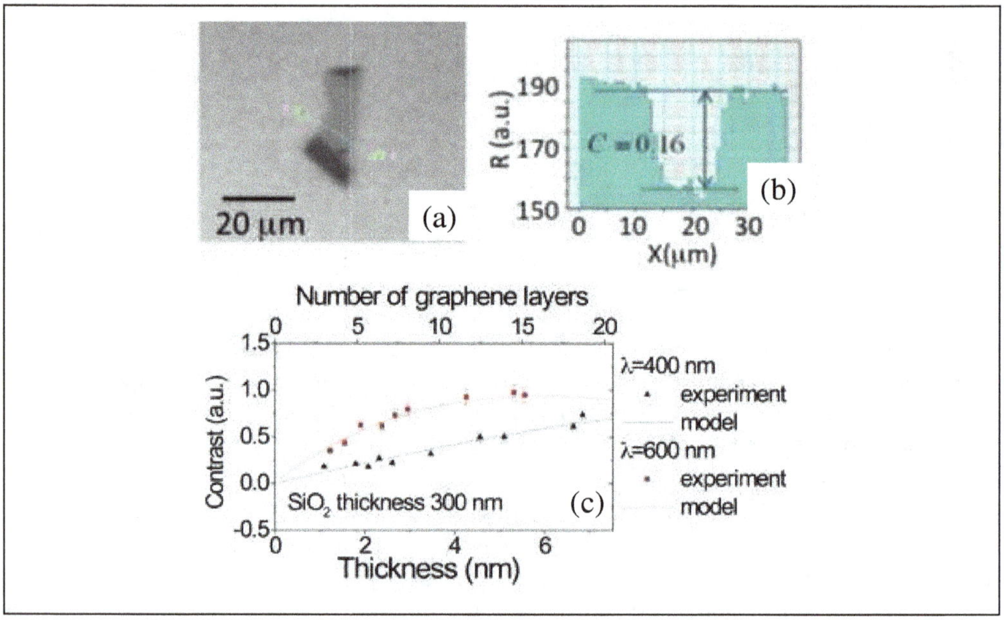

Fig. 5. Optical image of a SLG (a), reflected intensity profile along the dashed line (b) and optical contrast versus the number of graphene layers for two different incident light wavelengths (c) (Data from Ref. [24]).

layers) for samples deposited on 300 nm SiO_2 and illuminated with light of 400 nm (black triangles) and 600 nm (red squares) wavelengths. The lines represent the calculated contrast as a function of the number of layers. The reflected light intensities R_{sub} and R_{gr} are calculated using the Fresnel equations. The literature values of the complex refraction index of Si and SiO_2 as a function of the wavelength were used.

For graphene, the real and imaginary part of the refraction index n_{gr} were calculated starting from its noticeable absorption properties in the wavelength range of visible light. In fact, recently, it has been shown that a single layer of graphene absorbs a fixed percentage ($\pi\alpha$= 2.3%, where α is the fine structure constant) of the intensity of incident visible light [27]. Moreover, for few layers of graphene, the absorbance increases linearly with the number of layers. Based on these properties, refraction index values of graphene for the two considered wavelengths were calculated to be:

$$n_{gr}(400nm) = 2.57 - i0.81 \quad (3.a)$$

$$n_{gr}(600nm) = 2.53 - i1.21 \quad (3.b)$$

2.2 Epitaxial growth of graphene of SiC

Graphene growth during annealing of SiC relies on the interplay of two different mechanisms: (i) the preferential Si sublimation from SiC surface, which leaves an excess of C atoms in the near surface region, and (ii) the diffusion of these C atoms on the SiC surface and their reorganization in the two dimensional graphene lattice structure. Both mechanisms depend in a complex way both on the annealing conditions (temperature T_{gr}, ramp rate, pressure in the chamber) and on the initial surface conditions of SiC wafers, e.g. the termination (Si or C -face), the miscut angle, the defectivity. It has been shown that the rate of Si evaporation can be strongly reduced by performing the annealing in inert gas ambient at atmospheric pressure instead than in vacuum [28]. This results in the possibility to raise the growth temperature, from typical values of T_{gr}~1300°C in vacuum to T_{gr}>1600°C at atmospheric pressure [29]. This has, in turns, beneficial effects on the crystalline quality of the EG layers, since at higher temperature a higher diffusivity of the C atoms on the surface is achieved. It has been also observed that graphene growth starts from the kinks of terraces on the SiC surface or from defects (e.g. pits or threading dislocations) [30]. This means that, for fixed annealing conditions, a higher growth rate is expected on vicinal, i.e. off-axis, SiC surfaces than on on-axis ones, since the spacing between terraces kinks is reduced. Most of the studies on EG growth reported in the literature were carried out on on-axis hexagonal SiC substrates.

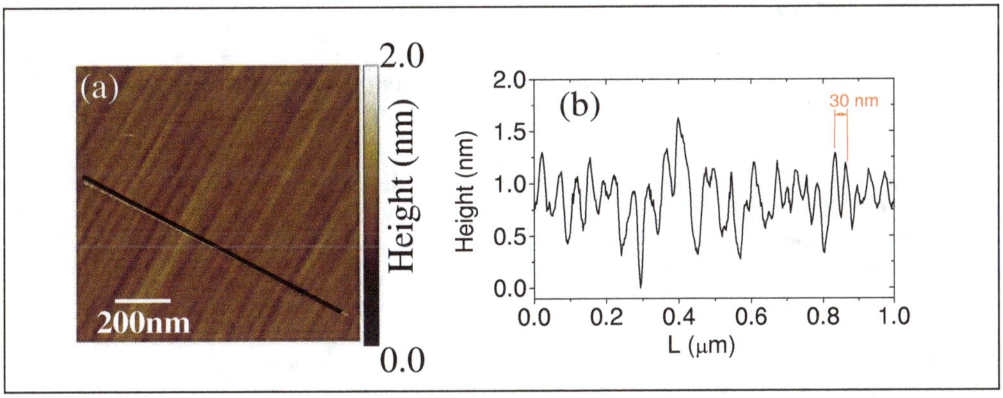

Fig. 6. AFM morphological image of a virgin 4H-SiC (0001), 8° off-axis surface (a) and linescan (b), showing the parallel terraces with a mean width of ~30 nm (Data from Ref. [31]).

The results here reported concern EG grown on the Si face of 8° off-axis 4H-SiC by annealing in inert gas (Ar) ambient in a wide range of temperatures (T_{gr} from 1600°C to 2000°C) [31]. A morphological AFM image of the virgin 4H-SiC (0001) surface is reported in Fig.6, showing the parallel terraces oriented in the <00-10> direction and with a mean width of ~30 nm. AFM images of the surface morphology for the samples annealed at 1600, 1700 and 2000 °C are reported in Fig.7(a), (b) and (c), whereas the corresponding phase maps on the same samples are reported in Fig.7(d), (e) and (f). Annealed samples show wide terraces (average width of ~150-200 nm) running parallel to the original steps in the virgin sample.

Such large terraces are the result of the step-bunching commonly observed on off-axis SiC substrates after thermal treatments at temperatures >1400°C. Interestingly, a network of nanometer wide linear features is superimposed to these large terraces. These features are particularly evident in the phase images of the surfaces.

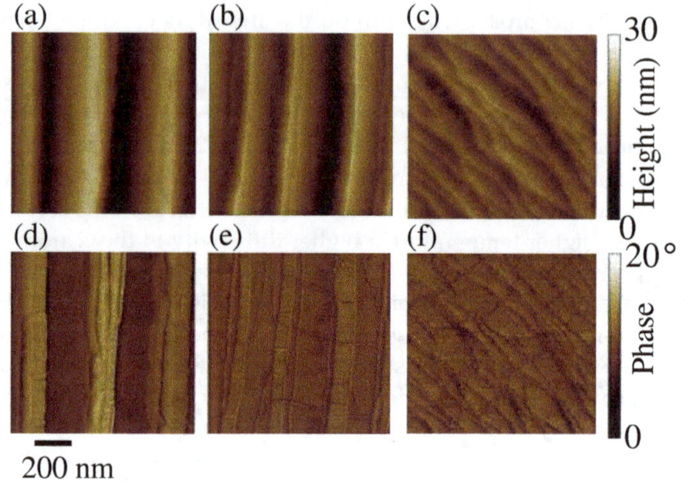

Fig. 7. AFM images of the surface morphology for the samples annealed at 1600 (a), 1700 (b) and 2000 °C (c). The corresponding phase maps on the same samples are reported in (d), (e) and (f). (Data from Ref. [31]).

Fig.8 shows two representative linescans taken in the direction orthogonal (b) and parallel (c) to the terraces length obtained from the morphology map of the sample annealed at 1700°C. In the orthogonal direction, it is worth noting very small steps with nm or sub-nm height, which are overlapped to the large terraces of the SiC substrate. The height of these steps is always a multiple of 0.35 nm, the height value corresponds the interlayer spacing between two stacked graphene planes in HOPG. As an example a ~0.35 nm and a ~1.1 high step are indicated in Fig.8(b), associated, respectively, to one and three graphene layers over the substrate or stacked over other graphene layers.

From the linescan in the direction parallel to the terraces length (c), peculiar corrugations with ~2 nm typical height can be observed.

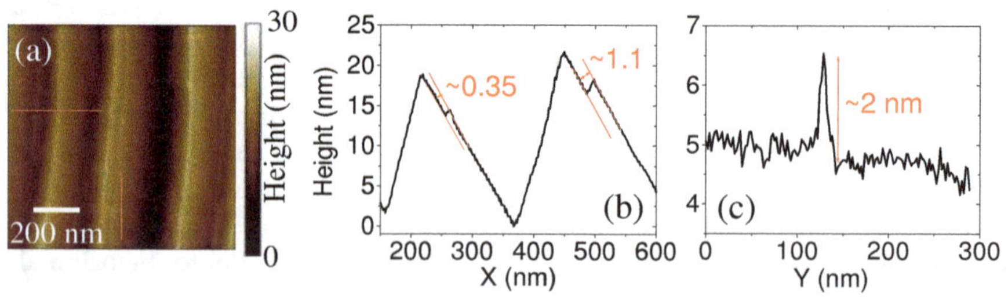

Fig. 8. Two representative linescans from the morphology map (a) of a sample annealed at 1700°C taken in the direction orthogonal (b) and parallel (c) to the steps. (Data from Ref. [31]).

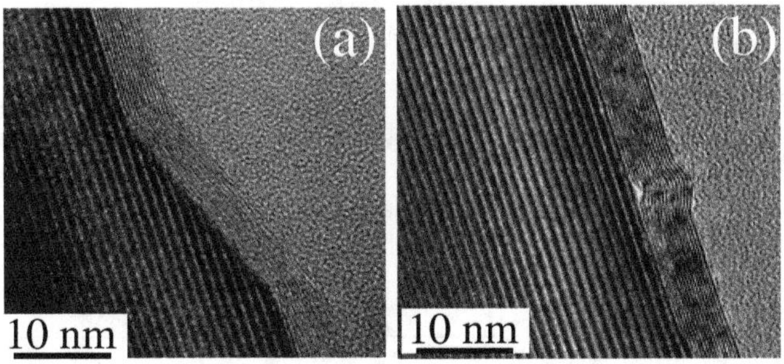

Fig. 9. Two representative HRXTEM images for 8° off-axis 4H-SiC (0001) annealed at 2000 °C. (Data from Ref. [31]).

HRXTEM analyses allowed to get further insight on the structural properties of the graphene films grown at the different temperatures and to clarify the nature of these peculiar corrugations.

In Fig.9(a) and (b) two representative HRXTEM images for a sample annealed at 2000 °C are reported. From Fig.9(a), it is evident that a graphene multilayer uniformly covers the SiC surface also on the terrace step edges. From Fig.9(b), the structure of one of the peculiar corrugations running orthogonal to the steps is reported. This cross sectional image unambiguously demonstrate that those features are wrinkles in the multilayer graphene film. These peculiar defects have been observed also by other authors in the case of few layers of graphene grown on the C face [32] or on the Si face [33] of hexagonal SiC, on-axis. However, in that case wrinkles do not exhibit any preferential orientation with respect to the steps, but form an isotropic mesh-like network on the surface, where wrinkles are interconnected into nodes (typically three wrinkles merge on a node and the angles subtended by the wrinkles are ~60° or 120 °C [32]). The formation of that mesh-like network of wrinkles was attributed to the release of the compressive strain which builds up in FLG during the sample cooling due to mismatch between the thermal expansion coefficients of graphene and the SiC substrate [32]. Comparing the morphology and phase images in Fig.7 for the samples annealed at the different temperatures, it is easy to observe that the density of wrinkles increases with temperature.

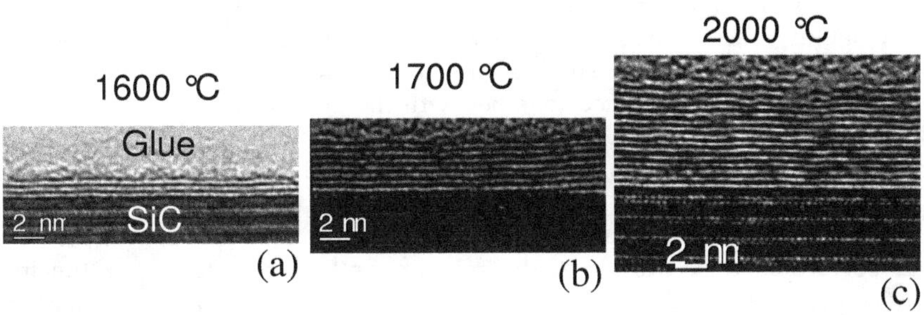

Fig. 10. HRXTEM images on the samples annealed at 1600 (a), 1700 (b) and 2000°C (c) are reported, showing 3, 8 and 18 layers on the surface of 4H-SiC. (Data from Ref. [31]).

In Fig.10(a), (b) and (c), HRXTEM images on the samples annealed at 1600, 1700 and 2000°C are reported, showing 3, 8 and 18 layers on the surface of 4H-SiC. These cross-sectional analyses give a direct measure of the number of grown layers, but only on a very local scale. Lateral variation of the FLG thickness on different sample positions cannot be determined by such a method. To get an estimation of the number of layers at selected surface positions and with higher statistics, AFM can be used to measure the depth of selectively etched stripes in FLG by O_2 plasma. This plasma treatment is known to remove efficiently carbonaceous species through a chemical reaction leading to the formation of CO_2.

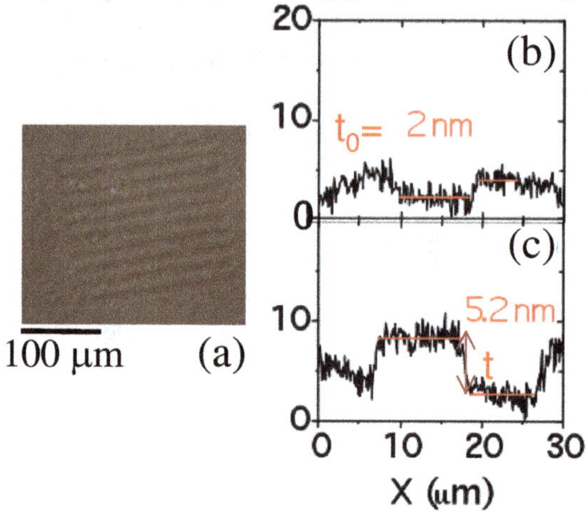

Fig. 11. Optical image of the etched graphene stripes by O_2 plasma treatment in multilayer graphene grown on SiC at 1700 °C (a). To estimate the graphene thickness, a lithographically patterned pristine SiC was simultaneously etched together with the graphitized SiC sample, and the etched SiC depth was measured. AFM Height profiles taken on a etched stripe on pristine SiC (b) and on the sample annealed at 1700 °C (c), respectively. (Data from Ref. [31]).

In Fig. 11(a), an optical image of the etched stripes in on the sample annealed at 1700 °C is reported. To obtain an accurate estimation, we checked if the SiC substrate is slightly etched by the used plasma processing. To this aim, a lithographically patterned pristine SiC substrate was simultaneously etched together with the graphitized SiC samples. Fig. 11(b) and (c) show the height profile taken on a stripe on pristine SiC and on the sample annealed at 1700 °C, respectively. From Fig. 11(b), it is clear that a thickness t0~2 nm of SiC is etched during the plasma treatment, due to the physical action of the plasma. This depth must be subtracted while evaluating the number of layers on graphitized SiC. Hence the number of layers can be estimated according to the relation n=(t-t0)/tgr, being tgr the interlayer separation between to stacked graphene layers (tgr≈0.35 nm). The average number of grown layers as a function of the growth temperature Tgr is reported in Fig.12, where the error bars represent the standard deviations obtained from a large statistics on the number of layers determined at several sample positions.

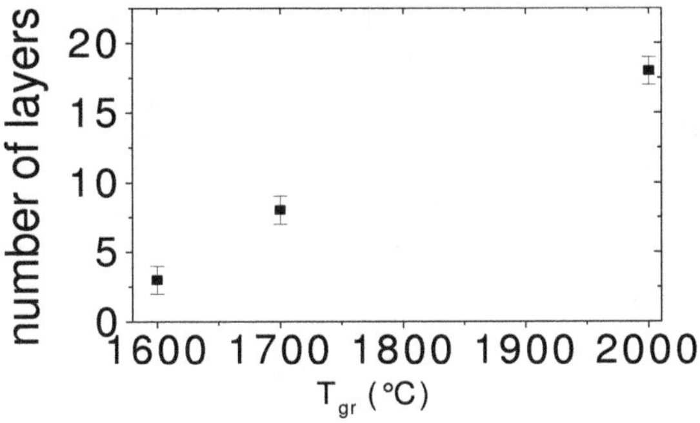

Fig. 12. Average number of grown layers on 4H-SiC (0001), 8° off, as a function of the growth temperature (T_{gr}). The error bars represent the standard deviations obtained from a large statistics on the number of layers at several sample positions. (Data from Ref. [31]).

3. Local transport properties of graphene by scanning probe microscopy

One of the main issues in the research on graphene is to understand how the various scattering mechanisms limit the transport properties, i.e. the electron mean free path and the mobility. Ideally, in a very clean graphene sheet (i.e. with no adsorbed impurities) sufficiently isolated from its environment to be considered free standing, charge carriers can exhibit a giant intrinsic mobility [34] and can travel for micrometers without scattering at room temperature. Indeed, very high values of mobility ($>2\times10^5$ $cm^2V^{-1}s^{-1}$) and electron mean free path have been observed in vacuum and at low temperature (5K) in 'suspended' graphene, obtained by locally etching the substrate under the sheet, after a cleaning by current-induced heating [35]. However, graphene for electronics applications is commonly supported by a dielectric substrate (typically SiO_2 or high-*k* dielectrics) or by semi-insulating SiC. The values of the electron mean free path and mobility observed in supported graphene layers are usually significantly lower than in suspended ones. So far, graphene on a substrate has shown field-effect mobility ranging from 10^2 to 2×10^4 $cm^2V^{-1}s^{-1}$, depending on the kind of substrate and on the graphene synthesis method.

Several factors that can affect the transport of carriers in graphene and degrade its electron mean free path and mobility have been identified and are still under active debate. Carrier scattering with charged impurities [36] (either adsorbed on graphene surface or trapped at the interface with the substrate) is typically indicated as one of the main mechanisms limiting mobility of graphene two-dimensional-electron-gas (2DEG). Since charged impurities interact with graphene 2DEG by a screened Coulomb potential, the strength of the interaction is expected to decrease significantly with increasing the permittivity of the substrate and/or of the dielectric layer deposited on graphene. Experimental results by different authors suggest that scattering by charged impurities is not the only mechanism that limits the mean free path attainable for substrate-supported graphene. Inelastic scattering by surface polar phonons (SPP) of the substrate [37] has been indicated as an additional mechanism limiting the carrier mobility in graphene. It has been shown theoretically that, due to the polar nature of the substrates commonly used for graphene (like SiO_2 and SiC), a long-range polarization field is associated to the thermally induced lattice vibrations at the surface of the substrate (i.e. the SPP). This field electrostatically couples with the 2DEG, resulting in a sizeable degradation of mobility at room temperature. Recently, the experimental evidence of such SPP scattering at room temperature has been reported, based on temperature dependent transport measurements performed on devices in graphene deposited on SiO_2 substrate [34]. It has also been predicted that for graphene on SiC, the SPP scattering has a weaker effect on the electron mobility than for graphene on SiO_2, due to weaker polarizability of SiC and relatively high phonon frequencies associated with the hard Si-C bonds [37].

As a matter of fact, charged impurities are randomly distributed on the graphene surface. The random distribution of these scattering sources causes local inhomogeneities in the transport properties, which adversely influence the reproducible operation of graphene nanodevices. Under this point of view, nanoscale resolution methods are required to probe the lateral homogeneity of the transport properties in graphene sheets.

3.1 Local electron mean free path

Recently, scanning capacitance spectroscopy (SCS) on graphene was applied to determine the "local" electron mean free path l (defined as the average of the distances traveled by electrons between "few" subsequent scattering events) and mobility μ [38,39]. A schematic representation of the SCS measuring setup is reported in Fig.13(a). The graphene/insulator/semiconductor backgate system can be described as a metal-insulator-semiconductor (MIS) capacitor. Under electric fields, graphene does not behave as a "classical" metal film, but manifests itself as a capacitor, whose capacitance (the quantum capacitance C_q) adds in series to the insulator and semiconductor capacitance contributions [40].

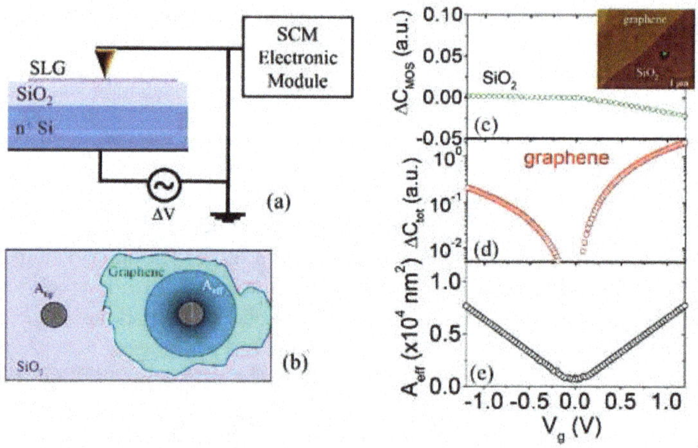

Fig. 13. Schematic representation of the scanning capacitance spectroscopy (SCS) measuring setup (a) and of the effective area in graphene (b). Representative □C-Vg characteristics measured with the tip in contact on 100 nm SiO2 on n+ Si (c) and with the tip on graphene coated SiO2 (d). Experimental dependence of the effective area on the gate bias (e), obtained from the ratio of the curves in (d) and (c). (Data from Ref. [39]).

Local measurements were carried out by an AFM, using Pt coated n^+-Si tips as probes. A modulating bias $\Delta V = \frac{V_g}{2} + \frac{V_g}{2}\sin(\omega t)$ with amplitude V_g and frequency ω=100 kHz was applied between the Si n^+ backgate and the probe. An ultra-high-sensitive (10^{-21} F/Hz$^{1/2}$) capacitance sensor connected to the conductive AFM tip allowed to measure, through a lock-in system, the capacitance variation $\Delta C(V_g)$ induced by the modulating bias.
The AFM tip in contact with bare SiO_2 forms a MOS capacitor with capacitance $A_{tip} \times C'_{MOS}$, where C'_{MOS} is the per unit area capacitance of the MOS system and $A_{tip} = \pi r_{tip}^2$ is the tip contact area with tip radius r_{tip}. When the tip is in contact with graphene, a positive gate bias causes the accumulation of electrons in graphene to screen the positively charged uncompensated ions in the depletion layer at the SiO_2/Si interface. On the contrary, a negative gate bias induces the accumulation of holes in graphene to screen the electrons at the SiO_2/Si interface. In both cases, the induced carriers in graphene are distributed over an effective area A_{eff} [40,41] This is schematically illustrated in Fig. 13(b). The total capacitance, when the tip is on graphene, is $C_{tot}=A_{eff} \times C'_{tot}$, where C'_{tot} is the series combination of C'_{MOS} and C'_q, the per unit area quantum capacitance of graphene. Since $C'_q >> C'_{MOS}$ for the typical thicknesses (~100 or ~300 nm) of the adopted SiO_2 layers, C_{tot} can be expressed as $C_{tot}=A_{eff} \times C'_{MOS}$. Hence, $C_{tot}/C_{MOS} = A_{eff}/A_{tip}$.
Fig.13(c) and 13(d) show two representative ΔC-V_g characteristics measured with the tip in contact on two distinct positions on a sample with graphene deposited on 100 nm SiO_2. The curve in Fig.13(c) was measured with the tip on bare SiO_2 and exhibits the typical behavior observed for a MIS capacitor on a n-type semiconductor. The curve in Fig.13(d), measured with the tip on graphene, has a minimum value close to V_g=0 and increases both for negative and positive V_g values. The experimental dependence of the effective area on the gate bias can be obtained from the ratio of the curves in Fig.13(d) and (c), i.e.

$$A_{eff} = A_{tip} \frac{|\Delta C_{tot}|}{|\Delta C_{MOS}|} \quad (4)$$

As shown in Fig.13 (e), A_{eff} exhibits a minimum at V_g~0 and increases linearly with V_g in a symmetric way for both bias polarities. $A_{eff}=\pi L_{eff}^2$ represents the effective area of the graphene capacitor responding to the AC bias, i.e. L_{eff} is the distance from the tip contact where the induced screening charges (electrons/holes) spread in graphene. A_{eff} is much smaller than the area of the entire graphene sheet (~100 μm² in this specific case).
In the following, it is shown that L_{eff} corresponds to the local electron mean free path in the probed graphene region. The induced charge carriers diffuse over a length L_{eff} from the tip contact, traveling at velocity v_F. Their diffusivity in graphene can be expressed as $D=v_F L_{eff}/2$. On the other hand, the diffusivity D can be related to electron mobility μ by the generalized Einstein relation $D/\mu=n/(q\partial n/\partial E_F)$, being q the electron charge, n the electron density and E_F the Fermi energy in graphene. The concepts of mobility and carrier density in graphene are meaningful when the Fermi level is far from the Dirac point but is still in the linear region of the dispersion relation. In these conditions, the density of states is linearly dependent on E_F. As a consequence, the carrier density can be expressed as $n=E_F^2/(\pi\hbar^2 v_F^2)$, and $D/\mu=E_F/2q$. Furthermore, far from the Dirac point, Boltzmann transport theory can be used to describe the electronic transport in graphene and μ can be expressed in terms of l as $\mu=qv_F l/E_F$. As a consequence, L_{eff} measured by SCS correponds to l. This continuum treatment, using the concepts of diffusivity and mobility and the Einstein relation, is

justified on the "mesoscopic" length scale investigated by SCS (from 10 to 100 nm). A quantum mechanical treatment of electron scattering phenomena, based on the interference of incident and reflected electron wavefunctions by a single scattering center, is required on a length scale of ~1 nm, as recently demonstrated in some investigations by scanning tunneling microscopy [42].

3.2 Role of graphene/substrate interface on the 2DEG transport properties

In this section, 'local' measurements of the electron mean free path are reported in graphene on most relevant substrates for electronic applications: (i) graphene exfoliated and deposited on 4H-SiC(0001) (DG-SiC), (ii) graphene epitaxially grown on 4H-SiC(0001) (EG-SiC), and (iii) graphene deposited on SiO_2 (DG-SiO_2) [43].

In the case of DG-SiO_2, the SiO_2 film works as the gate dielectric and the n^+ Si substrate works as the semiconductor of a metal-insulator-semiconductor (MIS) capacitor. Similarly, both in the case of DG-SiC and of EG-SiC, the very lowly doped SiC film works as the gate dielectric, whereas the n^+ SiC substrate works as the semiconductor back gate of the capacitor. Bias amplitude V_g was varied from 0 to 10V at bias frequency of ω = 100kHz.

In Fig.14 are reported representative capacitance-voltage characteristics obtained on DG-SiC (a), EG-SiC (b) and DG-SiO_2 (c) on arrays of 5×5 tip positions with an inter-step distance of 1μm×1μm. For reference, SCS measurements were carried also on bare SiO_2 and SiC regions of the samples.

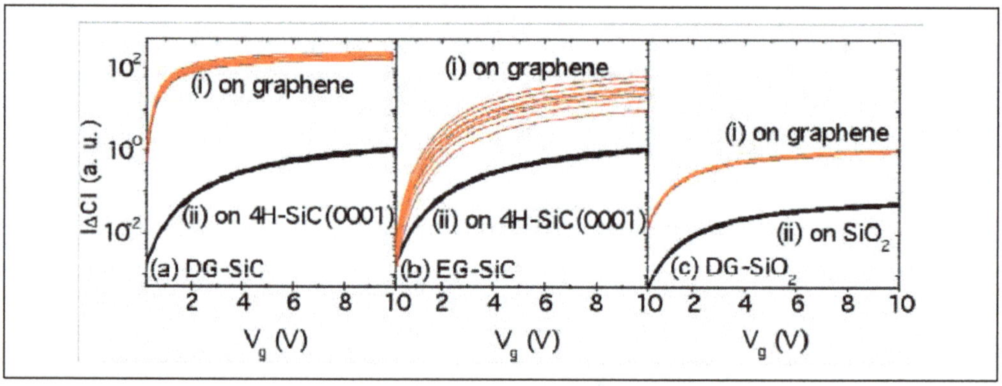

Fig. 14. Representative capacitance-voltage characteristics obtained on DG-SiC (a), EG-SiC (b) and DG-SiO_2 (c) on arrays of 5×5 tip positions with an inter-step distance of 1μm×1μm. (Data from Ref. [43]).

Fig.15(a) shows the evaluated l_{gr} for DG-SiC, EG-SiC and DG-SiO_2. l_{gr} is reported versus n_{Vg}-n_0, being n_{Vg} the carrier density induced in graphene by the gate bias V_g and n_0 the carrier density at V_g=0. The values of n_{Vg}-n_0 are obtained as

$$n_{Vg} - n_0 = e_0 e_{ins} V_g / qt_{ins} \tag{5}$$

where ε_{ins} and t_{ins} are the relative dielectric constant and the thickness of the insulating layer under graphene and ε_0 is the vacuum absolute dielectric constant. The histograms of the l_{gr} values at a fixed value of n_{Vg}-n_0=1.5×10^{11}cm^{-2} are reported in Fig.15(b) for DG-SiC (i), EG-SiC (ii) and DG-SiO_2 (iii).

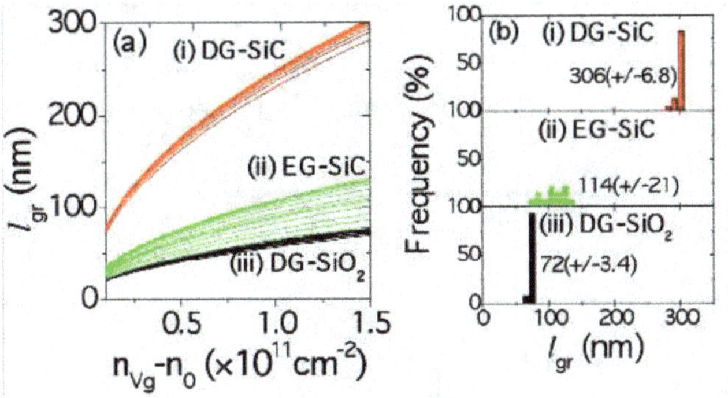

Fig. 15. Evaluated l_{gr} for graphene exfoliated and deposited on 4H-SiC(0001) (DG-SiC), for graphene epitaxially grown on 4H-SiC(0001) (EG-SiC), and for graphene deposited on SiO_2 (DG-SiO_2) (a). l_{gr} is reported versus n_{Vg}-n_0, being n_{Vg} the carrier density induced in graphene by the gate bias V_g and n_0 the carrier density at V_g=0.

It is worth noting that l_{gr} in EG-SiC is on average 37% of l_{gr} in DG-SiC, but the spread of the l_{gr} values in EG-SiC is much larger than in DG-SiC. These differences can be explained in terms of the peculiar structure of EG/4H-SiC (0001) interface, as discussed in the following section.

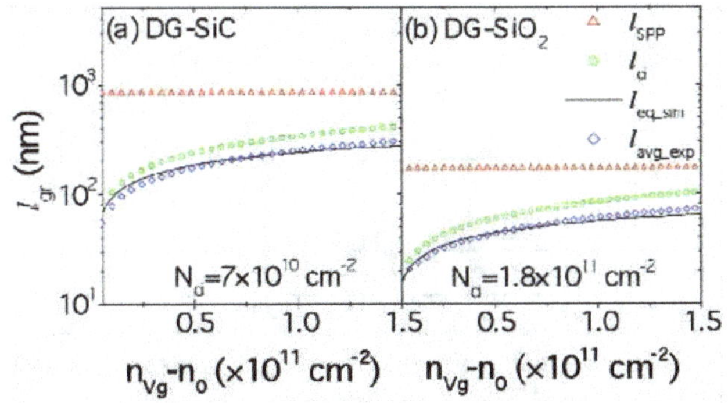

Fig. 16. Average of the l_{gr} vs n_{Vg}-n_0 curves measured on different tip positions on DG-SiC (a) and DG-SiO_2 (b). Data from Ref. [43].

It is also worth noting that l_{gr} on DG/SiC is on average ~4× than on DG/SiO_2 and the spread of the l_{gr} values are comparable in the two cases. This difference can be explained in terms of the higher permittivity of SiC (ε_{SiC}=9.7) than SiO_2 (ε_{SiO2}=3.9) and of the lower coupling of the 2DEG with surface polar phonons in SiC than in SiO_2. This is better discussed in the following. The electron mean-free path limited by scattering on charged impurities (l_{ci}) in graphene could be expressed as a function of the carrier density as [44],

$$l_{ci}(n) = \frac{16\varepsilon_0^2\varepsilon^2\hbar^2 v_F^2}{Z^2 q^4 N_{ci}}\left(1+\frac{q^2}{\pi\hbar v_F \varepsilon_0 \varepsilon}\right)^2 \sqrt{\pi n} \tag{5}$$

where Z is the net charge of the impurity (assumed to be 1 for this study), N_{ci} is the impurity density and ε is the average between the relative permittivity of the substrate (ε_{ins}) and of vacuum permittivity (ε_{vac}=1). The electron mean free path limited by scattering with a SPP phonon mode of characteristic frequency ω_v can be expressed as [45]

$$l_{SPP,\upsilon} = \sqrt{\frac{\beta}{\hbar\omega_\upsilon}\frac{\hbar v_F 4\pi\varepsilon_0}{q^2}\frac{q v_F}{F^2_\upsilon}\frac{\exp(k_0 z_0)}{N_{SPO,\upsilon}}\frac{\hbar\sqrt{\pi}}{q}} \quad (6)$$

where

$$k_0 \approx \sqrt{\left(\frac{2\omega_{SO,\upsilon}}{v_F}\right)^2 + \chi n} \quad (7)$$

$\chi \approx 10.5$, $\beta \approx 0.153\times10^{-4}$ eV and $z_0 \approx 0.35$ nm is the separation between the polar substrate and graphene flake. $N_{SPP,\upsilon}$ is SPP phonon occupation number. The magnitude of the polarization field is given by the Fröhlich coupling constants, F^2_υ [46].

In Fig.16(a) and (b) the average of the I_{gr} vs n_{Vg}-n_0 curves measured on different tip positions on DG-SiC and DG-SiO_2 are fitted with the equivalent mean free path obtained by,

$$l_{gr\_eq}^{-1} = l_{gr\_ci}^{-1} + \sum_\upsilon l_{gr\_SPO,\upsilon}^{-1} \quad (8)$$

The characteristic SPP frequencies and the corresponding Fröhlich coupling constants for SiO_2 and 4H-SiC substrate are listed in Table 1 .

	SiO_2	4H-SiC
$\hbar\omega_{SO1}$ (meV)	58.9	116.0
$\hbar\omega_{SO2}$ (meV)	156.4	-
F^2_1 (meV)	0.237	0.735
F^2_2 (meV)	1.612	-

Table 1. Characteristic surface polar phonons frequencies and corresponding Fröhlich coupling constants for SiO_2 and 4H-SiC substrate.

The only fitting-parameter is N_{ci}. The charged-impurity density limiting the mean free path was found to be ~7×10^{10} cm^{-2} for DG-SiC, and ~1.8×10^{11} cm^{-2} for DG-SiO_2 [43]. The calculated l_{SPP} and l_{ci} versus n_{Vg}-n_0 curves are also reported in both cases. It is worth noting that l_{SPP} for DG-SiC is more than five times l_{SPP} for DG-SiO_2. As a result, scattering by charged impurities is the limiting scattering mechanism in DG-SiC (see Fig.16(a)), whereas a significant contribution is played by scattering with SPP in the case of DG-SiO_2, especially at higher carrier densities (see Fig. 16(b)).

3.3 Electronic properties of the graphene/4H-SiC(0001) interface.

As shown in Fig.15(a), the interface between epitaxial graphene and 4H-SiC (0001) strongly affects the overall electronic and transport properties of the 2DEG in EG.

Both experimental and theoretical studies have shown that EG synthesis on the (0001) face of SiC, occurs through a series of complex surface reconstructions [47]. This starts from an initial (Si-rich) ($\sqrt{3}\times\sqrt{3}$)$R30°$ phase which rapidly converts into a second (C-rich) ($6\sqrt{3}\times6\sqrt{3}$)$R30°$ reconstruction when the temperature increases. This layer may be more or less defective with more or less dangling bonds at the interface with the Si face. Angle resolved photoemission spectroscopy (ARPES) measurements indicate that its electronic structure is not that of graphene single layer, i.e. no linear dispersion is observed. This layer is simply an intermediate (buried) buffer layer (zero layer, ZL) with a large percentage of sp^2 hybridization. The first graphene layer which forms on the ZL (see schematic in Fig.17(a)) exhibits, instead, a linear dispersion relation. The presence of the ZL makes EG n-type doped and causes an overall degradation of the carrier mobility.

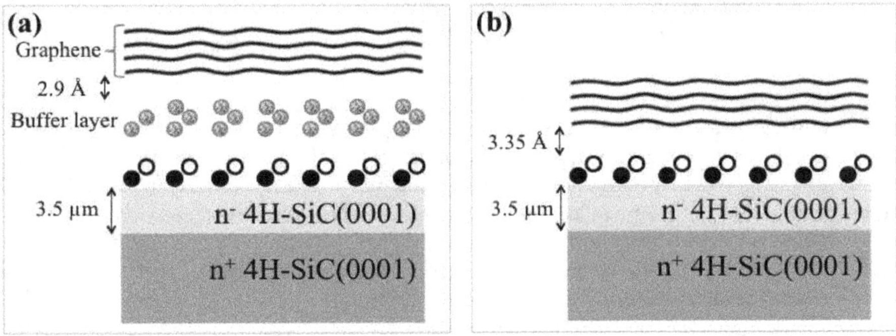

Fig. 17. Schematic representation of the interface between epitaxial graphene grown 4H-SiC (0001) and the substrate (a) and of the deposited graphene/4H-SiC (0001) interface (b).

Recently, the properties of the interface between EG and the zero layer have been investigated with nanoscale lateral resolution and compared with those of graphene mechanically exfoliated from HOPG and deposited on 4H-SiC (0001) [48].

In fact, in the case of deposited graphene (DG) no zero layer is present at the interface. EG was grown on n^+-doped (0001) *4H*-SiC substrate, 8° off axis, with a weakly doped n^--epitaxial layer on top, in an inductively heated reactor at a temperature of 2000 °C and under 1 atm Ar pressure [29]. A piece of the same *4H*-SiC wafer (not subjected to annealing) was used as the substrate for DG. This local investigation has been performed by measuring the current flowing across the interface using torsion resonance conductive atomic force microscopy (TR-CAFM) and scanning current spectroscopy (SCS).

TR-CAFM (see schematic in Fig.18(a)) is a dynamic scanning probe method which allows nondestructive electrical measurements from a conductive tip oscillating in a torsional or twisting mode in close proximity to the sample surface (0.3–3 nm). When a bias is applied between the tip and the sample backside, a map of the current flowing from the tip to sample surface is acquired. This noncontact method has a distinct advantage over the conventional conductive atomic force microscopy performed in contact mode because of the absence of shear forces that can damage the graphene sheets. Commercially available Si n^+-doped probes with platinum (Pt) coating were used, with a typical radius of the apex of 10–20 nm. TR-CAFM was used to measure, simultaneously, topographic and current maps on the sample surface. Fig.18(b) shows a TR-CAFM current map taken on DG at tip bias of 1 V. The brighter portions in the current map (region carrying a higher current) correspond to graphene. After identifying the SiC regions coated with DG, local *I-V* measurements were

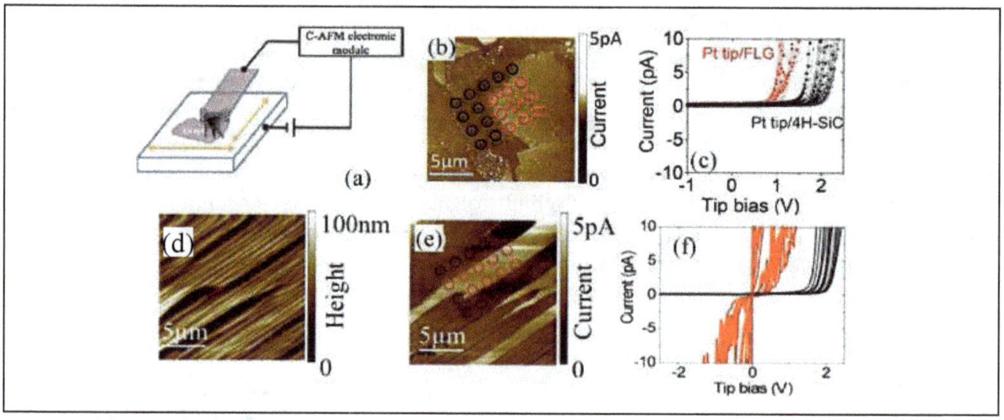

Fig. 18. Schematic representation of the Torsion Resonant–Conductive Atomic Force Microscopy (TR-CAFM) setup (a). TR-CAFM current map taken at tip bias of 1 V on deposited graphene on SiC (b). The brighter portions in the current map (region carrying a higher current) correspond to graphene coated regions. Local *I-V* curves measured on the indicated tip positions (c). TR-CAFM morphology (d) and current (e) maps collected on EG on SiC. The investigated sample region is only partially covered with a single layer of epitaxial graphene. Local *I-V* characteristics collected in the same sample region (f). (Data from Ref. [48]).

carried out by scanning current spectroscopy (SCS). In SCS the conductive tip is placed in contact on a discrete array of positions on the sample surface non invasively, lifting the tip by 20 nm at every interval. A swift transition between TR-CAFM and SCS is possible without altering the sample position. The representative positions of the tip during the SCS scan are depicted in Fig.18(b) and the recorded *I-V* curves are shown in Fig.18(c). All *I-V* curves exhibit a rectifying behavior, with a low leakage current under reverse bias and a sudden increase in the current for positive bias higher than a threshold voltage. It is worth noting that the acquired curves can be divided in two groups, which correspond to the tip positions on DG and bare SiC substrate. It is clear that DG forms a Schottky contact on top of the 4*H*-SiC layer.
Based on these results on DG, TR-CAFM was used to get local information on the graphene coverage on EG. Fig.18(e) shows a typical TR-CAFM current map along with the morphology collected on EG (Fig.18(d)). By comparing the morphology and current maps, brighter regions in the current map appear evident, corresponding to areas covered by a few MLs of graphene. The *I-V* characteristics collected by SCS in the same sample region are reported in Fig.18(f).
Similarly to the sample with DG, the *I-V* curves on the sample with EG can be divided into two distinct families: the curves acquired on an area covered with graphene and those acquired on an area free of graphene. The *I-V* characteristics on the graphene-free regions remained similar to those measured on the bare 4*H*-SiC surface in the sample not subjected to thermal treatment (see Fig.18(c)). The curves obtained on the regions coated with EG show completely different *I-V* characteristics. The leakage current is now very large with, in some cases, a clear Ohmic behavior. This is a direct evidence that the Schottky barrier height (SBH) at the EG/4*H*-SiC interface has been reduced with respect to DG/4*H*-SiC with, in some cases, a possibility of direct tunnel injection.
These SBH values have been evaluated for each of the *I-V* curves in Figs.18 (c) and (f), using the procedure of Refs. [49,50].
The histograms of the corresponding values have been reported in Figs.19(a) and (b), respectively. While the SBHs for Pt-tip/4*H*-SiC remain similar for both samples, the average

SBH value for EG on 4*H*-SiC appears very much reduced (Φ_{EG}=0.36±0.1 eV) with respect to DG (Φ_{DG}=0.85±0.06 eV). This fact is especially interesting considering that the SBHs on positions uncovered with graphene remain similar. The SBH difference must then be explained in terms of the band diagrams schematically drawn in Figs.19(c) and (d). Here, E_C is the energy of the conduction band edge, E_F is the Fermi energy at the bulk 4*H*-SiC, $E_{F,DG}$ is the Fermi energy of DG, and $E_{F,EG}$ the Fermi energy of EG. Due to the weak (van der Waals-like) interaction with the substrate, DG can be considered neutral (or close to neutral) in the absence of external bias. In this case, $E_{F,gr}$ should (more or less) coincide with the Dirac point energy (E_{Dirac}). Hence, in the first approximation, the SBH between DG and the *n*-type 4*H*-SiC epilayer should be given by the difference between the neutral graphene work function (W_{gr}≈4.5 eV) and the 4*H*-SiC electron affinity ($\chi_{4H\text{-}SiC}$≈3.7 eV). The value of Φ_{DG} is in good agreement with the value predicted by the simple model, i.e. 0.8 eV. The SBH in the case of EG can be explained in terms of a pinning of $E_{F,EG}$ at an average value ΔE_F≈0.49 eV higher than $E_{F,DG}$, as shown in Fig.19(d). This pinning is a direct evidence of the presence of positively charged states located at the interface between the Si face of 4*H*-SiC and the zero layer. The spread of SBH values for EG/4*H*-SiC is an evidence of a non uniform density of the interface states.

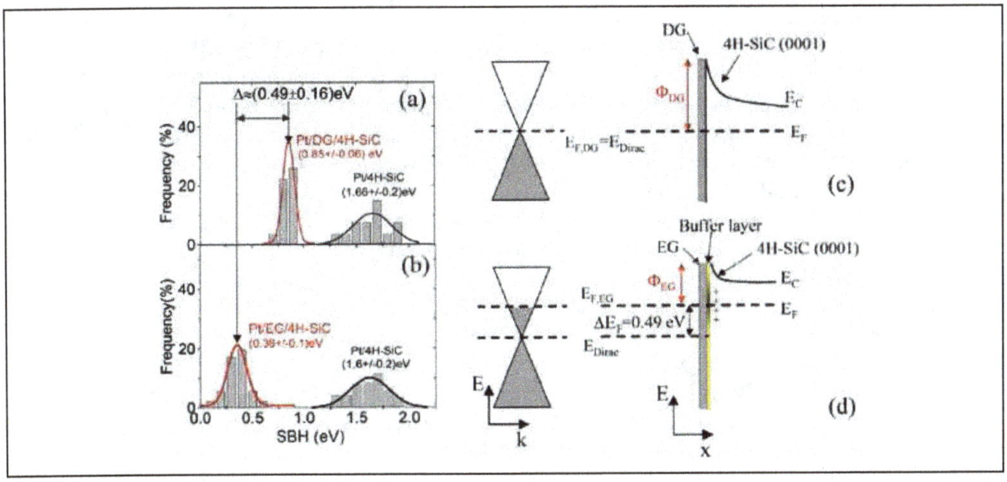

Fig. 19. Histograms of the local Schottky barrier height at DG/4H-SiC (a) and at EG/4H-SiC (b). The SBH difference is explained in terms of the band diagrams schematically drawn in (c) and (d). (Data from Ref. [48]).

3. Summary

After discussing two of the main graphene synthesis routes (mechanichal exfoliation of HOPG and thermal decomposition of SiC), the local characterization of graphene electronic properties by recently demonstrated scanning probe microscopy techniques has been presented. In particular, the lateral inhomogeneity of the electron mean free path has been determined, in relation to the graphene/subtrate interface and to the substrate dielectric properties. The peculiar electrical properties of epitaxial graphene/4H-SiC (0001) interface have been discussed.

The results of these measurements will allow to explain some peculiar aspects of the electrical behavior of macroscopic devices fabricated on graphene.

4. Acknowledgment

This work has been supported, in part, by the European Science Foundation (ESF) under the EUROCORE program EuroGRAPHENE, within GRAPHIC-RF coordinated project.

5. References

[1] Novoselov, K. S.; et al. (2004). Electric Field Effect in Atomically Thin Carbon Films. Science, Vol. 306, pp. 666.
[2] Andreoni, W., 2000, The Physics of Fullerene-Based and Fullerene-Related Materials (Springer, Berlin).
[3] Saito, R., G. Dresselhaus, and M. S. Dresselhaus, 1998, Physical Properties of Carbon Nanotubes, Imperial College Press, London.
[4] Charlier, J.-C.; Blase, X. & S. Roche. (2007). *Rev. Mod. Phys.* Vol. 79, pp. 677.
[5] Novoselov, K. S.; Geim, A. K.; Morozov, S. V.; Jiang, D.; Zhang, Y.; Katsnelson, M. I.; Grigorieva, I. V.; Dubonos, S. V. & A. A. Firsov. (2005). Two-dimensional gas of massless Dirac fermions in graphene, *Nature* Vol. 438, pp. 197- 200.
[6] Zhang, Y.; Tan, Y.-W.; Stormer, H. L. & P. Kim. (2005). Experimental observation of the quantum hall effect and Berry's phase in graphene, *Nature* Vol. 438, pp. 201–204.
[7] Novoselov, K. S.; Jiang, Z.; Zhang, Y.; Morozov, S. V.; Stormer, H. L.; Zeitler, U.; Maan, J. C.; Boebinger, G. S.; Kim, P.; and & Geim, A. K. (2007). Room-Temperature Quantum Hall Effect in Graphene. *Science* Vol. 315, pp. 1379–.
[8] Lin, Y.-M.; Dimitrakopoulos, C.; Jenkins, K. A.; Farmer, D. B.; Chiu, H.-Y.; Grill, A.; Avouris, Ph. (2010). 100-GHz Transistors from Wafer-Scale Epitaxial Graphene. Science, Vol. 327, pp. 662.
[9] Wang, X.; Zhi, L.; Tsao, N.; Tomovic, Z.; Li, J.; Mullen, K. (2008). Transparent Carbon Films as Electrodes in Organic Solar Cells. *Angew. Chem.*, Vol. 120, pp. 3032-3034.
[10] Elias, D. C.; Nair, R. R.; Mohiuddin, T. M. G.; Morozov, S. V.; Blake, P.; Halsall, M. P.; Ferrari, A. C.; Boukhvalov, D. W.; Katsnelson, M. I.; Geim, A. K.; Novoselov, K. S. (2009). Control of Graphene's Properties by Reversible Hydrogenation: Evidence for Graphane. Science, Vol. 323, pp. 610-613.
[11] Ni, Z. H.; Yu, T.; Lu, Y. H.; Wang, Y. Y.; Feng, Y. P.; & Shen, Z. X. (2008). Uniaxial Strain on Graphene: Raman Spectroscopy Study and Band-Gap Opening, *ACS Nano*, Vol. 2, pp. 2301–2305.
[12] Zhang, Y.; Tang, T.-T.; Girit, C.; Hao, Z; Martin, M.C.; Zettl, A.; Crommie, M. F.; Shen, R. Y. & Wang, F. (2009). Direct observation of a widely tunable bandgap in bilayer graphene. *Nature Letters*, Vol. 459, pp.820-823.
[13] Rotenberg, E.; Bostwick, A.; Ohta, T.; McChesney, J. L.; Seyller, T. & Horn, K. (2008). Origin of the energy bandgap in epitaxial graphene, Nature Materials, Vol. 7, pp. 258 – 259.
[14] Han, M. Y. ; Ozyilmaz, B.; Zhang, Y.; Kim, P. (2007). Energy Band-Gap Engineering of Graphene Nanoribbons. *Phys. Rev. Lett.* 98, 206805.
[15] Wang, X.; Li, X.; Zhang, L.; Yoon, Y.; Weber, P. K.; Wang, H. & Dai, H. (2009). N-Doping of Graphene Through Electrothermal Reactions with Ammonia. *Science*, Vol. 324, pp. 768-771.
[16] Lee, E. J. H.; Balusubramanian, K.; Weitz, R. T.; Burghard, M. & Kern, K. (2008). Contact and edge effects in graphene devices. *Nat. Nanotechnol.* Vol. 3, pp. 486-490.

[17] Casiraghi, C.; Pisana, S.; Novoselov, K. S.; Geim, A. K.; & Ferrari, A. C. (2007). Raman fingerprint of charged impurities in graphene. *Appl. Phys. Lett.* Vol. 91, 233108.
[18] Martin, J.; Akerman, N.; Ulbricht, G.; Lohmann, T.; Smet, J.; Klitzing, K. V. & Yacoby, A. (2008). Observation of electron–hole puddles in graphene using a scanning single-electron transistor. *Nature Phys.* Vol. 4, pp. 144-148.
[19] Zhang, Y.; Brar, V. W.; Girit, C.; Zett, A.; Crommie, M. F. (2009). Origin of spatial charge inhomogeneity in graphene. *Nature Phys.* Vol. 5, pp. 722-726.
[20] Hernandez, Y.; Nicolosi, V.; Lotya, M.; Blighe, F. M.; Sun, Z.; De, S.; McGovern, I. T.; Holland, B.; Byrne, M.; Gun'Ko, Y. K.; Boland, J. J.; Niraj, P.; Duesberg, G.; Krishnamurthy, S.; Goodhue, R.; Hutchison, J.; Scardaci, V.; Ferrari, A. C. & Coleman, J. N. (2008). High-yield production of graphene by liquid-phase exfoliation of graphite, Nature Nanotechnology, *Vol.* 3, pp. 563 - 568.
[21] Li, X.; et al. (2009). Large-Area Synthesis of High-Quality and Uniform Graphene Films on Copper Foils. *Science,* Vol. 324, pp. 1312.
[22] Lee, Y.; Bae, S.; Jang, H.; Jang, S.; Zhu, S.-E.; Sim, S. H.; Il Song, Y.; Hong, B. H.; Ahn, J.-H. (2010). Wafer-Scale Synthesis and Transfer of Graphene Films, *Nano Lett.* Vol. 10, pp. 490.
[23] Berger, C.; Song, Z.; Li, X.; Wu, X.; Brown, N.; Naud, C.; Mayou, D.; Li, T.; Hass, J.; Marchenkov, A. N.; Conrad, E. H.; First, P. N. & de Heer, W. A. (2006). Electronic Confinement and Coherence in Patterned Epitaxial Graphene, *Science* Vol. 312, pp. 1191.
[24] Giannazzo, F.; Sonde, S.; Raineri, V.; Patanè, G.; Compagnini, G.; Aliotta, F.; Ponterio, R.; & Rimini, E. (2010). Optical, morphological and spectroscopic characterization of graphene on SiO_2, *Phys. Status Solidi* C Vol. 7, pp. 1251.
[25] Ferrari, A. C.; Meyer, J. C.; Scardaci, V.; Casiraghi, C.; Lazzeri, M.; Mauri, F.; Piscanec, S.; Jiang, D.; Novoselov, K. S.; Roth, S.; & A. K. Geim. Raman Spectrum of Graphene and Graphene Layers, (2006). *Phys. Rev. Lett.* Vol. 97, pp. 187401
[26] Blake, P.; Hill, E. W.; Castro Neto, A. H.; Novoselov, K. S.; Jiang, D.; Yang, R.; Booth, T. J.; & Geim, A. K. (2007). Making graphene visible, Appl. Phys. Lett. Vol. 91, pp. 063124.
[27] Nair, R. R.; Blake, P.; Grigorenko, A. N.; Novoselov, K. S.; Booth, T. J.; Stauber, T.; Peres, N. M. R.; Geim, A. K. (2008). Fine Structure Constant Defines Visual Transparency of Graphene. *Science* Vol. 320, 1308.
[28] Emtsev, K. V.; Bostwick, A.; Horn, K.; Jobst, J.; Kellogg, G. L.; Ley, L.; McChesney, J. L.; Ohta, T.; Reshanov, S. A.; Rohrl, J.; Rotenberg, E.; Schmid, A. K.; Waldmann, D.; Weber, H. B.; & Seyller, T. (2009). Towards wafer-size graphene layers by atmospheric pressure graphitization of SiC(0001). *Nature Materials.* Vol. 8, pp. 203.
[29] Virojanadara, C.; Syvajarvi, M.; Yakimova, R.; Johansson, L.; Zakharov, I. & T. Balasubramanian. (2008). Homogeneous large-area graphene layer growth on 6H - SiC(0001). *Phys. Rev. B* Vol. 78, 245403.
[30] Bolen, M. L.; Harrison, S. E.; Biedermann, L. B.; & Capano, M. A. (2009). Graphene formation mechanisms on 4*H*-SiC(0001). *Phys. Rev. B* Vol. 80, pp. 115433.
[31] Vecchio, C.; Giannazzo, F.; Sonde, S. ; Bongiorno, C. ; Rambach, M.; Yakimova, R.; Rimini, E.; Raineri, V. (2010). Nanoscale structural characterization of epitaxial graphene grown on off-axis 4H-SiC (0001). *Nanoscale Research Letters*, in press.
[32] Prakash, G.; Capano, M. A.; Bolen, M. L.; Zemlyanov, D.; Reifenberger, R. G. (2010). AFM study of ridges in few-layer epitaxial graphene grown on the carbon-face of 4H–SiC (000-1). *Carbon.* Vol. 48, pp. 2383-2393.

[33] Sun, G. F.; Jia, J. F.; Xue, Q. K. & Li, L. (2009). Atomic-scale imaging and manipulation of ridges on epitaxial graphene on 6H-SiC(0001). *Nanotechnology*, Vol. 20, pp. 355701.

[34] Chen, J.H.; Jang, C.; Xiao, S.; Ishigami, M. & Fuhrer, M.S. (2008). Intrinsic and extrinsic performance limits of graphene devices on SiO_2. *Nature Nanotechnology*, Vol. 3, pp. 206-209.

[35] Bolotin, K. I.; Sikes, K. J.; Jiang, Z.; Klima, M.; Fudenberg, G.; Hone, J.; Kim P. & Stormer, H. L. (2008). Ultrahigh electron mobility in suspended graphene. *Solid State Commun.* Vol. 146, pp. 351-355.

[36] Hwang, E. H.; Adam, S.; & Das Sarma, S. (2007). Carrier Transport in Two-Dimensional Graphene Layers. *Phys. Rev. Lett.* Vol. 98, pp. 186806.

[37] Fratini, S.; & Guinea, F. (2008). Substrate-limited electron dynamics in graphene. *Phys. Rev. B* Vol. 77, pp. 195415.

[38] Giannazzo, F.; Sonde, S.; Raineri, V.; & Rimini, E. (2009). Irradiation damage in graphene on SiO_2 probed by local mobility measurements, *Appl. Phys. Lett.* Vol. 95, pp. 263109.

[39] Giannazzo, F.; Sonde, S.; Rimini, E. & Raineri, V. (2010). Lateral homogeneity of the electronic properties in pristine and ion irradiated graphene probed by scanning capacitance spectroscopy, *Nanoscale Research Letters*, in press.

[40] Giannazzo, F., Sonde, S., Raineri, V., & Rimini, E. (2009). Screening Length and Quantum Capacitance in Graphene by Scanning Probe Microscopy. *Nano Letters* Vol. 9, pp. 23

[41] Sonde, S.; Giannazzo, F.; Raineri, V.; & Rimini, E. (2009). Dielectric thickness dependence of capacitive behavior in graphene deposited on silicon dioxide. *J. Vac. Sci. Technol. B*. Vol. 27, pp. 868

[42] Rutter, G. M.; Crain, J. N.; Guisinger, N. P.; Li, T.; First, P. N.; & Stroscio, J. A. (2007). Scattering and Interference in Epitaxial Graphene. *Science* .Vol. 317, 219-222.

[43] Sonde, S.; Giannazzo, F.; Vecchio, C.; Yakimova, R.; Rimini, E.; & Raineri, V. (2010). Role of graphene/substrate interface on the local transport properties of the two-dimensional electron gas, *Appl. Phys. Lett.* Vol. 97, pp. 132101

[44] Stauber, T.; Peres, N. M. R.; & Guinea, F. (2007). Electronic transport in graphene: A semiclassical approach including midgap states, *Phys. Rev. B* Vol. 76, pp. 205423

[45] Perebeinos, V.; & Avouris, Ph. (2010). Inelastic scattering and current saturation in graphene, *Phys. Rev. B* Vol. 81, pp. 195442.

[46] Wang, S. Q.; & Mahan, G. D. (1972). Electron Scattering from Surface Excitations. *Phys. Rev. B*. Vol. 6, pp. 4517-4524.

[47] Emtsev, K. V.; Speck, F.; Seyller, Th.; Ley, L.; Riley, J. D. (2008). Interaction, growth, and ordering of epitaxial graphene on SiC{0001} surfaces: A comparative photoelectron spectroscopy study. *Phys. Rev. B* Vol. 77, pp. 155303.

[48] Sonde, S.; Giannazzo, F.; Raineri, V.; Yakimova, R.; Huntzinger, J.-R.; Tiberj, A. & Camassel, J. (2009). Electrical properties of the graphene/4H-SiC (0001) interface probed by scanning current spectroscopy, *Phys. Rev. B* Vol. 80, pp. 241406(R).

[49] Giannazzo, F.; Roccaforte, F.; Raineri, V. & Liotta, S. F. (2006). Transport localization in heterogeneous Schottky barriers of quantum-defined metal films, *EuroPhysics Letters* Vol. 74, pp. 686.

[50] Giannazzo, F.; Roccaforte, F.; Iucolano, F.; Raineri, V.; Ruffino, F. & Grimaldi, M. G. (2009). Nanoscale current transport through Schottky contacts on wide bandgap semiconductors, *J. Vac. Sci. Technol.* B 27, pp. 789.

8

Intercalation of Graphene Films on Metals with Atoms and Molecules

E.V.Rut'kov and N.R.Gall
Ioffe Physico-Technical Institute of Russian Academy of Sciences, Russia

1. Introduction

Our studies [1--3] resulted in a first observation of intercalation of graphene films on metal surfaces, which consists essentially in penetration (diffusion) of foreign atoms and molecules between the surface of a metal and the graphene film. The fact of intercalation itself implies a weak binding of graphene to the metal surface (§ 6) and finds a reasonable explanation in the graphene layer being valence saturated. Intercalation is usually observed to occur in layered solids (graphite, MoS_2) whose interlayer space may become occupied by some penetrating atoms and molecules. For instance, exposing a heated graphite crystal to cesium vapor initiates spontaneous penetration of cesium atoms between the graphite layers to form the C_8Cs intercalation compound, in which graphite layers with concentration N_c = 3.56 10^{15} cm^{-2} alternate with cesium layers with N_{Cs} = 5 10^{14} cm^{-2}, and the distance between neighboring graphite layers increases from 3.35 Å in a graphite single crystal to 5.94 Å in C_8Cs [4, 5]. Interestingly, intercalation leaves the structure and lattice constant of the graphite layer and, hence, its individual features unchanged.
Our research group has studied in considerable detail the various aspects of intercalation of graphene films on metals with a range of atoms, more specifically, K, Cs, Na, Ba, Sr, Pt, Si, C, Ag, Al, Ir, Cu, Mo, as well as with molecules of the C_{60} fullerenes [1--3].

2. Intercalation with atoms of alkali metals (K, Cs, Na) [1--3, 6]

2.1 Adsorption of potassium atoms on graphene-coated iridium

Potassium atoms were adsorbed at 300 K on a graphene film atop iridium to saturation, i.e., until a reverse flux of potassium atoms desorbing from the surface became equal to the incident flux. Next the surface was flash-heated. Three phases were observed in the thermal desorption spectrum (Fig. 1). The potassium atoms in the α phase desorbing completely at T < 850 K are produced by desorption of potassium from graphene; indeed, the temperature of potassium desorption from the α phase matches nicely with the lifetime of potassium adatoms on graphene. The β phase having a low adsorption capacity (~10^{11} at/cm^2) is attributed to the potassium adatoms decorating defects in graphene films (for a more detailed account of the β phase, see [2]). The adsorption capacity of the β phase grows strongly with removal of a part of carbon atoms from the graphene film [7]. Particularly interesting is the γ phase, with potassium desorbing from this state at $T \geq 2000$ K (sic!) when the graphene film breaks up. This desorption phase is associated with the release of

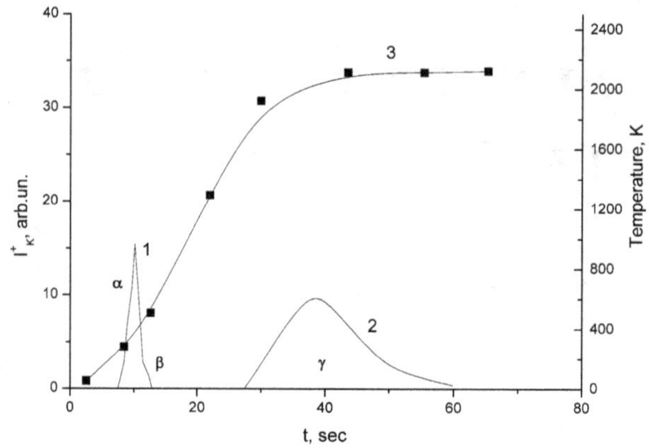

1, 2 - α- and γ- are desorption phases, 3 - the temperature - time dependence T(t).

Fig. 1. Thermal desorption spectra of K+ ions in heating of a graphene layer on Ir(111). K was deposited at 300 K with the flux $\nu_K = 1.0 * 10^{11}$ cm^{-2}*s^{-1} for 40 s.

potassium atoms buried under the graphene film. Figure 2 illustrates filling of the γ phase with potassium for $\upsilon_K = 6\ 10^{11}$ cm^{-2}s^{-1} reached in 10 min for different substrate temperatures in the $300 \leq T \leq 900$ K region, and the kinetics of this process in the 300--670 K interval is plotted in Fig. 3.

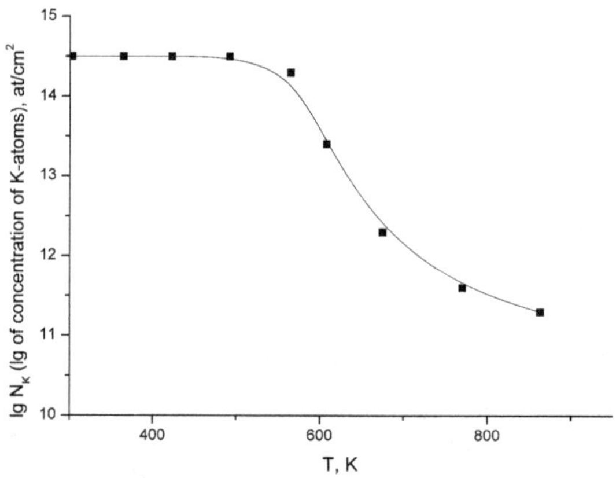

Fig. 2. K atomic concentration in logarithmic scale for K stored behind the graphene layer on Ir(111). K was deposited at various temperatures with the flux $\nu_K = 6.0 * 10^{11}$ cm^{-2}*s^{-1} for 10 min at the each temperature.

In the $300 \leq T \leq 550$-K region, the γ phase was filled in 10 min to its maximum possible level of $N_\gamma \approx$ (2--3) 10^{14} at/cm^2, which did not change during subsequent exposure to a potassium atom flux. There is more than one experimental verification of the potassium in the γ phase residing indeed under the graphene film [6]. Consider only two of them.

Verified by the work function. Potassium present with a concentration $N_K \sim$ (2--3) 10^{14} at/cm^2 on the surface of graphene was experimentally shown by CPD to reduce the surface work function by close to 2 eV. The potassium in the γ phase adsorbed to the same concentration reduces the work function by 0.3 eV only.

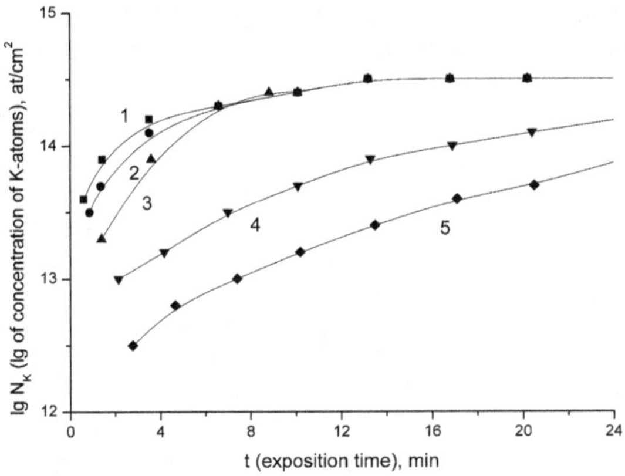

Fig. 3. K atomic concentration for K stored behind the graphene layer on Ir(111). K was deposited versus exposure time at various temperatures T (K): 1 - 300; 2 - 530; 3 - 590; 4 - 630; 5 - 670. The K flux was $\nu_K = 6.0 * 10^{11}$ cm^{-2}*s^{-1}.

Verified by AES. This is probably the most direct method of all, because it was found that filling of the γ phase with potassium to $N_\gamma \approx$ (2--3) 10^{14} at/cm^2 (following desorption of potassium from the α phase) reduces the intensity of the iridium Auger peak by a factor 1.4, although that of the carbon Auger peak does not change, thus implying that potassium is sandwiched between metal and the graphene film, i.e., resides in the intercalated state. The presence of potassium in the γ phase suggests that the graphene film is physisorbed on iridium, and that after intercalation with potassium it rose above iridium by a considerable distance, because the diameter of the potassium atom d_0 = 4.72 Å, and that of the potassium ion, d_+ = 2.66 Å [6].

It is common knowledge that graphite can be intercalated with atoms and molecules. Obviously enough, insertion of foreign atoms between layers should increase the interplanar distance *h*. This is evidenced by the data [4] listed below in Table 1.

Intercalant	Cs	K	Li	Rb	Br	HNO_3	HCl_3	$FeCl_3$
h, Å	5,94	5,35	3,71	5,65	7,04	7,84	7,94	9,37

Table 1.

Similar results are obtained in adsorption at 300 K of Cs and Na atoms on Ir(111)-graphene.

2.2 Adsorption of potassium atoms on iridium with graphene islands

The transition from a continuous film to graphene islands brings about appearance of a number of features in thermodesorption spectra. The first of them is the sharp increase of the capacity of the β phase, which should be assigned to potassium atoms decorating the edges of graphene islands.

Second, as the area of the islands S_0 decreases, the time lag t^* in filling of the γ phase after the beginning of potassium evaporation grows rapidly. Figure 4 displays graphs illustrating γ phase filling for different S_0. We readily see that t^* grows strongly with increasing area free of graphene islands.

Assuming all potassium atoms to roll off the graphene islands onto iridium, we find that for all S_0 filling of the γ phase starts at the same concentration of potassium on iridium, ~2.5 10^{14}

at/cm2. Significantly, in the case of a continuous graphene film on iridium, filling of the γ phase with potassium comes to completion after the concentration of atoms on the graphene surface has reached ~5 10^{13} at/cm2.

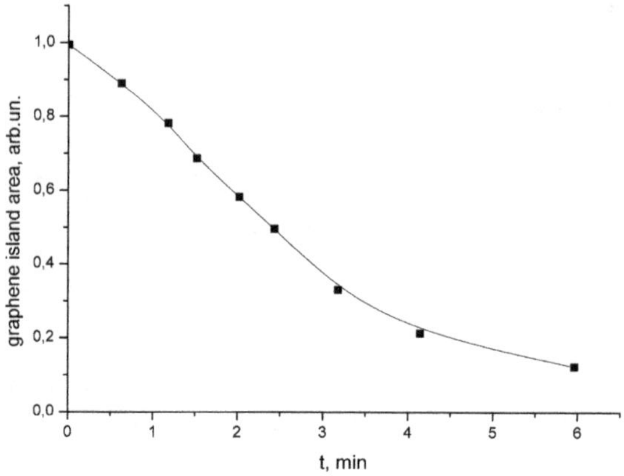

Fig. 4. Dependence of the initial time of K penetration into the γ-phase t* versus the relative graphene island area on iridium S_0. K was deposited at 300 K with the flux $\nu_K = 8.4 *10^{11} cm^{2*} s^{1}$.

Figure 5 plots the dependence of the maximum filling by potassium of the γ phase on the relative area of graphite islands. We see N_γ to scale directly proportional to S_0.

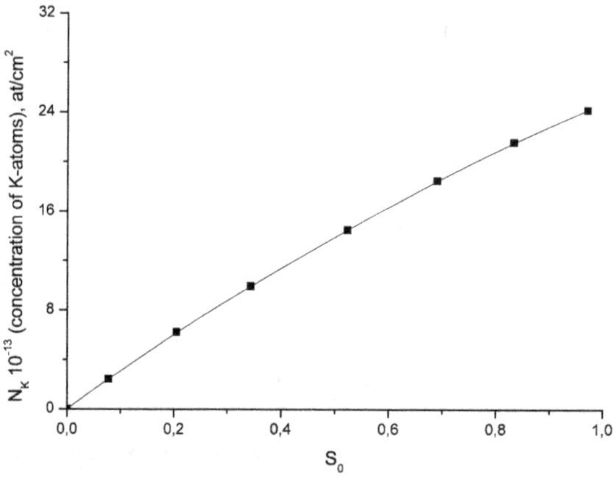

Fig. 5. Maximal K concentration under graphene (in the γ-phase) N_K versus the relative island area on Ir(111). K was deposited at 300 K with the flux $\nu_K = 2.0 *10^{12} cm^{-2} *s^{-1}$. N_K was determined by TDS method.

Consider now how could one explain these results. Obviously enough, the first centers to be filled by adsorbing potassium atoms should be those with strong binding; here they would be edges of the graphite islands (the β phase). The next to be filled are less energetically favorable centers; potassium adsorbs on iridium. For $\theta > 0.1$, a substantial part should be played by Coulomb repulsion among positively charged potassium adatoms, and it is apparently this repulsion that prevents complete filling of the potassium monolayer on iridium, so that for $N_K \geq 2.5\ 10^{14}$ at/cm^2 potassium atoms acted upon by Coulomb forces will start occupying the free surface of graphene islands. This is followed by effective filling of the γ phase. Just as in the case of a continuous graphene film, potassium from the γ phase for graphene islands desorbs from the surface at high temperatures ($T \geq 1900$ K). That potassium persists for a long time at high temperatures in the γ phase suggests that valence-non-saturated carbon atoms at the edges of graphene islands are chemically bound to atoms in the surface layer of iridium. It is the closed island edges that prevent potassium from escaping from under the islands ("bottom-up saucer" island, Fig. 6).

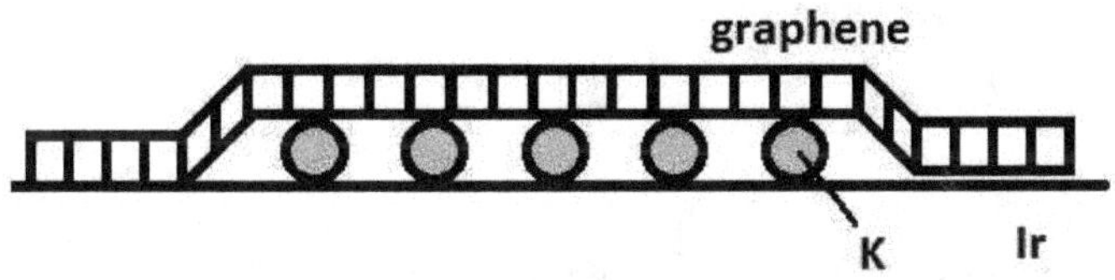

Fig. 6. Schematic graphene island on a metal with atomic K in the γ-phase.

5.3 Adsorption of cesium atoms on graphene atop rhenium [7--11]

One more example of a substrate metal, rhenium, may be in order here to demonstrate that graphene islands are physisorbed in this case too. A rhenium ribbon was carbonized in benzene vapor at T_c = 1700 K up to formation of a graphene film, after which the temperature was lowered to 300 K by turning off the heating current. Experiments showed that in this way one can preserve a graphene layer on iridium at temperatures of 300--1100 K, because volume diffusion becomes frozen at these temperatures. Thereafter, cesium atoms were adsorbed at 300 K on the graphene layer atop iridium in a constant flux $\upsilon_{Cs} = 2\ 10^{12}$ cm^{-2}s^{-1} to saturation, i.e., until the reverse flux became equal to the incident one. After this, the temperature was flashed (Fig. 7). We readily see that cesium desorption occurs in two phases, with the first of them (α phase) observed at $T \sim 900$ K, and the second (γ phase), at high temperatures, $T \geq 2100$ K (sic!). The temperature of cesium desorption from the α phase matches with the lifetime of cesium on a graphene film (§ 7) and can be identified with cesium escaping from the surface of the graphene film. The high temperatures of cesium desorption from the γ phase are associated with cesium becoming freed from under the graphene film when it breaks up. The maximum adsorption capacity of cesium in the γ phase, $N_\gamma \approx 8\ 10^{13}$ at/cm^2. Significantly, in these experiments the adsorption capacity of the β phase of cesium (similar to the potassium β phase for the iridium-graphene system), which is related with graphene defects, was at a level of $N_\beta \sim 10^{11}$ at/cm^2, and is not reflected in Fig. 7. Just as for the Ir-C system, we are going to present here two arguments in support of cesium in the γ phase being buried under the graphene film.

1. The work function of the surface of a graphene film decreases after maximum filling by cesium of the γ phase by ~0.25 eV, although the same amount of cesium present in the α phase causes a decrease of the work function by ~1.25 eV.
2. After desorption of the cesium α phase, the Auger peak intensity of carbon recovers to the level corresponding to a clean graphene film, whereas the rhenium peak intensity is found to be lower by a factor ~1.3 than that before the beginning of cesium adsorption. This implies that cesium in the γ phase is sandwiched between the graphene film and the metal.

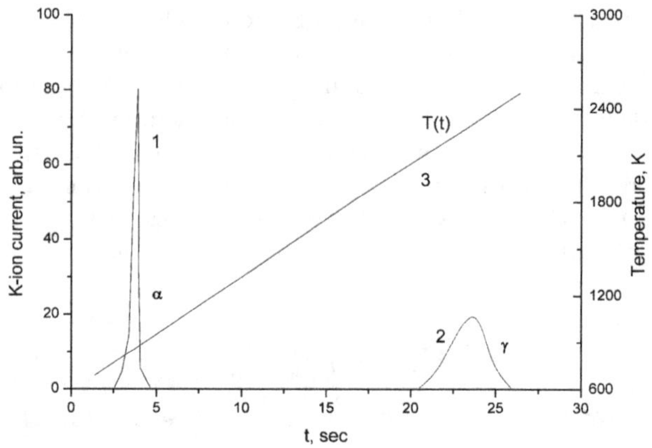

Fig. 7. Thermal desorption spectra for Cs+ ion after Cs deposition on a graphene layer on Re. Cs was deposited at 300 K with the flux $\nu_K = 8.5 \cdot 10^{11}$ cm^{-2}*s^{-1} for 15 min. 1, 2 - α- and γ-desorption phases, 3 the temperature - time dependence T(t).

Figure 8 illustrates filling of the γ phase by evaporation of cesium atoms for υ_{Cs} = 3.3 10^{12} cm^{-2}s^{-1} during 10 min, measured for different substrate temperatures within the interval $300 \leq T \leq 900$ K. Examining Fig. 8, we see that intensive filling of the γ phase by cesium starts at $T \leq 800$ K. Just as in the case of potassium atoms, the maximum adsorption capacity of the cesium γ phase does not depend on the sample exposure time to the flux of cesium atoms.

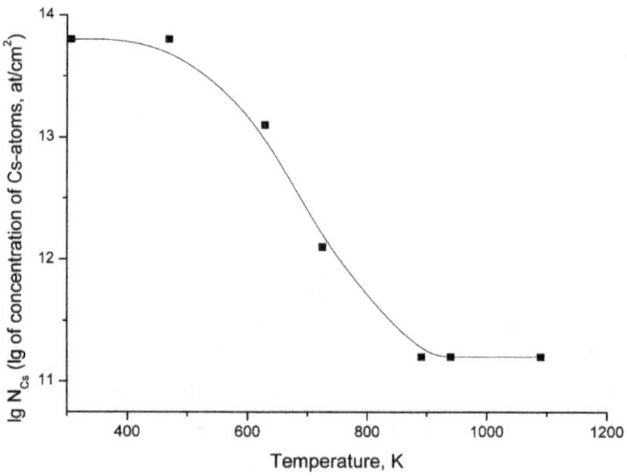

Fig. 8. Cs atomic concentration in logarithmic scale for Cs stored behind the graphene layer on Ir(111) versus temperature. Cs was deposited at various temperatures with the flux ν_{Cs} = 3.3 *10^{12} cm^{-2}*s^{-1} for 10 min at the each temperature.

Thus, the assumption of a graphene film on a metal being physisorbed is valid not only for iridium but for rhenium as well. Similar experiments on adsorption of K, Na, and Cs atoms performed on graphene grown atop other metals as well---rhodium, molybdenum, platinum, and nickel, provide convincing support for the general physisorptive character of graphene bonding with a metal surface.

2.4 Intercalation of graphene on a metal with alkali atoms [2, 3, 12, 13]

Application of TDS (in studies of adsorption of Cs atoms on graphene atop rhenium) involving detection of the ions released in surface ionization increased strongly the sensitivity of the method in the initial stages of adsorption, when the surface work function does not yet change and each desorbing Cs atom escapes in the form of the Cs^+ ion.

The following features observed at $T = 300$ K appear noteworthy.

1. At cesium concentrations on graphene atop rhenium below $N_{Cs} \approx 1\ 10^{12}$ at/cm^2, cesium atoms reside only in the α phase, i.e., on the surface of the graphene film.
2. For $N_{Cs} > 1\ 10^{12}$ at/cm^2, surface migration of cesium atoms sets in (rather of cesium ions, because cesium on the surface is positively charged), which is apparently initiated by Coulomb repulsion. In the process, cesium transfers to the back side of the ribbon to the α state, which is corroborated by the observation that the thermodesorption spectrum contains only the α phase with the cesium concentration $N_{Cs} = ½\upsilon_{Cs}t_{exp}$.
3. After the cesium concentration on both sides of the rhenium ribbon has reached the level $N_{Cs} \geq 1\ 10^{13}$ at/cm^2, cesium atoms start to diffuse actively into the space under graphene, signaling filling of the γ phase. The total concentration of cesium on the surface becomes $N_{Cs}(\text{tot}) = ½\upsilon_{Cs}t_{exp}$, because half of the material (cesium) escapes to the back side of the ribbon (this is why the α and β phases of cesium on the front side of the ribbon are easy to fill by evaporating it on the back side of the sample). It is apparently only at noticeable cesium concentrations (≥1 10^{13} at/cm^2) that the forces of electrostatic repulsion among cesium adatoms with a charge close to 1 become strong enough to overcome an activation barrier and start to diffuse under the graphene film. The nature of this activation barrier remains unclear, but it could be related, for instance, with the Coulomb repulsion of the cesium adatom from a positively charged defect in a graphene film, and it is possibly this defect that mediates efficiently the escape of charged adatoms migrating over the surface.

The process of filling of the γ phase by the other alkali atoms, sodium and potassium, exhibits basically the same features as that of the filling by cesium, which should be assigned to these alkali atoms having a noticeable positive charge on the graphene surface.

Cs, Na, and K atoms fill actively the γ phase in the 300--700-K temperature interval, where evaporation of atoms can result in building up on the surface of a certain "critical" concentration of adatoms which would initiate filling up of the γ phase. For $T > 700$ K, the lifetimes of alkali metals on graphene become too short (§7), which reduces the equilibrium surface concentration, with the result that the migration transfer under the graphite layer drops sharply (Figs. 2 and 8).

2.5 Why do Cs atoms find it easy to diffuse under the graphene film but hard to escape from under it?

There are presently no grounds to doubt that adatoms of alkali metals are "pushed" under the graphene film by their electrostatic repulsion. It was observed that for the intercalation of single-crystal graphite by alkali metals to begin, they should first build up to a certain concentration on the crystal faces [14]. This reminds one of the problem where one should force into a rubber tube balls of a slightly larger diameter; it can be solved by applying a certain force F (Fig. 9A). As the pushing "finger" acts in our case repulsion among positively charged adatoms (Fig. 9B). The above is supported by experiments aimed at studying Cs migration under the graphene islands; indeed, single Cs adatoms were found to practically not migrate under graphene at all [15].

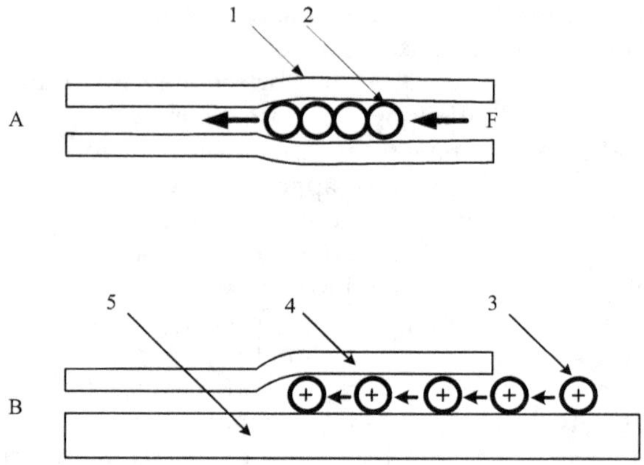

Fig. 9. **A**. Schematic structure of solid balls movement inside a ribbon tube under the action of an external force: 1 – ribbon tube, 2 – solid ball. **B.** Schematic structure for electropositive atomic metal penetration between the graphene layer and metal substrate: 3 - adsorbed Cs, 4 - graphene layer, 5 - metal substrate.

Now about the question of why it is easy (T = 300 K) for atoms of alkali metals to transfer under a graphene film but difficult (at $T \geq$ 2000 K!) to escape from under it, i.e., why the entry--escape ways are so strongly asymmetric.

Studies of the intercalation of graphene on Rh(111) with Cs atoms revealed that after desorption of the Cs α phase from the graphene surface, the cesium intercalant emerges partially from under the graphene film and desorbs likewise at low temperatures, 900--1000 K. The emergence of cesium from under graphene is signaled by the formation of diffusion "tails" in thermodesorption spectra (Fig. 10).

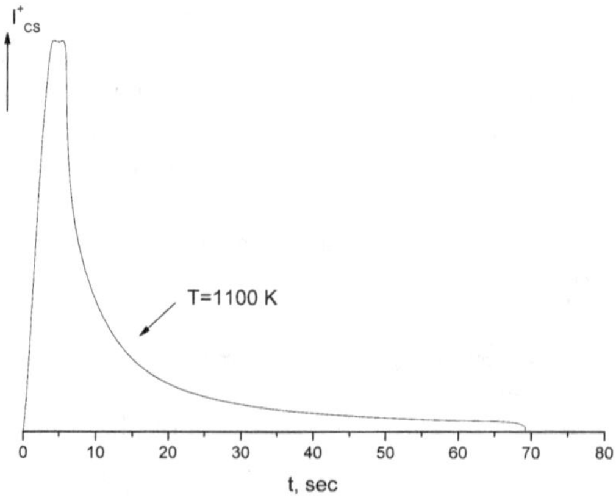

Fig. 10. Thermal desorption spectroscopy for Cs+ ions from the graphene layer on Rh(111) in the temperature increase form 300 to 1100 K. Cs was deposited at 300 K up to the concentration $N_{Cs} = \nu_{Cs}*t = 4.4 *10^{14}$ cm^{-2}.

Significantly, the larger was the amount of Cs adsorbed into the α phase, the deeper penetrated Cs atoms and in larger numbers under the graphene layer, and the longer was the diffusion "tail" in the TDS spectra.

Thus, as T increases, Cs emergence from under the layer is initially not blocked in any way, which makes it difficult to separate the α phase of the Cs emergence from the onset of the Cs escape from the γ phase. We are turning now to the second question of what forced adatoms to stay in intercalated state, so that they escape from under the graphene film only when it breaks up, and this may require quite high temperatures, often above 2000 K (sic!).

The answer to this question may be contained in Fig. 11. As T increases, Cs escapes from the α phase, i.e., from the graphene surface (1 in Fig. 11A), the next to desorb is Cs from the defects in the layer (2 in Fig. 11B), and it is through them that adatoms penetrate under the layer in the first place. The turn to emerge comes after that to Cs adatoms bound close to graphene island edges (3 in Fig. 11C). The Cs adatoms left far from the layer edge do not have time enough to migrate to the edge closing to the metal (4 in Fig. 11D). These adatoms remained trapped under the "bottom-up saucer" islands (Fig. 6).

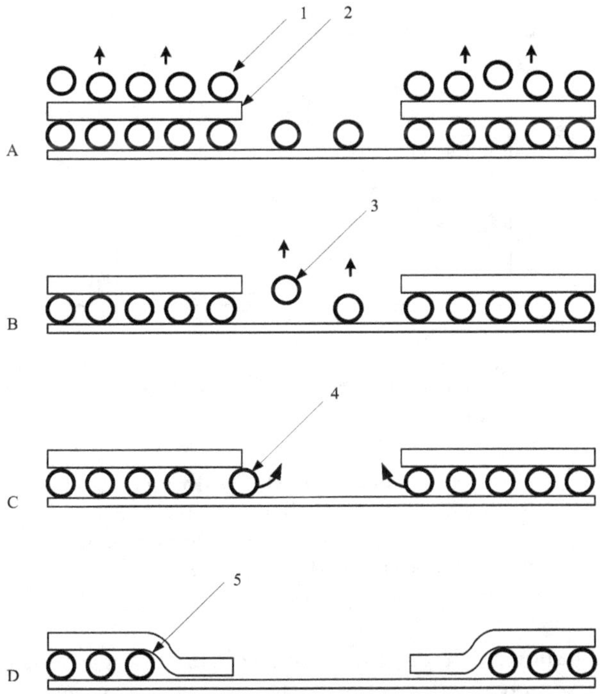

Fig. 11. Schematic model (A – D) showing the process of Cs liberation from the γ-phase and the island edge closure to the metal substrate after the temperature increase for the intercalation system Me-Cs-graphene. 1 - atomic Cs above the graphene layer, α- phase; 2 – graphene; 3 - atomic Cs desorbing from the layer defect; 4 - intercalated Cs near the island edge; 5 - 'locked' atomic Cs under the layer; graphene traps.

2.6 Excitation of edges of graphene islands

Under certain temperatures, the graphene island edges bound to metal should open periodically stimulated by thermal excitation. Rupture of the Me-C chemical bond may open the way for a part of the intercalant cesium to escape from under the graphene film and desorb. This accounts for a certain, albeit low-density, flux υ_{Cs} of Cs atoms leaving at T >

1000 K. For instance, for the Ir(111)-graphene system with a cesium concentration under graphene $N_\gamma \sim 2\ 10^{14}$ at/cm² and T = 1100 K, $\upsilon_{Cs} \sim 10^9$--10^{10} cm^{-2}s^{-1}. The Cs flux decreases with time, primarily because of the edge region becoming depleted in cesium, which makes it ever more difficult for the remaining adatoms to reach the island edge. Experiments demonstrate that removing Cs from under a graphene film without destroying it takes up several hours at $T \sim 1100$ K on Ir(111), although the lifetime of Cs, both on graphene and on iridium, at this temperature is only ~ 0.01 s (§ 7).

Figure 12 plots the variation of the flux of Cs atoms emerging from under graphene on Rh(111) with the substrate temperature varied from T_1 = 1300 K to T_2 = 1350 K, for the Cs concentration under the graphene islands $N_{Cs}(\gamma) = 5\ 10^{13}$ at/cm².

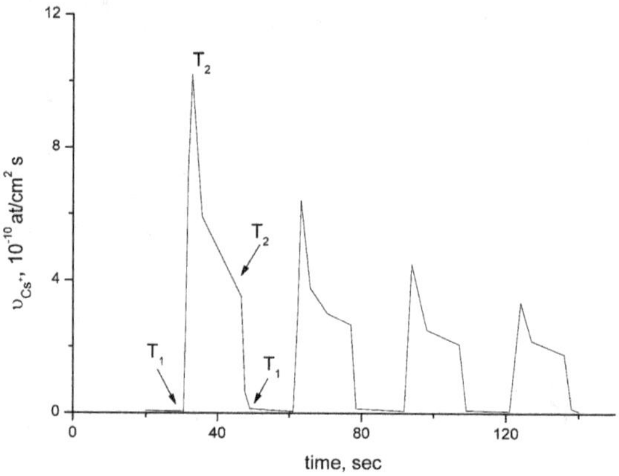

Fig. 12. The flux of Cs^+ ions out of graphene islands in their thermal excitation. The temperature was T_1 = 1300 K and T_2= 1350 K, relative island area was s_0~15%. The islands were grown on Rh(111), and Cs was intercalated to saturation at 300 K.

We immediately see a sharp rise of the cesium flux with increasing temperature associated with the increase of thermal excitation of graphene. As the temperature returns to its initial level T_1, the cesium flux density falls off rapidly too. Experiments with cyclic variation of temperature $T_1 \rightarrow T_2 \rightarrow T_1$ can be performed many times, and the results will be the same, except for the gradual decrease of the burst in the flux density of Cs atoms escaping from under the graphene (Fig. 12).

Knat'ko *et al.* [1989] showed that illumination with light or electron bombardment of a graphene film on Cs-intercalated iridium heated to ~900 K initiates emergence of these adatoms from under the graphene islands, which was visualized by monitoring the Cs^+ current. This effect was observed at photon energies varied from 2.3 to 1.5 eV, with the low-energy threshold remaining not quantified. It was assumed that the adatoms escape from under the graphene islands through some "valves" forming in the rupture of bonding of island-edge C atoms with iridium initiated by irradiation by light or electrons. The transport of particles taking place in the absence of external stimulating factors was explained as due to diffusion via defects present in the graphene film. The kinetics of these processes stimulated by light and electrons was studied, and a model reproducing them proposed [15]. It was shown that the burst of the Cs^+ ion current can be attributed to the emergence of adatoms buried under graphene close to the island edges. Filling of the near-edge region after its having being depleted by irradiation is a very slow process, which sets in probably following formation of a compact Cs cluster diffusing as a whole to the edge of an island from its central region. Thus Knat'ko *et al.* [1989] established an unusual pattern of migration of intercalated Cs atoms, in that they move in compact clusters rather than singly, with the rate of migration growing with increasing concentration of the intercalated atoms.

Besides the edges of graphene islands, thermal excitation may set in in any part of an island, accompanied by the appearance of a graphene hillock and increase of the graphene-metal distance at this point [16].

Even at T = 300 K, one may expect excitation of any part of graphene islands bound with the metal by weak van der Waals forces; indeed, such bubbles form and disappear all the time over the film. It appears reasonable to assume that from time to time, each carbon atom in the graphene network, is capable, while participating in thermal random motion, of building up on the bond with the surface an energy high enough to be able, in moving away from the surface, to drag along its nearest neighbors and form a graphene "bubble".

A graphene film could be said to remind one, as a suitable figure of speech, the surface of a rough sea, with crests and hollows coming to life and disappearing continuously, with the sea roughness becoming the stronger, the stronger is the gale (the temperature in the case of graphene).

Tontegode (1985) [16] resorted to the relation of Frenkel $\tau = \tau_0 \exp(E/kT)$ to estimate the kinetic energy E building up on a particle bond in random thermal motion in a given observation time τ at a temperature T, which is expended in formation of a "bubble". In this sense, the motion of a carbon atom over the network on the surface resembles largely the process of desorption, the only difference being that the graphene network itself, not unlike a spring, restores the carbon atom to its original position.

Estimates show that for a binding energy of a carbon atom with the metal in a graphene film of ~0.2 eV, a graphene "bubble" should accumulate four atoms in a time τ =1 s (for $\tau_0 = 10^{-13}$ s) at T = 300 K, while at T = 1500 K the "bubble" will be quite large and contain ~20 atoms. Formation of graphene "bubbles" facilitates migration of intercalated atoms.

2.7 Compression of a layer of Cs atoms under graphene islands

The first successful monitoring of the area of graphene islands with carbon atoms trapped underneath was made in the course of their destruction. The experiment started with intercalation of a graphene film on rhodium at 300 K with cesium atoms. Next the cesium was removed from the surface at ~900 K, after which the temperature was raised to T_1 > 1250 K, and the relative island area S_{01} was determined. Next the surface was flash heated, and the amount of cesium N_1 under the islands of area S_{01} deduced. The experiment was repeated from the very beginning, but now the temperature was raised to $T_2 > T_1$, the magnitude of $S_{02} < S_{01}$ found, and N_2 derived. The results obtained are summarized graphically in Fig. 13A, in which each data point is essentially a separate experiment.

We readily see that a decrease of the relative island area from 100% (continuous layer) to ~20% (curve 1 in Fig. 13A) does not actually entail a noticeable decrease of the total number of cesium atoms on the surface (curve 2 in Fig. 12A). This implies that while the islands may shrink markedly in area, they do not "let go" the cesium atoms buried under them (Fig. 13B). It is as obvious also that although the number of cesium atoms under the islands remains unchanged, their concentration grows through the decrease of islands in area (curve 3 in Fig. 13A for T = 1150--1270 K).

We note also that for T > 1300 K graphene islands become so small that they can possibly no longer keep trapped the cesium atoms confined under them, with the result that the average concentration of intercalated cesium will start to decrease. Apart from this, the concentration

A.

B.

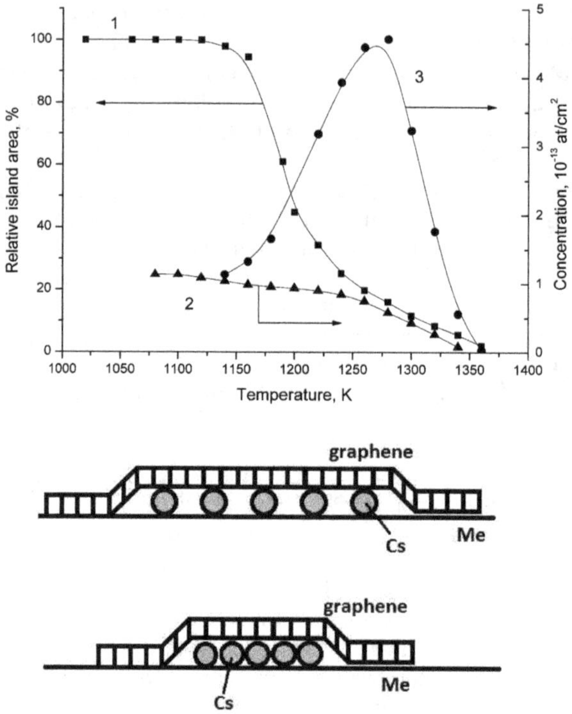

Fig. 13. Compression of graphene islands with intercalated Cs. **A**. Temperature dependences of: 1 - relative graphene island area; 2 - Cs concentration relative the total surface area; 3 - calculated Cs concentration relative to the actual area occupied by graphene. **B**. Schematic structure of the process:

of the graphene islands themselves will decrease as they break up with increasing temperature. Recalling the well known expression for particle lifetime on the surface, $\tau = \tau_0 \exp(E/kT)$ (τ_0 is the prefactor, E, the desorption energy, is in our case $E \sim 2$ eV [17]), we come readily to the estimate that graphene traps increased the lifetime of cesium adatoms on rhodium surface ~10^6 times.

Thus, we have revealed a new effect inherent in intercalation of an alkali metal under graphene films and consisting in a substantial increase of its concentration in the intercalated state under heating of the adlayer. While annealing reduces the graphene island area, their edge atoms bound with the metal by chemisorption forces do not let cesium emerge from under the graphene; this gives rise to compression of the cesium layer, and it is under the islands that its surface concentration exhibits a noticeable increase.

3. Intercalation with barium [18, 19]

It was found that Ba intercalates under graphene on iridium and rhenium, but the pattern of the processes involved depends essentially on the temperature of the substrate metal. We are going therefore to consider these processes separately for $T < 1000$ K and $T > 1000$ K.

3.1 Adsorption of Ba on Ir-C for $T < 1000$ K

In the case of Ba adsorbed on iridium at 300 K intercalation does not seem to occur, which can be deduced from a characteristic change in shape of the carbon C_{kvv} Auger peak (to be discussed in detail below), but because of strong Ba-Ba bonding on the graphene surface Ba undergoes condensation, which complicates monitoring the intercalation process. Adsorption of Ba on iridium at 300 K does not probably result in intercalation, which is suggested by a characteristic change in the shape of the C_{kvv} Auger peak of carbon (to be discussed in detail below), but strong Ba-Ba binding gives rise to condensation of Ba on graphene surface, which makes difficult monitoring the intercalation processes. These difficulties are sidestepped by invoking adsorption of Ba on heated Ir-C. Consider Fig. 14A,a displaying the dependence of the Ba, Ir, and C Auger peak intensities observed in Ba adsorption on Ir-C at 880 K. In the course of adsorption, the Auger peak of Ba grows in amplitude and that of iridium falls off, while the carbon Auger peak does not change. This implies that barium accumulates only under the graphene as intercalant (Fig. 14B). For $t \geq 150$ s, the concentration of intercalated Ba becomes as high as 5 10^{14} cm^{-2} (which is close to its monolayer concentration on the surface of transition metals) and changes no more. The efficiency of intercalation is ~0.3, in other words, 30% of the Ba flux striking the Ir-C surface becomes intercalated, with the remaining 70% desorbing. If we heat now this system, approximately 50% of the intercalated barium desorb at $T \leq 1300$ K, while the barium left under the graphene with the concentration of 2 10^{14} cm^{-2} leaves the surface only at $T > 2000$ K, when the islands break up. This experiment provides one more argument for the absence of any "blocking" of Ba escape from under graphene in the initial stage, as this was observed in the case of cesium.

A.

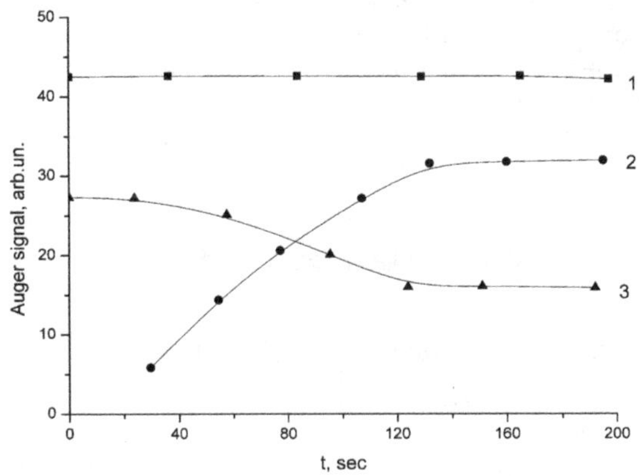

B.

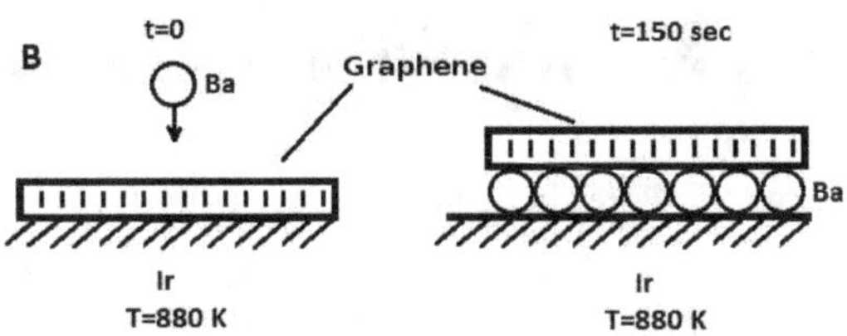

Fig. 14. Ba intercalation under a graphene layer on Ir(111). **A.** Auger signals of carbon (1), barium (2), and iridium (3) versus total amount of deposited Ba onto Ir-C at 880 K. The deposition flux is $\nu_{Ba} = 3.2 * 10^{12}$ cm^{-2}*s^{-1}. **B.** Schematic structure of the process:

3.2 Adsorption of barium on Ir-C at 1000 < *T* < 1500 K

Ba atoms with activation energy for desorption E = 1.9 eV desorb intensively from graphene on iridium at 880 K. Therefore observation of existence of barium islands on Ir-C at substantially higher temperatures, 1000 < T < 1500 K, came as a total surprise [19]. One studied the specific features of growth and dissolution of two-dimensional barium islands on Ir-C [17, 21]. Barium in these islands was bound strongly with the substrate with an energy of 3.4 eV and was present in a concentration of 2 10^{14} cm^{-2}, which means that the Ba islands were not close packed. In this temperature region, Ba did not intercalate with graphene. One had to account for the high binding energy of barium with the substrate in an island, 3.4 eV, which exceeds substantially the energy required for Ba sublimation, ~1.8 eV, and that for desorption of a Ba atom from Ir-C, which is 1.9 eV. We assumed that Ba adatoms are adsorbed at the centers of benzene rings in the graphene film, so that when the latter performs random thermal vibrations, they come closer to the surface atoms of iridium and compress slightly the graphene film to form an electron bond with iridium with a binding energy increased from 1.9 to 3.4 eV (Fig. 15). It appears actually a reasonable assumption, because the energy of desorption of a Ba atom from the iridium surface is high, 5.7 eV. If such a Ba monolayer is heated to T ~ 1600 K, about half of the barium will desorb, while the other half will migrate under the film to become an intercalant. If now we adsorb Ba again at 1000 < T < 1500 K on graphene confining intercalated barium in a concentration of 5 10^{13} cm^{-2}, strongly bound Ba islands do form, but only on a part of the surface [18]. It appears probable that the intercalated Ba is distributed nonuniformly in these experiments, so that regions with its maximum concentration of 5 10^{14} cm^{-2} border areas with no intercalated Ba present, but it is in these regions that strongly bound Ba islands form on graphene (Fig. 16). At concentrations of intercalated Ba in excess of 2 10^{14} cm^{-2}, strongly bound islands do not form at all. Neither do they form on the surface of a graphene film on iridium at film thicknesses above two monolayers. Similar results were obtained by us in studies of strontium adsorption on Ir-C [18--21].

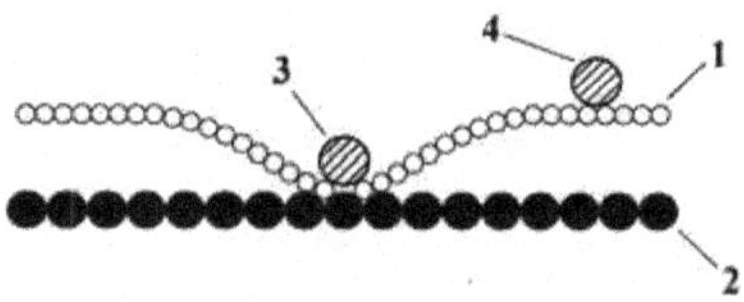

Fig. 15. Schematic structure of the graphene layer (1) on iridium (2) with strongly (3) and weakly (4) bonded atomic Ba.

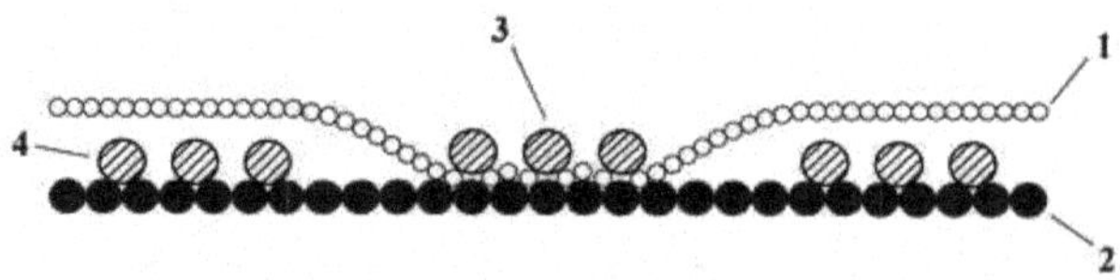

Fig. 16. Schematic structure of the graphene layer (1) on iridium (2) with 2D Ba island (3) and intercalated Ba (4).

4. Intercalation with platinum [22]

It was found that a graphene film on a metal is intercalated efficiently with atoms having low ionization potentials (Cs, K, Ba,...), which are positively charged on graphene and are strongly bound with it by mirror image forces. An intriguing question arises immediately of whether atoms with high ionization potentials, which are not charged on a graphene film and bound with it only weakly by van der Waals forces of polarization origin, would intercalate a graphene film on a metal. In an attempt at finding an answer to this question, we studied adsorption and intercalation of Ir-C with atoms of Pt, Si, and C.

Platinum was evaporated by sublimation from heated Pt ribbons. Pt adsorbed on Ir-C at 300 K grew to form a thick film on the graphene film. It was found that Pt atoms adsorbing on a graphene film atop iridium heated to 1000 K intercalate with Ir-C with an efficiency $\kappa = 1$. This intriguing result implies that *each* Pt atom incident at 1000 K on a graphene film penetrates under it to become an intercalant species. Figure 17A shows how the Auger peaks of carbon, platinum and iridium vary as a result of adsorption of Pt on Ir-C at a higher $T = 1200$ K, when the intercalation efficiency is slightly lower ($\kappa < 1$), and part of Pt atoms desorb from the graphene film. Interestingly, in contrast to Cs, K, and Ba atoms which form in intercalated state a monatomic film, Pt grows under the graphene in a thick, *multilayer* film, which, on the one hand, is strongly chemically bound with iridium, while on the other is coupled weakly, probably by van der Waals forces, to the graphene film (Fig. 17B). This is probably what accounts for our observation [2, 3] of the effect of *bi-intercalation,* in which consecutive adsorption first of Pt atoms, and subsequently, by atoms of Cs, brings about growth under the graphene film of a thick Pt film, with a Cs monolayer forming on top of it (Fig. 18).

A.

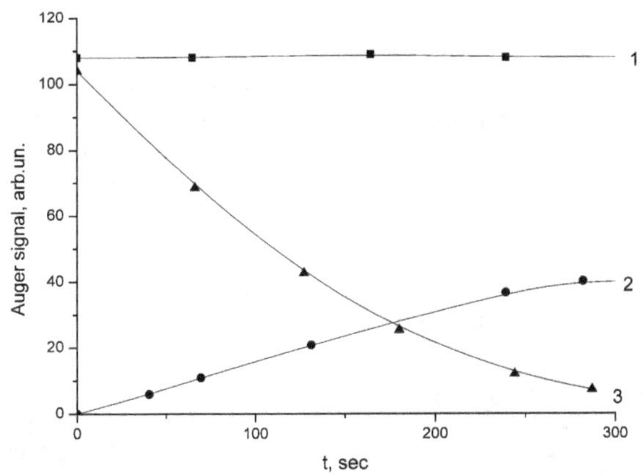

B.

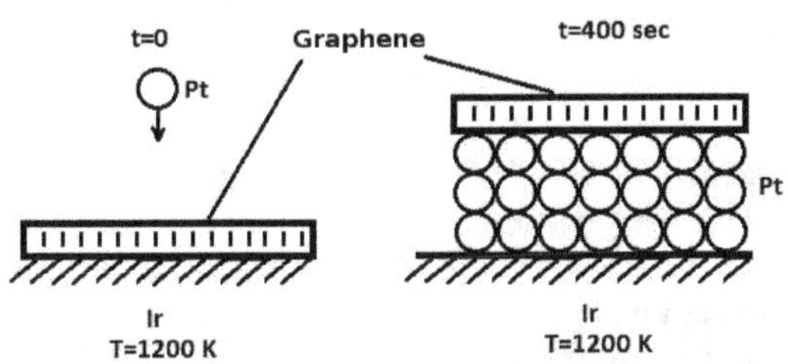

Fig. 17. Pt intercalation under a graphene layer on Ir(111). **A.** Auger signals of carbon (1), Pt (2), and iridium (3) versus Pt deposition time onto Ir-C at 1200 K. The deposition flux is $\nu_{Pt} = 1 \cdot 10^{13}$ cm$^{-2} \cdot$s^{-1}. **B.** Schematic structure of the process:

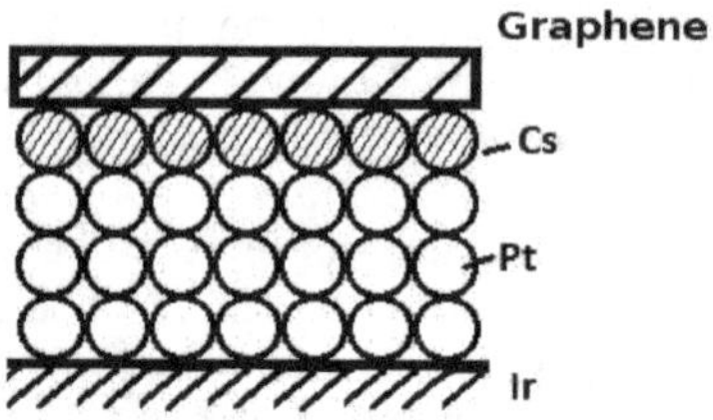

Fig. 18. Schematic structure of the layer in bi-intercalation: first Pt is deposited onto the graphene layer on Ir(111) at 1000 K; than Cs was deposited at 300 K and the system was annealed at 800 K to remove α-phase Cs.

The condition κ = 1 means that a Pt atom physisorbed on graphene can migrate in its lifetime against desorption from the center of the island to its edge, which is possible if migration over graphene is very efficient. For the process to be efficient enough, the graphene film should be smooth, and the potential relief for migration, fine; also, the graphene islands should not be very large. Because Pt adatoms on graphene residing atop iridium are electrically neutral, decoration of the graphene island edges by them leaves it neutral, as a result of which intercalation with Pt occurs without the concentration threshold characteristic of the intercalation with Cs, K, and Ba; this means that already the very first Pt atoms adsorbing on graphene with iridium heated to 1000 K penetrate under the film. Interesting data were obtained on the efficiency of intercalating Ir-C with Pt atoms at lower temperatures, 300 < T < 1000 K, as well. For instance, if platinum is evaporated on a graphene film on iridium at 300 K, then at room temperature no intercalation is observed; instead, Pt is adsorbed to form a monolayer (with $N_{Pt} \sim 1\ 10^{15}$ cm^{-2}, which reduces the Auger peak of carbon (with E = 272 eV) by a factor 1.5, and that of iridium (E = 54 eV), ~ 3 times. After this, the temperature was raised in steps of 100°, with exposure of the sample for 30 s at each temperature. As the temperature increased, the Auger peak of iridium was found not to change, while that of carbon grew by a factor 1.5. This means that as the temperature was raised, all of the adsorbed Pt monolayer became intercalated, apparently already at T < 1000 K, when Pt does not yet start to desorb.

5. Intercalation with silicon [2, 3, 22]

Silicon was deposited by sublimation from heated ribbons. Si adsorbed on Ir-C at 300 K grows to form a thick film which reduces the Auger peaks of carbon and iridium until they melt into the background. As T increases, the pattern of carbon film growth changes substantially, Figure 19A plots the variation of the intensities of Auger peaks of carbon, silicon, an iridium in the course of Si adsorption on Ir-C at 1000 K. The growth of the silicon Auger peak suggests that silicon builds up in the near-surface region, and the constancy of the carbon peak, that silicon accumulates under the graphene film. After 150 s of deposition, the Auger peaks of silicon and iridium stop to change, hence, stops to change also the composition of the near-surface region ≥(10--15) Å thick, which is accessible for AES. Studies of silicon adsorption on clean iridium [23] revealed that up to the concentration $N_{Si} \sim 3\ 10^{14}$ cm^{-2} all of the silicon incident on the surface accumulates in the form of a surface chemical compound, surface silicide. Silicon atoms arriving onto the surface after this state has been filled diffuse into the bulk of iridium to form in its near-surface region volume silicide (Fig. 19B). It was found that the curves plotting the variation of the silicon Auger peak intensities vs. deposition time of the same silicon flux on iridium and iridium coated by a graphene film at 1000 K practically coincide (if one accounts for the attenuation of the silicon Auger peak intensity by the graphene film). This means that silicon intercalates into Ir-C at 1000 K with an efficiency $\kappa \sim 1$. When Si adsorbs on Ir-C at higher temperatures $T \geq 1300$ K, a sizable part of it becomes bound in the intercalated state, with the remainder desorbing.

A.

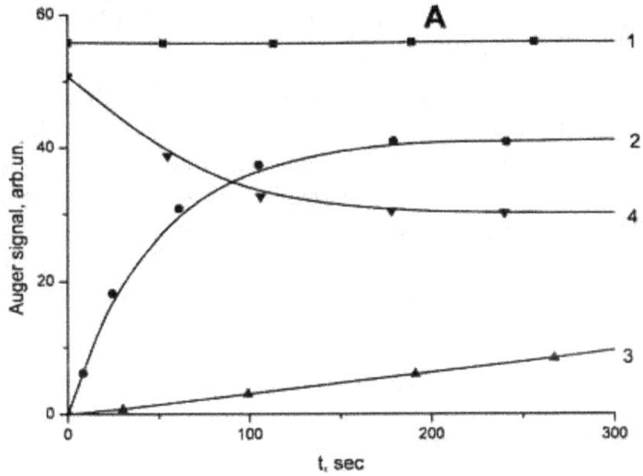

B.

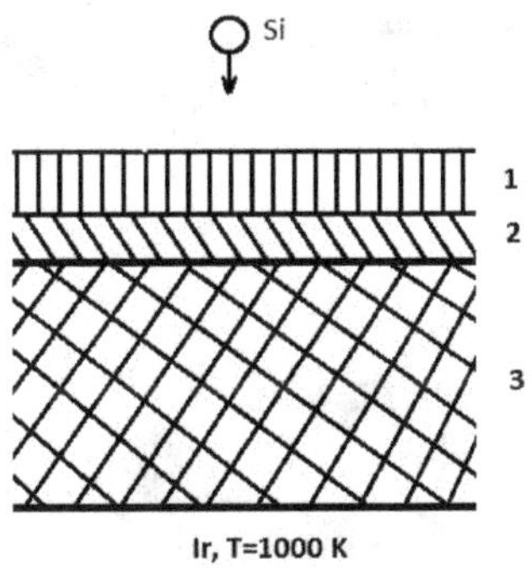

Fig. 19. Si intercalation under a graphene layer on Ir(111). **A.** Auger signals of carbon (1), Si (2), and iridium (4) versus Si deposition time onto Ir-C at 1000 K. (2) - Si Auger signal in its deposition onto Ir-C at 1500 K. The deposition flux is $\nu_{Si} = 1 * 10^{13}$ cm^{-2}*s^{-1}. **B.** Schematic structure of the process: 1 - graphene layer; 2 - surface silicide; 3 - bulk silicide

6. Intercalation with carbon [2, 3]

The experiments which demonstrated that atoms of Pt and Si intercalate with an efficiency $\kappa = 1$ under a graphene film adsorbed on iridium heated to $T = 1000$ K stimulated interest in looking for other atoms which could be intercalated with intrinsic carbon atoms. In these experiments, carbon atoms (emitted by a source developed by us and not producing C clusters [24]) were deposited at $300 < T < 1800$ K on a graphene film atop iridium, with the film thickness probed by AES. At all these temperatures, a thick carbon film was observed to grow, but its nature (graphitic or non-graphitic) and the growth mechanism involved were found to depend strongly on *T*.

Adsorption of C atoms on a graphene film atop iridium heated to $T \leq 700$ K produces a thick carbon film of non-graphitic structure (Fig. 20), which is evidenced by some of its characteristics. First, the shape of the C_{KVV} Auger peak becomes non-graphitic (the spectrum contains one peak at 247 eV, while graphite features two peaks at 247 and 263 eV). Second, adsorption of Cs atoms on such a film does not demonstrate a characteristic splitting of the C_{KVV} Auger peak (a point to be discussed below), which is observed when Cs atoms are adsorbed on a graphite film. Third, the work function increases from 4.45 eV, a figure typical of graphite, to 4.75 eV, an energy identified with chemisorbed carbon. And finally, fourth, the degree of dissociation of CsCl molecules increases from 10^{-4} on graphene to 10^{-1} in the case of a thick carbon film. From this we can infer that adsorption of C atoms on graphene atop iridium heated to 700 K translates into formation of a thick non-graphitic carbon film, which grows on graphene with no intercalation mechanism involved.

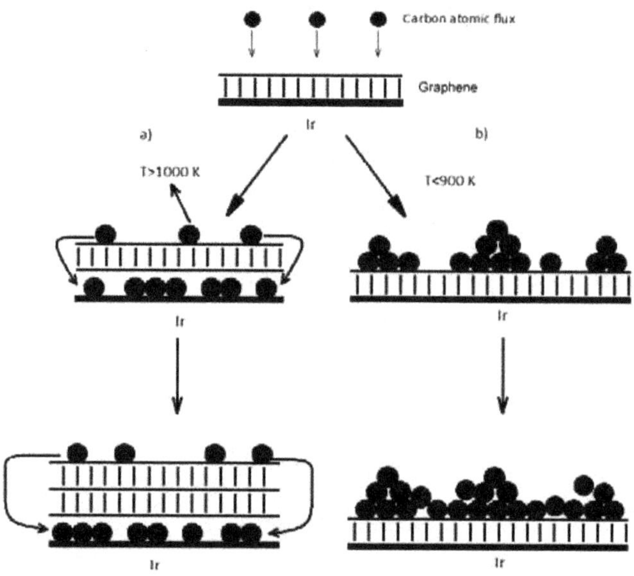

Fig. 20. A scheme of carbon film structure in atomic carbon deposition onto the graphene layer on Ir(111) at various temperatures: (a) - T > 1000 K; (b) - T < 900 K.

The situation is radically different for the formation of a carbon film when C atoms adsorb on graphene atop iridium heated to higher temperatures, $T \sim 1100$ K (see Fig. 20). Accumulation of carbon in the adlayer likewise was not observed, with all the C atoms striking the surface either intercalating under the graphene islands or desorbing. Intercalation started with formation under the graphene film of graphite islands, and after their coalescence, of a second graphene film above the first one. It was found that after the growth of the second graphene layer has come to completion, the rate of growth of the third graphene layer drops by a few times. This finds a ready explanation in that C atoms have to diffuse from the top of the film to its base, i.e., the iridium surface, where graphene islands of the next layer nucleate; this brings about loss of C atoms because of the increasing probability of their desorption and of decreasing film growth rate. We note that at $T \leq 700$ K the rate of film growth in thickness is constant for υ_C = const. Thus, in adsorption of C atoms on Ir-C at 1100 K carbon accumulates not in the adlayer but rather in intercalated state under the graphene film, where a thick carbon film with graphitic structure grows.

7. Intercalation with silver [25, 26]

Ag was evaporated by sublimation of Ag wires wound on a heated W spiral. Adsorption of silver onto graphene atop iridium did not result in intercalation at any temperature; indeed, at $T < 900$ K, a thick silver film grew, while at $T > 900$ K, silver atoms desorbed. It is possibly because of their tendency to aggregation that Ag atoms form at the edge of a graphite island, a place best suited for starting intercalation, Ag_2 and Ag_3 clusters, which interfere with penetration of silver under the island.

If, however, one intercalates Ir-C preliminarily with Cs atoms to $N \sim 2\,10^{14}$ cm^{-2}, and only after this deposits a thick silver film at 300 K and raises then the temperature to 1000 K, then silver will effectively roll off under the graphene in an amount of ~ 1 monolayer, the remaining silver desorbing. It appears reasonable to suggest that silver penetrates under the graphite islands when their edges are raised; it was found that it expels the intercalated cesium.

8. Intercalation with aluminum [27]

Evaporation of aluminum on graphene atop iridium or rhenium at room temperature brings about growth of three-dimensional aluminum islands on graphite. There is no noticeable intercalation of aluminum under the film, which is evidenced by the Auger signal intensities of carbon and the substrate being attenuated by the growing adsorbate film to the same extent.

The results obtained in annealing of three-dimensional aluminum islands ($N_{Al} \sim$ cm^{-2}) adsorbed on graphene atop iridium are presented in Fig. 21A. We readily see that up to ~700 K the system remains practically unchanged. At high temperatures, the Auger signal of aluminum begins to fall off, and that of carbon, to rise, with the substrate (iridium) signal remaining initially constant, to increase slightly afterwards. For $T > 1200$ K, all the three Auger signals reach constant levels remaining unchanged up to 1500 K. At high temperatures, the aluminum signal drops to zero, that of iridium grows almost twofold, and that of carbon, remains unchanged.

To gain understanding of the observed variation of Auger signals, consider Fig. 21B. At T = 700 K we see the onset, and at 1200 K, completion of a rearrangement of the adsorption layer, in which three-dimensional aluminum islands adsorbed on top of the graphene layer, dissolve, and the Al atoms released in the process partially evaporate, and partially intercalate under the graphene. For $T > 1200$ K, there is no aluminum on the outer side of the graphene, and the Auger signal of carbon has the same intensity as before the beginning of aluminum evaporation. By contrast, the Auger signal of the substrate became substantially smaller in amplitude as a result of its being screened both by the graphite layer and by the intercalated aluminum; indeed, its intensity is close to one half the initial (before Al evaporation) level (we used the iridium Auger signal with E = 54 eV).

With the heating continued ($T > 1500$ K), aluminum escapes from the adlayer, apparently through thermal desorption and, possibly, by partial dissolution in the bulk of the substrate, and its Auger peak drops to zero. By contrast, the Auger peak of iridium grows to its initial level, and that of carbon retains its initial amplitude (Fig. 21A).

A.

B.

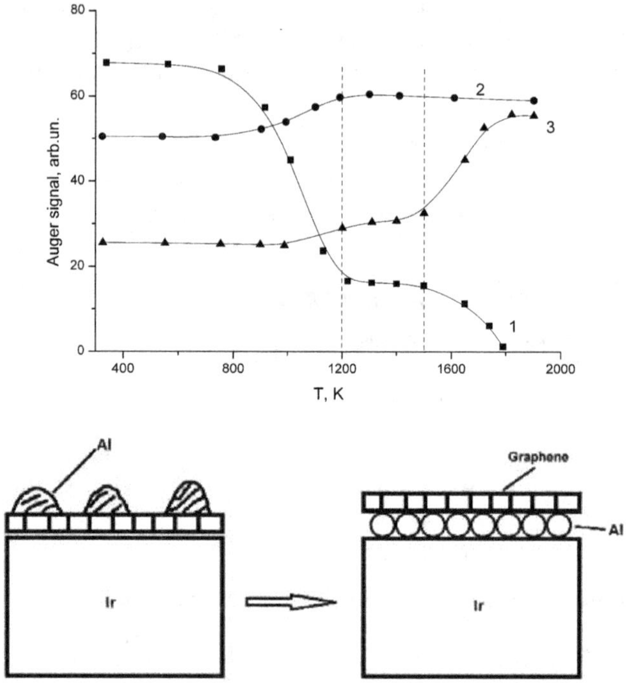

Fig. 21. Al intercalation under a graphene layer on Ir(111). **A.** Auger signals of Al (1), carbon (2), and iridium (3) versus temperature in a stepwise annealing of the Al film deposited onto Ir-C at 300 K. Initial Al surface concentration was ~$2*10^{15}$ cm^{-2}. **B.** Schematic structure of the process: at 300 K (left) and at 1200 K (right).

The absolute amount of aluminum intercalated under the film was found to be N_{Al} ~ 5 10^{14} cm^{-2} [27].

The net result of direct evaporation of aluminum on graphene at $T > 950$ K is that it ceases to accumulate on top of the graphene layer but intercalates under it instead. This is evidenced by Auger signal of carbon remaining unchanged whereas that of aluminum grows, and the signal of iridium, falls off, albeit insignificantly. This suggests that the aluminum entering the adlayer does not build up on top of the graphite layer and, hence, does not reduce its Auger signal but rather accumulates sandwiched between the graphite film and the metal surface.

An experiment was conducted [27] to measure the efficiency of aluminum intercalation under graphene on (111) iridium, i.e., the fraction of particles penetrating under the film relative to the total number of atoms incident on the surface. Estimates showed the intercalation efficiency to be ~14% at 1000 K, to fall off to 7% at 1100 K. The same figures are obtained when using rhenium as a substrate metal, thus supporting the earlier conclusion of the properties of graphene being practically independent of the nature of the metal it was formed on. The limiting concentration of aluminum accumulating in intercalated state on the surface of rhenium also matches with the composition of the surface aluminide ReAl.

The low efficiency of intercalation should apparently be assigned to desorption of the larger part of the aluminum incident on the surface of graphene. It appeared worthwhile to see whether the aluminum intercalation efficiency would change if graphene covered not all of the metal surface. For this purpose, we grew successively on a rhenium sample a graphene film of area *s* covering 40, 70, and 90% of the metal surface, and studied the process of intercalation. The fraction of the substrate surface coated by graphite was assumed to be proportional to the carbon Auger signal intensity from the graphene film.

As follows from these experiments, within experimental error the intercalation efficiency is practically independent of the amount of aluminum that has already penetrated under the film, while remaining dependent on the fraction of the surface coated by the island, to increase at $T = 1100$ K from $W \sim 0.35$--0.1 for $s = 90\%$ to $W \sim 1$ for $s < 70\%$. Significantly, the figures obtained exceed by far the efficiency of intercalation of aluminum atoms under a continuous graphene film.

While it would appear reasonable to assume that the quality of the graphene film itself does not depend on whether it is continuous or not, in the second case the number of island boundaries would certainly increase, as would possibly decrease their size. Indeed, the island breakup starts from the edges [3], which is not strange at all, particularly if we remember that the binding energy of a single carbon atom within the layer is ~9 eV [28], and at the island edges, substantially lower, ~3 eV on rhenium [10] and ~4.5 eV on iridium [3]. It may be that the noticeable difference between the efficiencies of intercalation under a monolayer and a submonolayer film suggests that it is the edges of graphene islands that are the defects involved in intercalation of particles under graphene, as was supposed in Ref. [2]. After the islands had coalesced to form a continuous monolayer, the penetration with respect to intercalating particles decreases. And conversely, the increase of the efficiency to a value approaching unity means that the islands have become close in size to the migration path lengths of aluminum atoms they can travel in their lifetimes over graphite surface at a given temperature. Scanning tunneling microscopy suggests that the island size is 3000--8000 Å, and, hence, the migration path lengths of aluminum atoms at 1000--1100 K should be of the same order of magnitude.

The above increase in the intercalation efficiency is very likely of a universal nature, so that it will be characteristic of many types of intercalating particles.

9. Intercalation with iridium [29]

Adsorption of iridium atoms on graphene. Iridium atoms were provided by a second evaporating iridium ribbon, which was freed thoroughly of impurities and heated to ~2400 K. The expected flux of Ir atoms subliming from the ribbon was ~10^{15} $cm^{-2}s^{-1}$ [30], a level high enough to provide formation of several layers of iridium atoms on graphene surface in a reasonable time.

The main difficulty of the experiment consisted in the need to discriminate by AES the evaporated iridium from the iridium of the substrate. To overcome it, formation of the graphene film was preceded by evaporation on iridium of molybdenum atoms to a concentration of ~0.1 of monatomic density (molybdenum is identified adequately by AES). Molybdenum atoms were provided by well cleaned molybdenum ribbons heated to high temperatures. Thus, already after formation of graphene, AES offered a possibility of probing carbon in the form of graphene ($E = 272$ eV), molybdenum ($E = 221$ eV), and iridium ($E = 176$ eV) (Fig. 22a). In these conditions, the intensities of the molybdenum and iridium Auger signals decreased ~1.6 times, a figure characteristic of screening by a graphene film [31]. Next, 2--3 layers of iridium were evaporated on graphene at 300 K; this resulted in a decrease of the Auger signal intensities of carbon and molybdenum by about a factor 3 (Fig. 22b).

a.

b.

c.

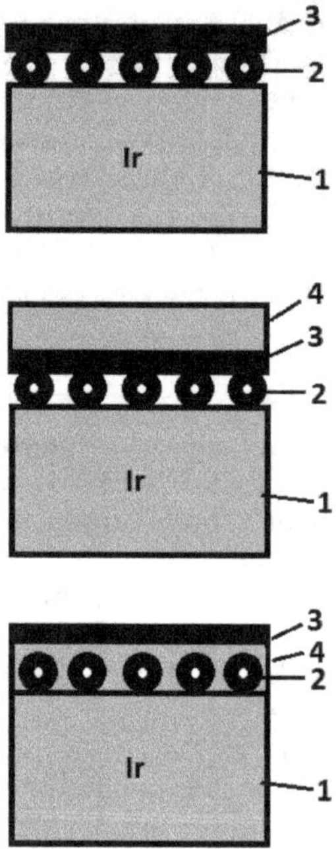

Fig. 22. A model scheme for the adsorption system Ir - graphene with intercalated atomic Mo as an inner standard: **a** - Ir substrate (1) with chemisorbed Mo (2) under graphene (3). Mo surface concentration is ~0.1 ML; **b** - the same system with Ir film of ~2 ML thickness (4) deposited above the graphene at 300 K; **c** - the result of the system annealing at 1200 K.

Annealing of the Ir(111)-Mo-graphene system. Figure 23 illustrates the results of annealing of an iridium film on graphene obtained by AES. The intensity of the carbon Auger signal at T = 1100 K was found to recover to the initial level corresponding to a clean graphene surface. At the same time, the molybdenum Auger signal intensity did not change. This can be understood by examining Fig. 22c. The evaporated iridium has diffused under heating of the sample under the graphene; the graphene surface became free of the adsorbate, with the intensity of the carbon Auger signal growing back to the original level (Fig. 23). At the same time, the iridium film confined under graphene continues to screen the layer of molybdenum, and its Auger signal intensity remains unchanged (Fig. 22c). The decrease of the iridium Auger signal in intensity should apparently be attributed to transition of Ir atoms to the intercalated state and to their screening by the graphene film (Fig. 23).

Direct experiments with evaporation of Ir atoms on graphene performed at moderate temperatures of 1100--1300 K suggest accumulation of iridium under the graphene; indeed, the molybdenum Auger signal intensity decreases, while the surface of graphene remains free of the adsorbate. For $T \geq 1300$ K, desorption of iridium atoms from the heated passive graphene surface begins to play a prominent part, with the result that buildup of Ir atoms in intercalated state decreases strongly.

Experiments with intercalation of graphene with iridium atoms provided supportive evidence for the overall pattern of this process for polyvalent atoms, namely, intensive

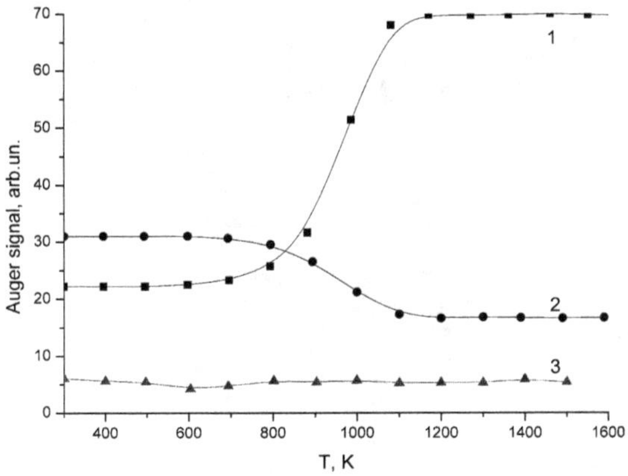

Fig. 23. Carbon (1), Ir (2), and Mo (3) Auger signals versus annealing temperature in a stepwise, within 100 K, heating of the Mo film initially deposited onto graphene at 300 K.

penetration of the adsorbate under the graphene is observed to occur at moderate and high temperatures (1000--1500 K). Recall that atoms with low ionization potentials (primarily the alkali metals) intercalate under graphene at low temperatures, $T \leq 700$ K; also, the lower the temperature, the more efficient is the diffusion of adatoms under the graphene film.

10. Intercalation with copper [32]

Evaporation and annealing of a copper film on iridium. Copper atoms were evaporated on an iridium surface at 300 K until the Auger signal intensity of the substrate dropped to one half its original level, which we estimated as corresponding to a copper concentration of about two monolayers. The main events occurring under stepped annealing of such a film are observed at $T \sim 1100$ K; more specifically, within a narrow temperature interval of the order of 50 K, the Auger signal of copper drops sharply in intensity to melt into the background noise, whereas that of iridium recovers to the original level identified with a clean surface. The most probable mechanism accounting for the surface cleaning is desorption of copper from the iridium surface. While this experiment does not permit us to exclude completely the possibility of copper dissolution in the bulk of the substrate, later experiments with graphene (see below) make this assumption highly unlikely.

Evaporation and annealing of a copper film on graphene atop iridium. Two copper monolayers were evaporated at 300 K on graphene (Fig. 24a), after which the system was annealed in steps at a number of temperatures. The variation of Auger signal intensity from the iridium substrate (E = 176 eV), copper (E = 60 eV), and carbon (E= 272 eV) is displayed in graphical form in Fig. 25. We readily see that at the transition from 1100 to 1200 K the Auger signal of carbon increased sharply to become equal to that from clean graphene surface. There still remains a noticeable Auger signal of copper, though; and since the Auger signal intensity of the substrate is substantially lower than that of the initial, copper-free "graphene on iridium" system, the only reasonable interpretation of the above observations appears to be that copper atoms escaped from the surface under the graphene film to become an intercalant (Fig. 24b). The graphene-involved decrease of the Auger signal of copper (E = 60 eV) to about one half its original value suggests that practically all of the

a.

b.

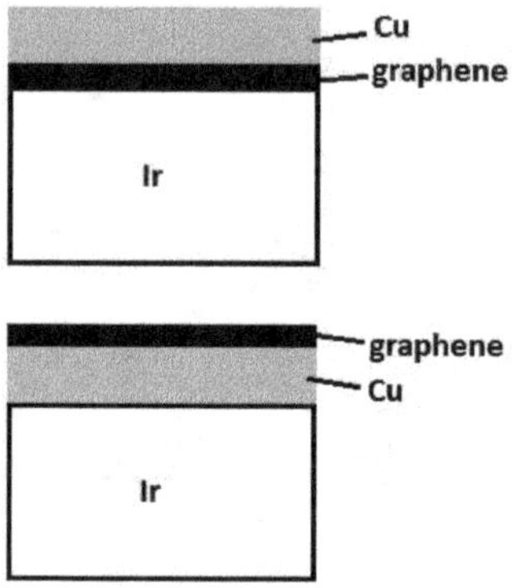

Fig. 24. Schematic structure of the layer transformation for the system Ir(111) - graphene - Cu: (**a**) - Cu film of ~2 ML thickness is deposited upon graphene on Ir(111) (T=3000K); (**b**) - the same layer after annealing at 1200 K.

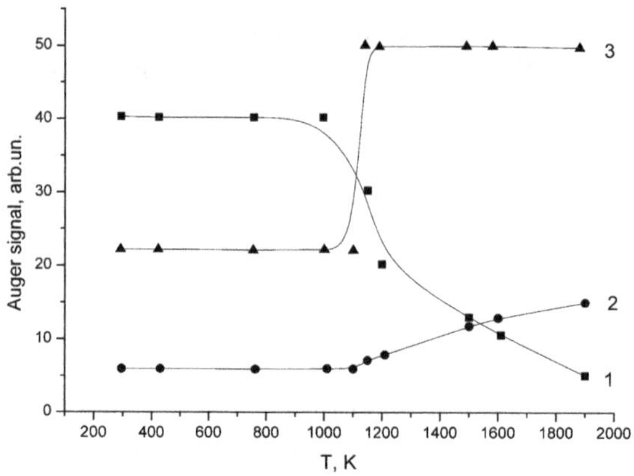

Fig. 25. Cu (1), Ir (2), and carbon (3) Auger signals versus annealing temperature in a stepwise, within 100 K, heating of the Cu film initially deposited onto graphene at 300 K. Time delay is 30 s at each temperature.

copper atoms initially adsorbed on graphene ended with being confined under it [32]. Further annealing brought about one more intriguing observation, namely, the Auger signal of copper is observed now under annealing up to ~1700 K, i.e., a temperature 500 K (sic!) higher than that in the case of clean iridium surface.

This seems to imply that in the first experiment copper atoms mostly desorbed from the clean surface of iridium, whereas in the second experiment the graphene "roof" interferes with this process, with copper atoms left confined to the surface up to high temperatures. Obviously enough, the channel of copper dissolution in the bulk of the metals (if it does exist) remains available in the presence of a graphene film too; moreover, the graphene "roof" which confines Cu atoms at high temperatures (≥1700 K) enhances the probability for adatoms to dissolve in the bulk of the substrate metal, as this was observed [33] to occur for cesium atoms intercalated in the same iridium-graphene system.

Thus, graphene adsorption on iridium surface acts as a kind of a trap for intercalated copper atoms by increasing substantially the temperature at which the sample is freed completely of the adsorbed atoms.

Graphene acts as a trap for atoms with both high and low ionization potentials, for instance, Cs, K, Na [2, 3, 32]. These two types of systems differ, however, radically from one another. First, atoms of alkali metals intercalate actively under graphene only at low temperatures, $T \leq 700$ K. Second, such atoms can be freed completely from under graphene only if all of the graphene islands break up under heating; the temperatures involved may, however, be quite high indeed---for instance, intercalated cesium escapes from under graphene islands on Re(10-10) at $T \sim 2200$ K (sic!) [2].

In the case of polyvalent atoms and atoms with high ionization potentials, breakup of graphene islands is not required altogether; indeed, the intercalant can escape from under a continuous graphene film through the same defects that were possibly instrumental in its penetration there in the first place (Fig. 26).

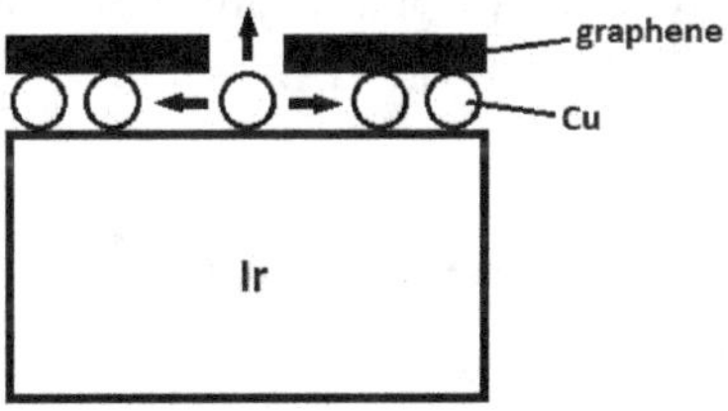

Fig. 26. A model scheme for atomic Cu thermal desorption from the intercalated state.

Now we are turning to the question of why atoms of copper or of other polyvalent high-ionization potential metals diffuse readily under a graphene film on a metal at $T \sim 1100$--1200 K, whereas their total removal from the intercalated state requires substantially higher temperatures. It might appear strange that the graphene film behaves in such a, say, anisotropic manner, namely, it is easy to penetrate when atoms transfer from the adsorbed to intercalated state, i.e., from the outer surface inward, but is difficult to cross in the opposite direction,

Let us turn to Fig. 26. As demonstrated by direct studies of migration of Si adatoms on Ir(111), silicon adatoms migrate freely both over a clean iridium surface by wandering from one side of the ribbon to the other (originally clean) in the 1200 K--1500 K temperature interval, and over iridium with the graphene layer forming [2, 3]; in other words, graphene does not interfere with the migration of adatoms over the metal surface. The same results can be expected with other polyvalent atoms as well (Pt, Cu, Al, etc.); indeed, they migrate easily and uniformly under the graphene film on a metal at temperatures of 1100--1200 K.

But then how can an intercalated atom leave an adsorption system? Adatoms in intercalated state get a chance to desorb when two events coincide. First---adatom occurs close to a defect in the graphene film, i.e., on the open surface, as it were, without a graphene roof overhead (Fig. 26). Second---if the adatom receives at exactly this instant a heat pulse from the substrate strong enough to desorb or transfer to the graphene film with subsequent desorption. Otherwise, the adatom will continue its migration further under the graphene film and, even if it is hit by a thermal pulse from the substrate strong enough to break the adsorption bond, will not leave the system, because it will be repelled back from the graphene film. Obviously enough, in this case desorption will become strongly complicated, so that atoms confined under the graphene will be forced to stay in the system up to very high temperatures. For illustration, Table 3 lists for the Ir(111)-graphene system the temperatures required for the adatoms to escape from a clean metal surface, T_1, and from the intercalated state, T_2.

Element	Cs	K	Al	Cu	Ag
T_1, K	900	900	1500	1250	1000
T_2, K	2000	2000	1900	1900	1900

Table 3.

Now atoms with low ionization potentials (Cs, K, Na) behave in a radically different way. Diffusion under the graphene film, migration under it and escape from under the graphene are essentially cooperative effects involving electrostatic repulsion of charged adatoms [2, 3]. It can be maintained with confidence that migration of single adatoms of alkali metals under a graphene film is all but suppressed [15].

Thus, the graphene "roof" reduces by many orders of magnitude the probability of desorption of adatoms from the surface compared with that from the open surface of a metal (Table 3).

11. Intercalation with fullerene molecules [34--36]

The density of the flux of C_{60} molecules was estimated from the time taken up by formation of a fullerene monolayer N_M on clean iridium at $T = 300$ K, when the Auger signal intensity of the substrate dropped to one fourth of its initial level, with $\upsilon_{C60} = N_M/t_M$, where $N_M \approx 2\ 10^{14}$ mol/cm^2.

If a fullerene monolayer is evaporated at 300 K on graphene atop iridium, the Auger signal intensity of the substrate drops practically down to background level. Heating such a film to $T = 800$ K brings about complete desorption of C_{60} molecules, because one observes recovery of the Auger signals of the graphene film and iridium. Thus, C_{60} molecules do not dissociate on the passive Ir-C surface and do not intercalate with it.

After this, a thick fullerite film (~ $3N_M$) was produced on the graphene (Fig. 27a). The intensity of the iridium Auger signal passes into background noise, while the carbon Auger peak took on "fullerene" shape with an energy of 269 eV. The results produced by heating of such a film to 800 K were as follows: only part of the C_{60} molecules desorbed, while a sizable part of C_{60} molecules diffused under the graphene film to become intercalated (Fig. 27b). The disappearance of the iridium Auger signal after the substrate had been heated to 800 K implies that C_{60} molecules had diffused under the graphene in an amount equivalent to not less than one monolayer. The Auger spectrum of carbon was in this case actually a superposition of the Auger spectra of graphene and a fullerene monolayer, with an energy of 271 eV. Experiments on adsorption of C_{60} on clean iridium demonstrated dissociation of C_{60} molecules within the $900 \leq T \leq 1200$ K [37] (Fig. 27C).

In the $1400 \leq T \leq 1600$ K interval, the carbon Auger signal falls off somewhat in intensity, and the substrate Auger signal becomes visible (Figs. 27 c and d). This temperature interval, 1400--1600 K, is associated with active migration of carbon over the surface involving transfer onto the back side of the ribbon (Fig. 27 d), which was supported by direct evidence in studies of migration on clean iridium. For $T > 1600$ K, the carbon film under the graphene undergoes intensive graphitization which changes the C_{kvv} Auger spectrum shape to "graphitic". At T = 1600--2100 K, the intensity of the Auger signal of carbon can be

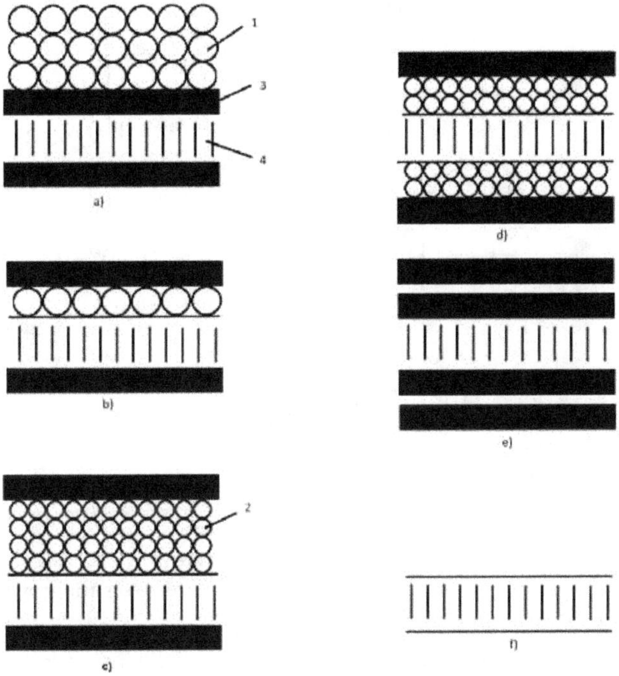

Fig. 27. Schematic representation for intercalation of the graphene layer on Ir(111) by molecular C_{60}. 1 - deposited polylayer film of molecular C_{60}; 2 - single carbon atom; 3 - graphene layer; 4 - iridium substrate. Temperature, K: (a) - 300; (b) - 800, (c) - 1200; (d) - 1600; (e) - 1900; (f) - 2200.

explained as due to screening by ~ two graphite monolayers (Fig. 27,e). Heating the substrate at temperatures above $T \sim 2200$ K cleaned efficiently the surface of carbon.

Thus, C_{60} molecules adsorbed in the form of a thick fullerite film (unlike a C_{60} monolayer) intercalate under the graphene atop iridium. This might be because the C_{60} molecules at the "bottom" of the fullerite film cannot desorb fast enough at $T = 800$ K and do not break up on the passive graphite surface, but rather transfer by diffusion to the intercalated state.

Graphene on rhenium can be intercalated with fullerene molecules just as this is done with iridium [36].

12. Electronic properties of graphene on metals with adsorbed and intercalated atoms [38--42]

As shown by calculations [43, 44] corroborated by experimental studies of the valence band of single-crystal graphite by photoelectron [45] and Auger electron [46] spectroscopy, the densities of both filled and empty electronic states of two-dimensional graphite drop dramatically close to the Fermi level.

In describing the interaction of particles with the metal surface, we shall follow the scheme suggested by Gurney [47]. As a single atom approaches the surface of a metal, interaction with it broadens the discrete level of the valence electron of the atom into a quasi-level and displaces it from its original position (Fig. 28). In this schematic description, one species of

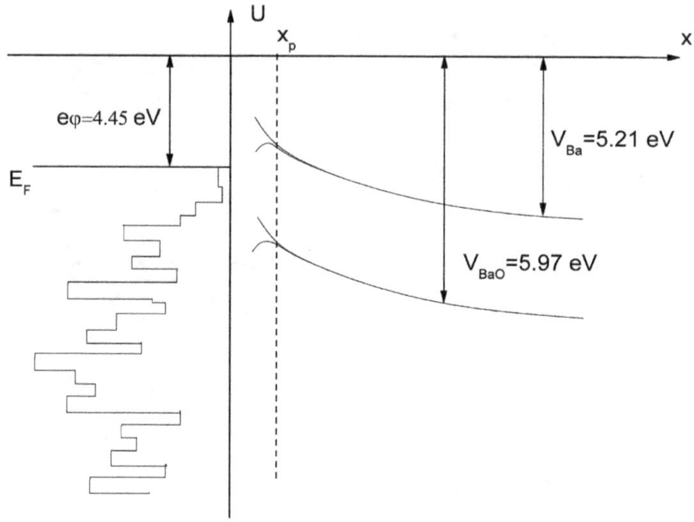

Fig. 28. Schematic transformation of the valence electron level of atomic Ba and molecular BaO in their moving towards the graphene layer on a metal. eφ is the graphene work function, V - atomic particle ionization potential.

adatoms differs from another in the shape and position of the quasi-level only. The adatom charge is determined both by the position of the center of the quasi-level relative to the Fermi level and its width. For the transfer of valence electrons from Cs and Ba adatoms to the valence band of graphite to be efficient, a sizable part of the quasi-level should be located in both cases above the Fermi level. Adsorption and intercalation with both Cs and Ba adatoms brings about filling of the empty band in the valence band of graphite lying above the Fermi level by depersonalized electrons. It thus appears reasonable to expect that the C_{kvv} Auger transition in two-dimensional graphene involving two electrons from the valence band should be sensitive to its filling. Discussion of the electronic properties of Re-C and Ir-C containing Cs and Ba can be found in Refs. [38--42].

Graphene residing on a metal is made up of atoms of two kinds, adsorbed and intercalated ones, albeit in somewhat different conditions. We start with experiments [28] in which adsorption of Ba atoms on graphene atop iridium heated to 900 K was performed by high-resolution ($\Delta E/E \approx 0.1\%$) AES. It was found that Ba fills only the intercalated state up to high concentrations, ~5 10^{14} cm^{-2}, at saturation. As evident from Fig. 29, before the intercalation the C_{kvv} Auger peak has a characteristic graphitic shape with energies 247 and 263 eV in the positive part of the spectrum, but no carbidic peak at the energy 255 eV (spectrum 1). We see that intercalation with Ba does not affect the "graphitic", positive part of the spectrum, while in its negative part, as the Ba concentration increases, first a peak at 276.5 eV (spectra 2 and 5), and after that, a second peak with an energy of 281 eV (spectra 3 and 4) appears on the high-energy side. One may thus conclude that the intercalated Ba is in the charged state, with its valence electron donated to the valence band of graphene to fill the levels adjoining the Fermi level, which is evidenced by the change in the high-energy part of the C_{kvv} Auger peak, because the transition in carbon is the *kvv*-type, and it involves two electrons of the graphite valence band. A similar pattern was observed in intercalation of HOPG graphite

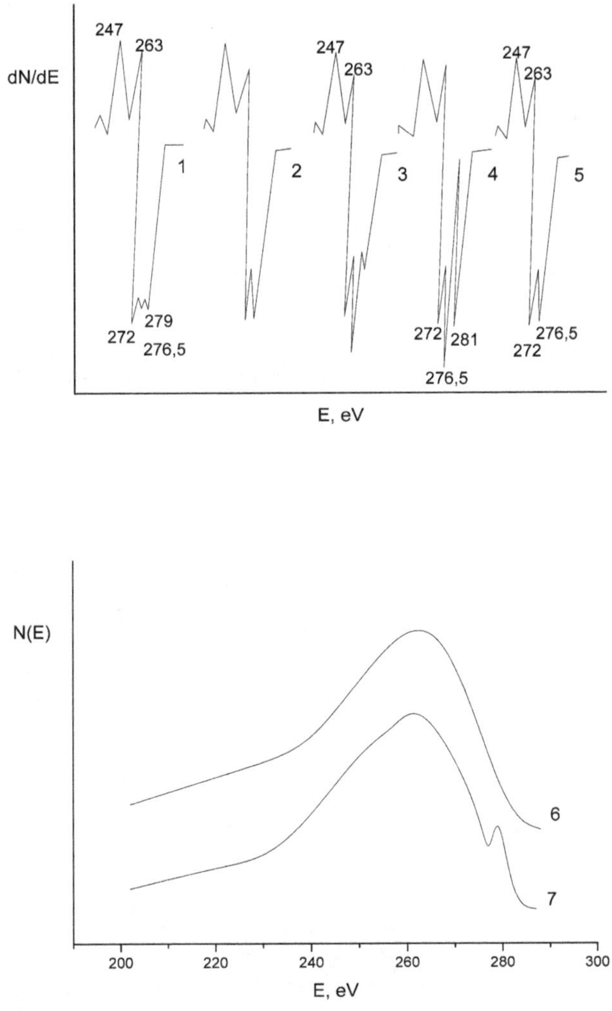

Fig. 29. Carbon KVV differential Auger spectra for sequential stages of Ba adsorption on the graphene layer on Ir(111) at 900 K. Ba concentration under the layer, 10^{14} cm^{-2}: 1 - 0; 2 - 1.5; 3 - 2.5; 4 - 5.0 (saturation). 5 - C_{KVV} Auger spectrum after system annealing at 1600 K (N_{Ba} ~ 2*10^{14} cm^{-2}). 6; 7 - integral C_{KVV} Auger spectra for pure grapene (6) and for the state 4: Ba adsorption to saturation (7).

with atoms of alkali metals [46, 48]. Heating an Ir-C ribbon to 1600 K forces part of the intercalated Ba to desorb, with the remaining Ba with a concentration of 2 10^{14} cm^{--2} being identifiable with the Auger peak at 276.5 eV (spectrum 5).

Figure 30 [41] presents C_{kvv} Auger spectra obtained in adsorption of Cs on Re-C. This process is paralleled by simultaneous filling of the adsorption (to a concentration N_α) and intercalation (to a concentration N_γ) states, and at the completion of adsorption, to saturation $N_\alpha \sim N_\gamma \sim 4$

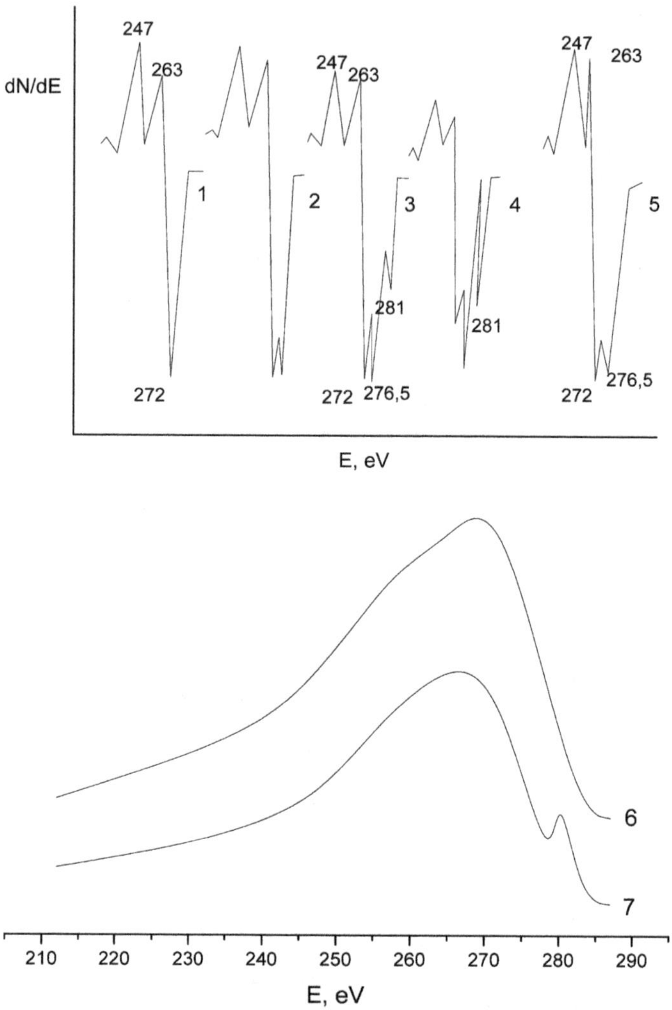

Fig. 30. Carbon KVV differential Auger spectra for sequential stages of Cs adsorption on the graphene layer on Re(10-10) at 300 K. Total Cs concentration over and under the layer, 10^{14} cm^{-2}: 1 - 0; 2 - 1.0; 3 - 2.0; 4 - 7.5 (saturation). 5 - C_{KVV} Auger spectrum after system annealing at 850 K (N_{Cs} ~ $1*10^{14}$ cm^{-2}). 6; 7 - integral C_{KVV} Auger spectra for pure grapene on Re (6) and for the state 4 - Cs adsorption to saturation (7).

10^{14} cm^{-2}. We readily see that the valence-band electrons of Cs fill the same levels in the graphite valence band as in adsorption of Ba, with peaks at 276.5 eV and 281 eV appearing successively, besides the one at 272 eV, in the negative part of the spectrum.

Interesting relevant information on the behavior of intercalated Cs and Ba with oxygen admitted to the instrument can be found in Ref. [42]. In this experiment, Re-C containing only intercalated Cs, and adsorbed and intercalated Cs was exposed at 300 K to O_2 (with a dose of 10^{-6} Torr x 60 s = 60 Langmuirs), and C_{kvv} Auger spectra were recorded (Fig. 31). It turned out that at 300 K the concentration of surface-adsorbed carbon on graphene is as low as to be undetectable by AES (O_2 is only physisorbed on the basal plane of crystalline graphite). Exposure to oxygen did not affect the shape of the C_{kvv} Auger peak of intercalated Cs, wherefrom we can infer that oxygen does not diffuse under the graphite islands on rhenium. Interestingly, the shape of the C_{kvv} peak of adsorbed Cs did change; indeed, the peak at 281 eV disappeared, which implies that as a result of oxidation part of the Cs valence electrons returns from the valence band of graphite into the oxide molecule. Figure 28 illustrates this in the particular example of Ba, in which case oxidation of Ba adatoms on graphene results in formation of a BaO admolecule with its inherent system of electronic levels. It is a safe guess that because of the high ionization potential of the BaO molecule (V_{BaO} = 6.97 eV, to be contrasted with V_{Ba} = 5.21 eV for the Ba atom) the quasi-level of the valence electron in the BaO admolecule lies below the graphite Fermi level and, thus, will be filled in oxidation by electrons leaving the valence band of graphite.

A.

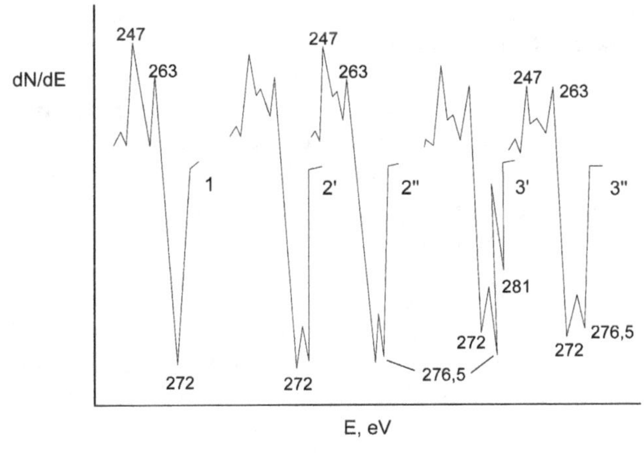

B.

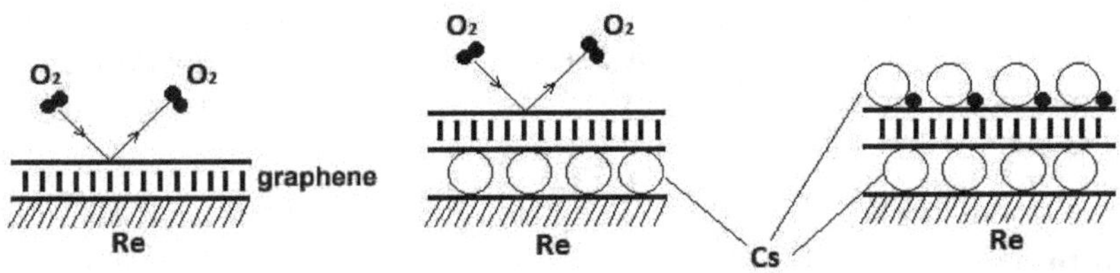

Fig. 31. Cs adsorption on the graphene layer on Re(10-10) at 300 K with subsequent oxidation at $P_{O2} = 1*10^{-5}$ Torr. **A.** Carbon KVV differential Auger spectra. Total Cs concentration over and under the layer, 10^{14} cm^{-2}: 1 - 0; 2′ and 2′′ - 1.5; 3′ and 3′′ - 7.0; (saturation). The spectra 2′ and 3′ are taken before exposure to oxygen; the spectra 2′′ and 3′′ - after the exposure. **B.** Schematic representation of the process.

An interesting study [42] compared the possibilities of intercalating with cesium the surface of iridium with graphene and a monolayer of chemisorbed carbon with a concentration $N_C \sim$ 3.9 10^{15} cm^{-2} produced by adsorption of C atoms on iridium at 300 K. It was found that, in contrast to graphite, Cs does not intercalate under the monolayer of chemisorbed carbon; indeed, there is no characteristic splitting of the C_{kvv} Auger peak and no high-temperature ($T \geq 2000$ K) γ phase in the TD spectrum.

Another study [41] covered intercalation not only of graphene films on metals with Cs and Ba, but that of thick graphite films on rhenium with cesium at 300 K. Figure 31 presents Auger spectra of graphite on rhenium taken before (spectrum 1) and after (spectrum 2-4a) intercalation with cesium. We readily see that intercalation of both graphene and graphite with cesium atoms initiates similar changes in the shape of the carbon Auger spectrum. As already pointed out, in the case of graphene the strong changes in the shape of carbon Auger peak stem from its unique electronic properties, namely, the layer being two-dimensional, there are many electrons in the valence band, and the levels are widely separated. The same explanation can apparently be assigned to a thick graphite film, because graphene layers are not coupled by electron exchange, and they are bound by van der Waals forces. Incidentally, studying an Auger spectrum of carbon of a similar shape in the case of intercalation into a thick graphite film required a substantially longer exposure time than in the case of one graphene layer. There is nothing strange in it, because as the number of graphite layers increases, the adsorption capacity of the system with respect to cesium which intercalates between all graphite layers increases proportionally. Significantly, heating a graphite film intercalated with cesium to $T \sim 1000$ K removes completely the intercalated cesium (except for the first graphene layer on the metal). We may infer that this heating temperature is high enough to remove cesium both from the surface and from the bulk of graphite, because the edges of graphite "plates" in a thick film are not closed and, thus, do not interfere with diffusion of Cs adatoms with their subsequent desorption.

If a thick, cesium-intercalated graphite film (its Auger spectrum is 4a in Fig. 31) is exposed to oxygen, the Auger peak of cesium decreases by about a factor 5 (from 5a to 5b in Fig, 31), and a high Auger peak of oxygen (6 in Fig. 31) appears in the spectrum. Explanation to this interesting observation can be readily found in that the charged cesium adatoms residing on the outer side of the graphite film become neutralized in the course of oxidation, and, thus, will not interfere with the release of cesium sandwiched inside the graphite film, where cesium adatoms are charged and, hence, repel one another. Thus, a layer of oxide of the alkali metal grows on the surface of the graphite film, whose thickness is large enough to bring about an increase of the Auger peaks of cesium and oxygen and a decrease of that of oxygen. Experiments showed that the thickness of the oxide layer grows with that of the graphite film intercalated with cesium.

13. References

E.V. Rut'kov, A.Ya. Tontegode Pisma Zhur. Tekh. Fiz., 7(18), 1122, (1981)

A.Ya. Tontegode, E.V. Rut'kov. Usp. Fiz. Nauk, 163(11), 57, (1993)

N.R. Gall, E.V. Rut'kov, A.Ya. Tontegode. Modern Phys. B., 11(16), 1765, (1997)

Graphite Intercalation compounds eds. H.Zabel, S.A.Solin, Springer Ser.Mater.Sci., 1990, v.14; 1992, v.18.

N.A. Kholin, E.V. Rut'kov, A.Ya. Tontegode. Surf. Sci., 139, 155, (1984)

J.E.Fischer, R. Thomson, Phys.Today, 1978, v.31, N.7, p.363.

N.R. Gall, E.V. Rut'kov, S.N. Mikhailov, A.Ya. Tontegode Fiz. Tverd. Tela. 27(8), 2351, (1985)

N.R. Gall, S.N. Mikhailov, E.V. Rut'kov, A.Ya. Tontegode Zhur. Tekn. Fiz., 56, 732, (1986)
N.R. Gall, S.N. Mikhailov, E.V. Rut'kov, A.Ya. Tontegode Surf. Sci, 191, 185, (1987)
N.R. Gall, S.N. Mikhailov, E.V. Rut'kov, A.Ya. Tontegode Fiz. Tverd. Tela, 28, 2521, (1986)
N.R. Gall, S.N. Mikhailov, E.V. Rut'kov, A.Ya. Tontegode Synthetic Metals, 34, 497, (1989)
E.V. Rut'kov, N.R. Gall, Fiz. Tverd. Tela 51(8), 1639 (2009)
E.V. Rut'kov, N.R. Gall. Pisma Zhur. Eksp. Tekn. Fiz. 88(4), 308, (2008)
M.S. Dresselhaus, G. Dresselhaus. Advan. Phys. 30, 139, (1981)
M.V. Knat'ko, V.I. Paleev, N.D. Potehina Zhur. Tekn. Fiz. 59, 154, (1989)
A. Ya. Tontegode Pisma Zhur. Tekn. Fiz., 15, 57, (1989)
E.V. Rut'kov, A.Ya. Tontegode Zhur. Tekn. Fiz., 45(9), 1884, (1975)
E.V. Rut'kov, A.Ya. Tontegode Fiz. Tverd. Tela, 32(10), 2960, (1990)
E.V. Rut'kov, A.Ya. Tontegode Materials Science Forum, (91-93), 709, (1992)
E.Y. Zandberg, E.V. Rut'kov, A.Ya. Tontegode Pisma Zhur. Tekn. Fiz., 1(16), 741, (1975)
E.Y. Zandberg, E.V. Rut'kov, A.Ya. Tontegode, N.D. Potehina Fiz. Tverd, Tela, 19(2), 1665, (1997)
N.R. Gall, E.V. Rut'kov, A.Ya. Tontegode, Pisma Zhur. Tekn. Fiz. 14(6), 527, (1988)
N.R. Gall, E.V. Rut'kov, A.Ya. Tontegode Povrkhnost, 10, 47, (1989)
N.R. Gall, E.V. Rut'kov, A.Ya. Tontegode J. Chemical Vapor Deposition, July, 6(1), 72, (1997)
N.R. Gall, E.V. Rut'kov, A.Ya. Tontegode Pisma Zhur. Eksp. Tekn. Fiz., 75(1), 28, (2002)
N.R. Gall, E.V. Rut'kov, A.Ya. Tontegode Phys Solid State, 46(2), 371, (2004)
N.R. Gall, E.V. Rut'kov, A.Ya. Tontegode Semiconductors. 36(3), 276, (2002)
A.R.Ubbelohde, F.A.Lewis, “Graphite and its Crystal compounds”, Oxford, Clarendon Press, 1960.
E.V. Rut'kov, N.R. Gall Pisma Zhur. Tekn. Fiz.35 (16), 1 (2009)
N.A.Nesmeianov "Vapor pressure of chemical elements", Academia Nauk Publ., Moscow, 1961. (in Russian)
V.N. Ageev, E.V. Rut'kov, A.Ya. Tontegode, N.A. Kholin Fiz. Tverd. Tela, 23(8), 2248, (1981)
E.V. Rut'kov, N.R. Gall Fiz. Tekn. Poluprovodnikov, 43(10), 57 (2009)
A.Ya. Tontegode, F.K. Yusifov Fiz. Tverd. Tela, 35, 987, (1993)
E.V. Rut'kov, A.Ya. Tontegode, M.M. Usufov, Yu.S. Grushko Mol. Mat., 4, 217, (1994)
E.V. Rut'kov, A.Ya. Tontegode, M.M. Usufov Phys. Rev. betters 74(5), 758, (1995)
N.R. Gall, E.V. Rut'kov, A.Ya. Tontegode, M.M. Usufov, Mol. Mat., 7, 187, (1996)
E.V. Rut'kov, A.Ya. Tontegode, Yu.S. Grushko Pisma Zhur. Eksp. Tekn. Fiz. 57(11), 712, (1993)
N.R.Gall, S.N. Mikhailov, E.V. Rut'kov, A.Ya. Tontegode Poverkhnost, 12, 14, (1986)
N.R.Gall, S.N. Mikhailov, E.V. Rut'kov, A.Ya. Tontegode Fiz. Tverd. Tela, 28(8), 2521, (1986)
N.R.Gall, S.N. Mikhailov, E.V. Rut'kov, A.Ya. Tontegode Poverkhnost, 4, 22, (1987)
N.R.Gall, S.N. Mikhailov, E.V. Rut'kov, A.Ya. Tontegode Synthetic Metals, 34(1-3), 447, (1989)
N.R.Gall, S.N. Mikhailov, E.V. Rut'kov, A.Ya. Tontegode Surf. Sci., 226, 381, (1990)
R.F. Willis, B.E. Fitton, G.S. Painter Phys. Rev., 9, 1926, (1974)
G.S. Painter, D.E. Ellis Phys. Rev. 1, 4797, (1970)
P. Murchand, G. Fretigny, M. Lagies and al. Phys. Rev., 30, 4788, (1984)
P. Oelhafen, P. Pfluger, H.I. Grüntherodt Solid State Commun., 32, 885, (1975)
R. Gurney Phys. Rev., 47, 479, (1935)

9

Chemical and Biosensing Applications Based on Graphene Field-Effect Transistors

Yasuhide Ohno, Kenzo Maehashi and Kazuhiko Matsumoto
The Institute of Scientific and Industrial Research, Osaka University
Japan

1. Introduction

Label-free electrical detection of biomolecule based on nano-meter size materials has attracted in many fields such as clinical diagnosis for health care, life science and practical pharmacy because they are expected for rapid and easy detection of various biological species at home. Chemical and biological sensors using silicon nanowires(Cui et al., 2001; Li et al., 2005; Zheng et al., 2005) and carbon nanotubes (CNTs)(Besteman et al., 2003; Chen et al., 2003; Ishikawa et al., 2010; Maehashi et al., 2007; Maehashi, Matsumoto, Kerman, Takamura & Tamiya, 2004; Martnez et al., 2009; Star et al., 2003) have been developed for the past decades. Especially, CNT field-effect transistors (CNT-FETs) are one of the strongest candidates for biosensing applications due to their high aspect ratio, high mechanical strength, large surface areas and outstanding electrical characteristics. These superior characteristics make CNTs ideal for nanoscale devices(Saito et al., 1998). There have been many reports about chemical and biological sensors using CNT-FETs such as proteins, glucose, DNA hybridizations and immunosensors(Besteman et al., 2003; Chen et al., 2003; Ishikawa et al., 2010; Maehashi et al., 2007; Maehashi, Matsumoto, Kerman, Takamura & Tamiya, 2004; Martnez et al., 2009; Star et al., 2003).

Although the CNT-FET based sensors have high potential, their electrical characteristics strongly depends on their chirality (diameter) and work function of the contact metal(Chen et al., 2005). Since the chirality control growth of CNTs has not been achieved, the reproducibility and stability of the CNT-FET based sensors have been major problem(Maehashi, Ohno, Inoue & Matsumoto, 2004). And the typical absolute value of the drain current (I_D) of a CNT-FET with one CNT channel in solution is several nA to several 10 nA because only small drain and gate voltages can be applied in solution due to the avoidance of solution electrolysis and oxidization of the electrodes, channel and analyte. To resolve these problems, aligned CNTs have been studied in recent years(Palaniappan et al., 2010). However, the separation between semiconductor and metallic CNTs is also very difficult problem.

Graphene, single layer hexagonal network of carbon atom, can modify these problems. Since they are ideal two-dimensional crystal with extremely high carrier mobility at room temperature without any sophisticated doping process and very stable materials, graphene field-effect transistors (G-FETs) have been expected for the next-generation practical devices(Geim & Novoselov, 2007; Novoselov et al., 2004). In recent years, sensing applications using graphene and graphene-like materials have been intensively studied because their electrical characteristics are very sensitive for the environmental conditions and

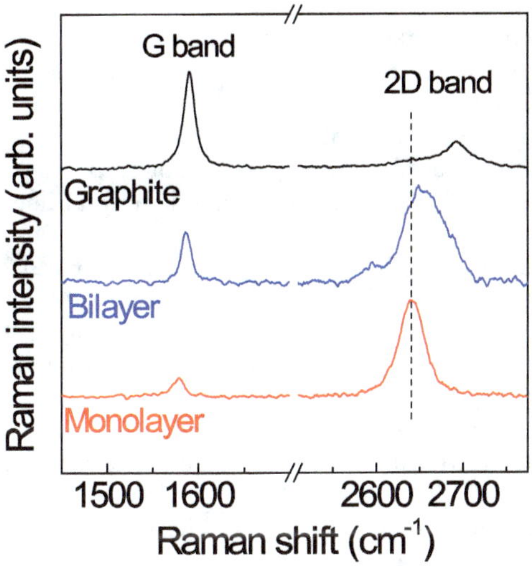

Fig. 1. Raman spectra of G- and 2D-bands for 1 and 2 layer(s) graphene and HOPG.

the surface-analyte or analyte-analyte bindings occur very close to the graphene channel(Yang et al., 2010). The mechanism of action for G-FET sensors is that chemical species adsorbed on the surface of the graphene act as electron donors or acceptors, resulting in conductance changes. Most of the graphene sensors have been used to detect gas molecules(Arsat et al., 2009; Dan et al., 2009; Fowler et al., 2009; Lu et al., 2009; Qazi et al., 2007; Robinson et al., 2008; Schedin et al., 2007). In addition, research on electrical detection of biomolecule detection using graphene has gradually increased over the last few years. Electrochemical detections of glucose and proteins have been investigated(Alwarappan et al., 2009; Lu et al., 2007; Shan et al., 2009; Shang et al., 2008; Wu et al., 2009). Field-effect transistors based on reduced graphene from graphene oxide or graphene amine have detected DNA hybridizations and negatively charged bacteria(Mohanty & Berry, 2008). In graphene, the highest carrier mobility is only achieved in single-layer graphene due to its linear energy dispersion at K point(Nagashio et al., 2008). Since the sensitivity using G-FET-based chemical and biological sensors depend on their transconductance ($= \partial I_D/\partial V_G = C_G V_D \mu$; where C_G, V_D and μ are the gate capacitance, drain voltage and field-effect mobility, respectively), G-FETs with single-layer graphene are thought to be suitable for the sensing devices.

In this chapter, we describe our attempts to apply G-FETs with single-layer graphene to chemical and biological sensors(Ohno et al., 2009). We investigated behavior of G-FETs immersed in an electrolyte and show that they have very good transfer characteristics compared with their characteristics in vacuum. They also exhibit clear pH-dependent I_D characteristics and could electrically detect surface-protein adsorption. Moreover, we demonstrate the achievement of electrical detection of biomolecules and their charge types by G-FETs.

2. Experimental

The G-FETs used in this study were fabricated on a 285-nm-thick thermally grown SiO_2 layer on a heavily p-doped silicon substrate ($\rho < 0.01\ \Omega\cdot$cm). Single-layer graphene flakes were obtained by micromechanical exfoliation using natural graphite and clear adhesive tape. The graphene flakes were searched by an optical microscope after slowly peeling off the tape from the substrate. As a result of this procedure, various types of graphene layers (or thick graphite) were identified on the surface SiO_2/Si substrate. Single-layer graphene flakes were identified by analyzing the shift in green intensity under optical microscope observation and by Raman spectroscopy. Figure 1 compares the 632.8 nm Raman spectra of 1 and 2 layer(s) graphene and highly oriented pyrolytic graphite (HOPG). Two strong peaks with a G band at $\sim$1580 cm^{-1} and a 2D band at $\sim$2650 cm^{-1} could be observed. A single peak of the 2D can be observed for the single-layer graphene, while a much broader peak with some shoulders, shifted to a high-frequency, could be observed for the bilayer graphene. As shown in Figure 1, a single peak in the 2D band in the Raman spectrum is direct evidence of single-layer graphene(Ferrari et al., 2006). Ti (5 nm)/Au (30 nm) source and drain electrodes were formed by electron beam lithography and lift-off method. The degenerately doped silicon substrate was also used for the back gate. Typical optical microscope image of G-FET was shown in Fig. 2.

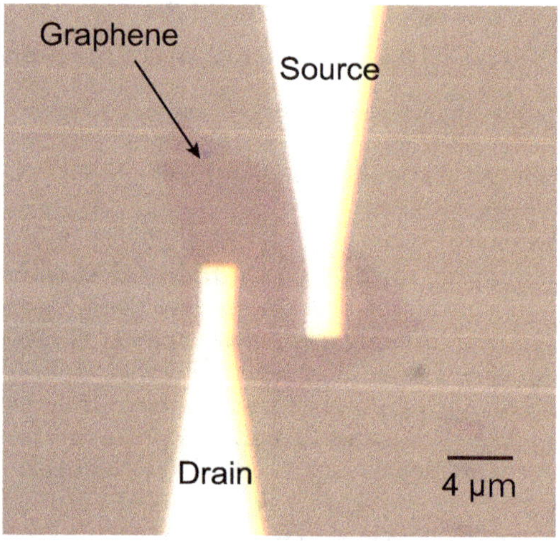

Fig. 2. Optical micrograph of a typical G-FET.

To measure the solution pH value and protein adsorption, the device was surrounded by a silicone rubber pool attached to the substrate. Then, the Ag/AgCl reference electrode was immersed into the solution contained within a silicone rubber barrier as shown in Fig. 3 and Fig. 4. The Ag/AgCl reference electrode was used as the top-gate electrode to minimize environmental effects(Minot et al., 2007). The electrical characteristics of the G-FETs were measured by a semiconductor parameter analyzer (4156C; Agilent technologies Inc., Santa Clara, CA), using two-terminal measurement. In the experiments, we drove the G-FETs under low voltage ($\leq$ 0.2 V) due to the avoidance of oxidization of electrodes and graphene channel. Natural graphite used in this work was kindly provided by Nippon Graphite Industries Ltd. (Shiga, Japan). 10 mM phthalate buffer solution of pH 4.0, phosphate buffer solution of pH 6.8 and 10 mM borate buffer solution at pH 9.3 were purchased from Horiba Ltd. (Kyoto, Japan).

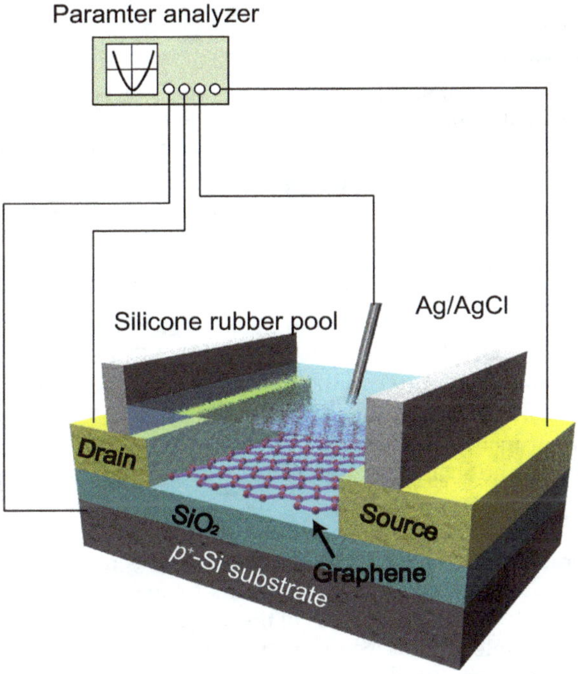

Fig. 3. Schematic illustration of experimental setup with G-FETs.

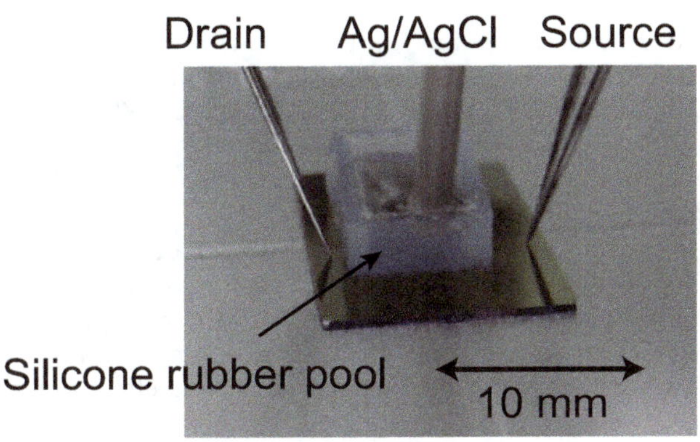

Fig. 4. Photograph of experimental setup with a G-FET.

Solutions of various pH values were prepared by mixing a 10 mM phthalate buffer solution of pH 4.0, a 10 mM phosphate buffer solution of pH 6.8 and a 10 mM borate buffer solution at pH 9.3. Bovine serum albumin (BSA), which was used as the target protein, was purchased from Sigma Aldrich Inc. (St Louis, MO).

3. Results and discussion

3.1 Transport characteristics in solution

At first, we investigated the transport characteristics of G-FETs in solution. In a solution or electrolyte, the electrical-double layer acts as a top-gate insulator, and the thickness of electrical-double layer is generally defined by the Debye-Hückel equation, which depends on the ionic strength and temperature, which is as small as 5 nm in the buffer solution with several ten mM. This value is very thin compared with the SiO_2 layer used as the back-gate insulator. For graphene devices, only 90 or 290 nm-thick SiO_2 can be used because single or a few layer graphene flakes on the 90 or 290 nm can be observed by a conventional optical microscope due to the refractive index of the graphene(Blake et al., 2007). Because of this thin gate insulator, the field-effect applied by the top-gate electrodes is very effective compared from the back gate. In fact, electrolyte-gated CNT-FETs and G-FETs have shown good electrical characteristics as thin top-gate insulators with high dielectric constants in ionic solutions(Das et al., 2008; Rosenblatt et al., 2002).

A typical plot of I_D as a function of back-gate voltage under reduced pressure (1×10^{-3} Pa) is shown in Fig. 5. Transconductance ($= \partial I_D / \partial V_G$) was estimated to be 0.13 μS and 30 μS for back-gate and top-gate operations, respectively. The transconductance was for top-gate operation was more than 200 times better than that of back-gate operation; thus, a thin electrical-double layer is formed on the graphene surface. This result shows that G-FETs can operate even in solution, and their electrical characteristics in solution are excellent.

To clarify the difference between the g_m values of the devices, back- and top-gate capacitances were estimated using a simple model. Assuming that graphene is equivalent to a metal disk placed on the insulator, the gate capacitance can be expressed by the equation below, (Gelmont et al., 1995)

$$C_G = \frac{2\pi\varepsilon_0(\varepsilon_r + 1)r}{\tan^{-1}\left[\dfrac{2h(\varepsilon_r + 1)}{r\varepsilon_r}\right]}, \tag{1}$$

where, r, h, ε_r and ε_0 are the radius of the metal disk (graphene), the thickness of the gate insulator, the relative permittivity and vacuum permittivity, respectively. For the electrolyte-gated G-FETs (with r=1.5 μm, h=1 nm and ε_r (H_2O)=80), the estimated electrostatic gate capacitance was $C_{G_EL} \sim$500 nF cm^{-2}. In this case, if total top-gate capacitance (C_{TG}) is considered as a series of the C_{G_EL} and the quantum capacitance (C_Q), (Meric et al., 2008; Rosenblatt et al., 2002) then C_{TG}=$C_{G\_EL}C_Q/(C_{G\_EL} + C_Q)$. Because the C_Q of the graphene channel is approximately 2 μF cm^{-2}, (Meric et al., 2008) the C_{TG} is 400 nF cm^{-2}. This value is more than three orders of magnitude larger than the back-gate capacitance of 85 pF cm^{-2} (with r=1.5 μm, h=285 nm and ε_r (SiO_2)=3.9), given by the above equation. It can be concluded that this larger top-gated capacitance gave rise to better transfer characteristics of G-FETs in the electrolyte, indicating their high potentials for use in field-effect sensing applications.

3.2 pH dependence

The dependence of the transfer characteristics and conductance of G-FETs on pH were evaluated. Figure 6 shows I_D plotted as a function of V_{TGS} for a G-FET in various electrolytes at various pH values from 4.04 to 8.16. The Dirac points of the G-FET shifted in a positive detection with increasing pH. This behavior indicates that G-FETs can detect the pH value by the electrical characteristics. A plot of the time-dependent I_D for a G-FET at V_{TGS} =−80 mV in pH values from 4.0 to 8.2. is shown in Figure 7. Every 10 min, either a 10-mM phosphate buffer solution at pH 6.8 or a 10-mM borate buffer solution at pH 9.3 were added to increase

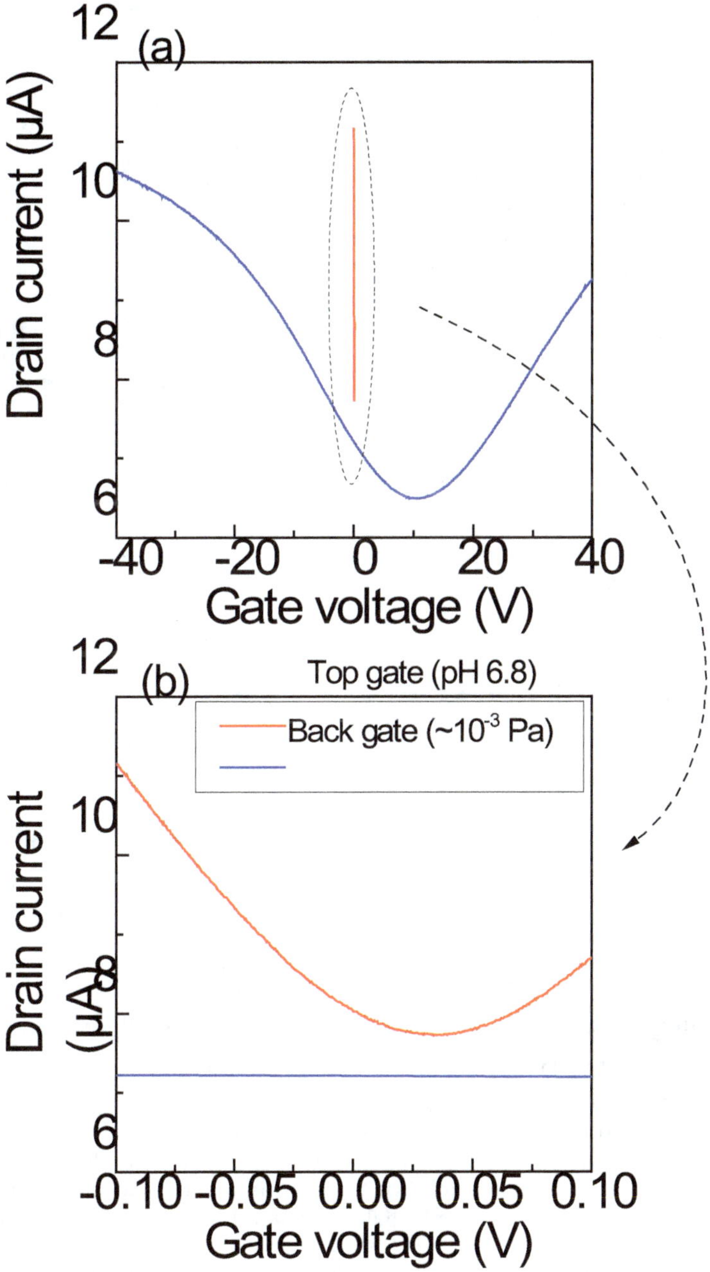

Fig. 5. (a) I_D as a function of back-gate voltage at 10^{-3} Pa and top-gate voltage in an solution at pH 6.8. (b) Enlarged view of I_D as a function of gate voltage.

the pH. The I_D increased stepwise with pH from 4.0 to 8.2, and the I_D at each pH value was virtually constant. The plot of the average I_D against pH values indicates that the relationship between pH and I_D is linear over the range from 4.0 to 8.2, as shown in Figure 8. Similar changes in gate transfer characteristics have been observed for a G-FET exposed to NO_2 and pH dependence of G-FETs with few-layer epitaxial graphene. It is concluded that the increased I_D can be attributed to the increased negative charge around the graphene channel, because the hole is the carrier in this condition. The origin of the increase in current is not clear at present. We speculate that hydrogen or hydroxide ions have some effect on the graphene surface. In case of the CNT-FETs, hydroxide ions bound to the carbon nanotube surface act as acceptors(Lee et al., 2007; Pan et al., 2004). As a nano-carbon material, graphene may exhibit chemical reactivity similar to carbon nanotubes. Further investigation is necessary to clarify this point.

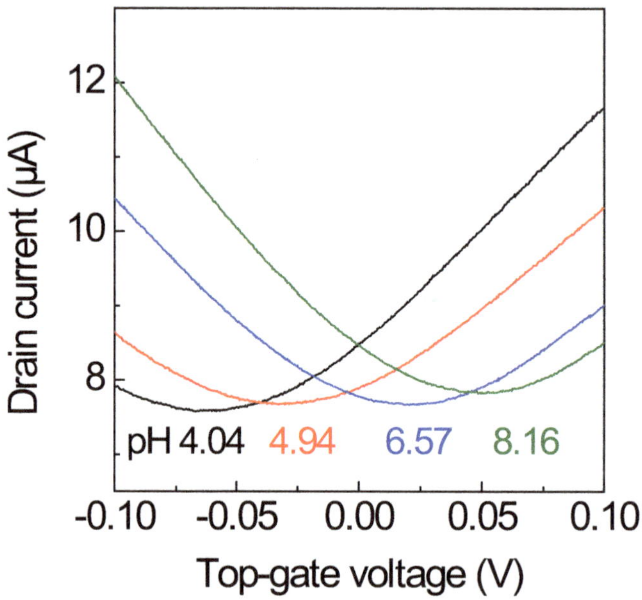

Fig. 6. I_D as a function of top-gate voltage of a G-FET at pH 4.04, 4.94, 6.57 and 8.16.

It should be noted that the Dirac point of G-FETs at a pH solution was not constant. Although the Dirac point = 0 V was realized at pH 5.8 in this device, other devices showed the Dirac point = 0 V at different pH values. One possible reason of the instability is due to charged impurities such as residual EB resist, defects or underlying SiO_2 and Si. Indeed, the solution pH slightly influenced the carrier mobility. Charged impurity scattering is major subject in the graphene technology. These uncontrollable charged impurities may lead the Dirac point instability.

The detection limit (=resolution, signal/noise=3) for changes in pH was estimated to be 0.025 in the pH range from 4.04 to 8.16. In this work, the signal and noise were defined by the average and standard deviation of the data, respectively. On the other hand, a pH sensor based on carbon nanotube FETs showed the detection limit of 0.67 due to the small drain current in

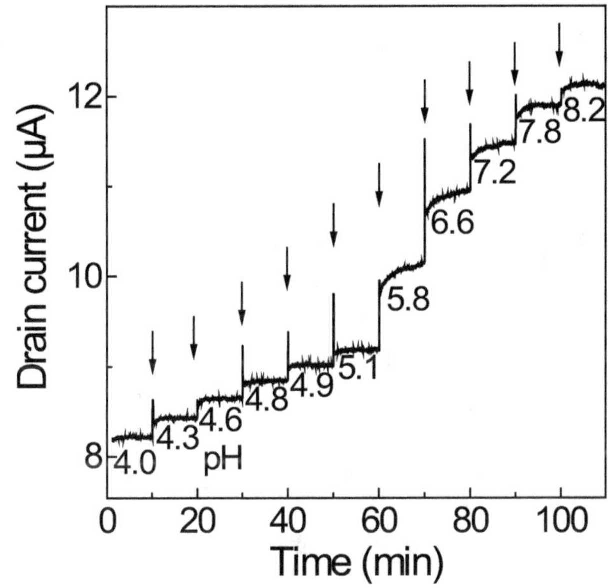

Fig. 7. I_D versus time data of a G-FET for pH values from 4.04 to 8.16. Arrows indicate the adding points of solution with different pH.

solution ($\sim$ 10 nA)(Yamamoto et al., 2009). Therefore, the G-FETs are useful as pH sensors for their stability. The pH dependence of the drain current was similar to that of few-layer epitaxial graphene.

3.3 Detection of protein adsorption

Finally, we demonstrated label-free biomolecule detection, based on electrolyte-gated G-FETs. A 10-mM phosphate buffer electrolyte solution at pH 6.8 was used, with a bovine serum albumin (BSA; Sigma Aldrich Inc., St Louis, MO) target biomolecule. The isoelectric point of BSA is 5.3, indicating that BSA molecules are negatively charged at the pH used for this measurement. Figure 9 shows the evolution of I_D of a G-FET for electronic monitoring of BSA adsorption on a graphene channel at $V_{TGS} = -0.1$ V and $V_{BGS} = 0$ V. Under this condition, the I_D was expected to be increased by the hole carrier when (negatively charged) BSA was attached to the graphene. Measurement began with a 10-mM pure phosphate buffer solution. After 10 min, further pure buffer solution was added and I_D was virtually unchanged. After 20 min, BSA concentration dependence of I_D was shown. Arrows in Figure 9 mark the points where solutions with various concentration of BSA were injected. The I_D clearly increased when BSA was introduced, indicating adsorption on the graphene surface. The drain current change (ΔI_D) is shown as a function of BSA concentration (C_{BSA}) in Figure 10. ΔI_D increased linearly at low concentrations and was saturated at higher concentrations. This result indicates that the adsorption of BSA molecules onto the graphene surface follows the Langmuir adsorption isotherm given by

$$\frac{C_{BSA}}{\Delta I_D} = \frac{C_{BSA}}{\Delta I_{DMax}} + \frac{K_d}{\Delta I_{DMax}}, \tag{2}$$

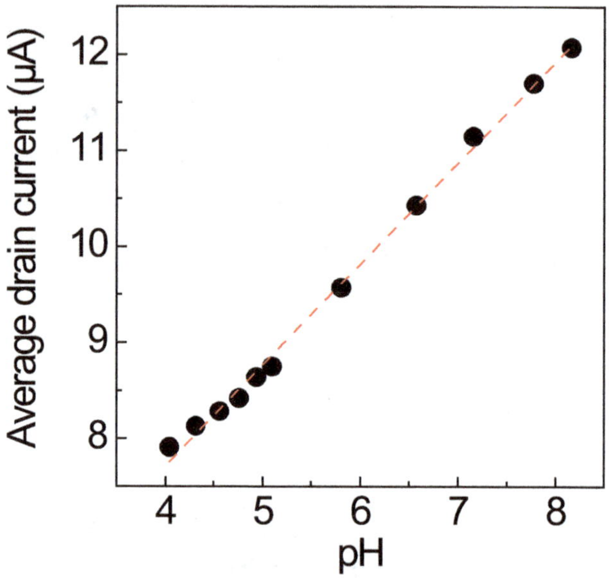

Fig. 8. Average I_D as a function of pH. The dashed line is a linear fit to the data points.

where K_d is the dissociation constant of the interaction between BSA molecules and graphene and ΔI_{DMax} is the drain current at saturation. The Langmuir adsorption isotherm fitted the experimental results well, as shown in Figure 10 (red dashed line). The K_d was estimated to be 1.5×10^{-8} M, which was comparable with the values obtained for Si nanowire(Cui et al., 2001) and CNT-FET biosensors(Abe et al., 2008; Maehashi et al., 2009) using antibody-antigen interactions, despite the fact that BSA molecules were considered to be directly adsorbed onto the graphene surface in this work rather than through a protein-protein interaction. Further experiments are needed to confirm the suitability of the Langmuir adsorption isotherm for these experimental results.

Moreover, adsorption of proteins with different charge types onto the graphene surface was detected. The isoelectric point of BSA is approximately 5.3; accordingly, BSA molecules are positively (negatively) charged in the 10-mM phthalate (phosphate) buffer solution. Four G-FETs were used in this experiment. Two of them were used in the phthalate buffer solution and others in the phosphate buffer solution. Figures 11 show the time course of normalized I_D at V_D and V_{TGS} of 0.1 and −0.1 V, respectively. The drain current decreased when 80 (red lines) and 100 (blue lines) nM BSA in 10 mM phthalate buffer solution (pH 4.0) was added, indicating that the graphene channel detected the positive charge. Conversely the drain current increased after adding negatively charged BSA in 10 mM phosphate buffer solution (pH 6.8). And the I_D changes seem to depend on the concentration of BSA. It should be noted that some I_D trends are unstable after adding the BSA as shown in Fig. 11. The origin of these unstable features may come from remote impurity in SiO_2 layer under the graphene channel. Very recent research showed that the existence of SiO_2 layer influenced the sensing properties for G-FETs.

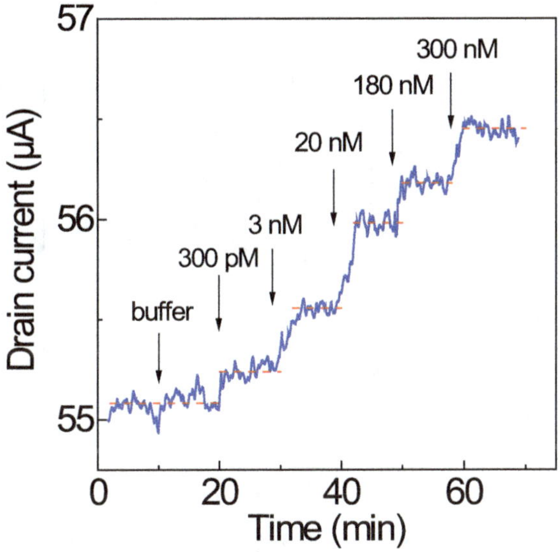

Fig. 9. I_D versus time for electrical monitoring of exposure to various BSA concentrations. Dashed lines indicate the average I_D.

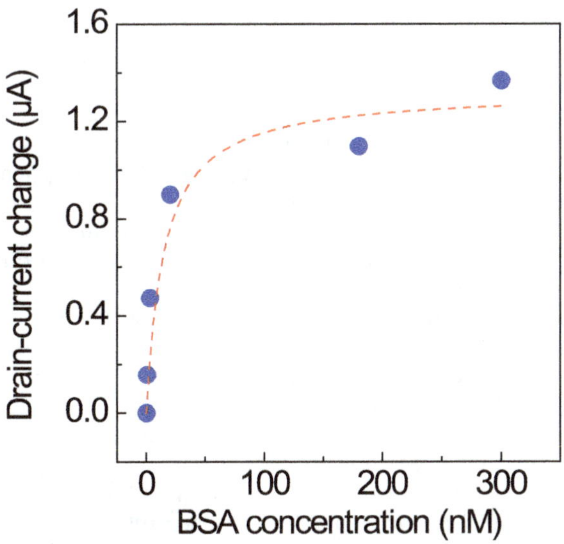

Fig. 10. Plot of the net I_D change of a G-FET versus BSA concentration. The dashed line is a fit to the Langmuir adsorption isotherm.

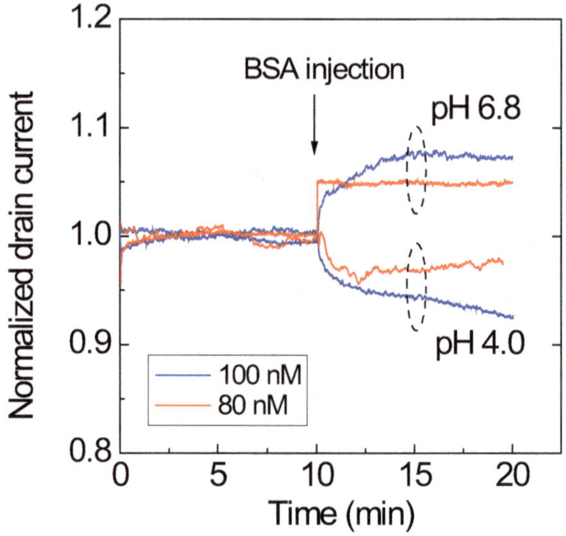

Fig. 11. Time course of normalized I_D for G-FETs at V_D and V_{TGS} of 0.1 and −0.1 V, respectively, in 10-mM phthalate and phosphate buffer solution. Red (blue) lines indicate that 80 (100) nM BSA was added at 10 min.

These results indicate that the graphene channel could detect the charge type of the adsorbed biomolecules. Under the experimental conditions, the carriers in the graphene channel were holes. Therefore, the decreased (increased) I_D observed for BSA in the phthalate (phosphate) buffer solution was a consistent result, because the hole carrier in the graphene channel decreased (increased) as a result of exposure to the positively (negatively) charged proteins. For biosensors based on CNT-FETs, only positive charged biomolecules can be detected, owing to their *p*-type semiconductor band characteristics(Yamamoto et al., 2010). In contrast, the G-FETs can detect positively and negatively charged biomolecules because a Schottky barrier does not form at the interface between the electrodes and graphene, owing to its the zero-gap semiconductor characteristics. Furthermore, the absolute value of the drain current is larger than that of carbon nanotube devices, indicating their robustness to noise.

In this work, the conductance changes by BSA adsorption were quite small. Three possible interpretations can be considered for the small conductance changes. One interpretation is due to the electrode-graphene contact resistance. Although two-terminal measurement was used in this experiment, four-terminal measurement is needed to ignore the contact resistance fluctuations. Another interpretation is due to desorption of the BSA molecules. Protein sensing using specific protein detection such as antigen-antibody effect is efficiency to interrupt the BSA desorption. The other interpretation is due to the difference between isoelectric point of the BSA (=5.6) and solution pH (=6.8) is relatively small. This small difference may lead some uncharged amino acid of the BSA molecules. Moreover, it is important to clarify where the charge transfer occurs and surface area dependence of the protein adsorption. These subjects should be investigated to develop the biomolecule detector using G-FETs.

The G-FETs can reuse after measuring of protein adsorption. Figure 12 shows the optical micrographs of a G-FET after exposing the several μM BSA (a) and after washing the proteins

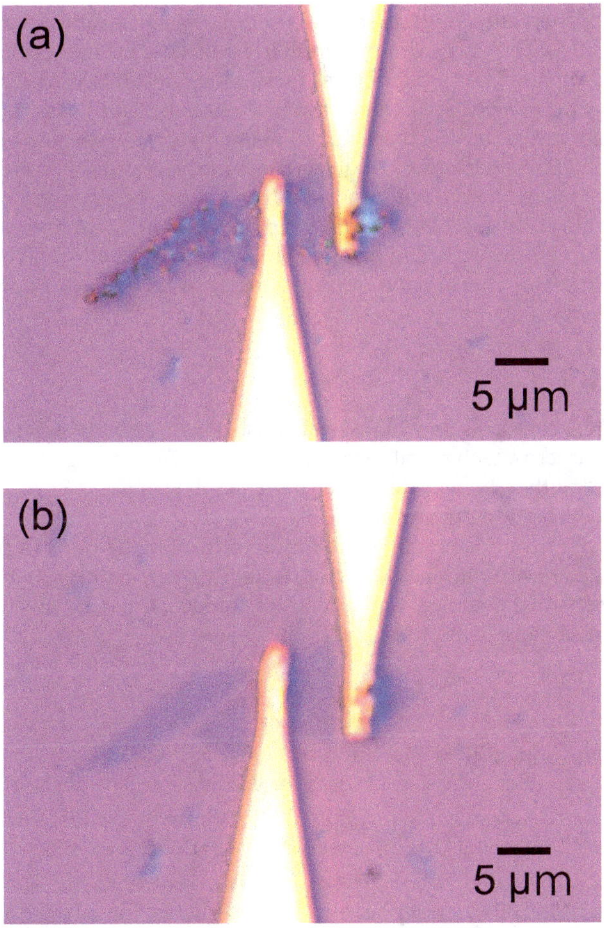

Fig. 12. Optical micrograph of a G-FET (a) after exposing proteins and (b) washing the proteins by acid and alkali solutions.

by acid and alkali solutions (b). Many bright points indicate the adsorbed proteins onto the graphene surface, and only these proteins disappeared after washing by sulfuric acid and sodium hypochlorite solution. The graphene flakes have not been broken by these solutions, indicating their high stability. And after the washing process, the G-FET can reuse as a sensor. These results indicate that the G-FETs have high potentials for the chemical and biological sensors.

3.4 Comparison of the CNT-FETs and G-FETs

Finally, comparison of the sensing characteristics of CNT-FETs and G-FETs are briefly described. The important requirements of materials for the sensing device are the charge sensitivity and stability. Since the charge sensitivity is strongly depends on the surface to volume ratio, it can be considered that the CNT-FETs have better charge sensitivity than G-FETs. On the contrary, the stability of G-FETs is superior to the CNT-FETs. Their absolute value of the drain current is as high as several ten μA, which is more than 1000 times larger than those of CNT-FETs. This large drain current indicates the robustness for the noise during the sensing measurements. The stable devices carry more credibility. And chirality-controlled CNT growth technique, which leads the stable FET characteristics, has not been achieved at present. This problem makes it difficult to fabricate the integrated sensors based on CNT-FETs. In the case of G-FETs, the development of growth technique of graphene is the key technology. Very recent reports says the roll-to-roll production of 30-inch graphene films can be grown by CVD system(Bae et al., 2010). Such large-scale graphene growth technique will lead large-scale integrated and multiple sensors based on G-FETs. Because both CNT-FETs and G-FETs have superior characteristics for the sensing applications each other, it is important to use them according to the situations.

4. Conclusion

We have investigated chemical and biological sensors using G-FETs. Single-layer graphene was obtained by a micro-mechanical cleavage method. Changes in the solution pH were electrically detected with a lowest detection limit (signal/noise = 3) of the 0.025. Their I_D showed protein-concentration dependence and their I_D changes with BSA concentration were fitted well by the Langmuir adsorption isotherm. In addition, the G-FETs clearly detected the different charge types of a biomolecule owing to its isoelectric point. G-FETs are promising devices for highly sensitive chemical and biological sensors. In near future, we try to evaluate the selective protein sensing using the G-FETs.

5. References

Abe, M., Murata, K., Ataka, T. & Matsumoto, K. (2008). Calibration method for a carbon nanotube field-effect transistor biosensor, *Nanotechnology* 19: 045505.

Alwarappan, S., Erdem, A., Liu, C. & Li, C.-Z. (2009). Probing the Electrochemical Properties of Graphene Nanosheets for Biosensing Applications, *J. Phys. Chem. C* 113: 8853–8857.

Arsat, R., Breedon, M., Shafiei, M., Spizziri, P., Gilje, S., Kaner, R., Kalantar-zadeh, K. & Wlodarski, W. (2009). Graphene-like nano-sheets for surface acoustic wave gas sensor applications, *Chem. Phys. Lett.* 467: 344–347.

Bae, S., Kim, H., Lee, Y., Xu, X., Zheng, J.-S. P. Y., Balakrishnan, J., Lei, T., Kim, H. R., Song, Y. I., Kim, Y.-J., Kim, K. S., Özyilmaz, B., Ahn, J.-H., Hong, B. H. & Iijima, S. (2010). Roll-to-roll production of 30-inch graphene films for transparent electrodes, *Nat. Nanotechnol.* 5: 574–578.

Besteman, K., Lee, J.-O., Wiertz, F. G. M., Heering, H. A. & Dekker, C. (2003). Enzyme-Coated Carbon Nanotubes as Single-Molecule Biosensors, *Nano Lett.* 3: 727–730.

Blake, P., Hill, E. W., Neto, A. H. C., Novoselov, K. S., Jiang, D., Yang, R., Booth, T. J. & Geim, A. K. (2007). Making graphene visible, *Appl. Phys. Lett.* 91: 063124.

Chen, R. J., Bangsaruntip, S., Drouvalakis, K. A., Kam, N. W. S., Shim, M., Li, Y., Kim, W., Utz, P. J. & Dai, H. (2003). Noncovalent functionalization of carbon nanotubes for highly specific electronic biosensors, *Proc. Natl. Acad. Sci.* 100: 4984–4989.

Chen, Z., Appenzeller, J., Knoch, J., lin, Y. & Avouris, P. (2005). The Role of Metal-Nanotube Contact in the Performance of Carbon Nanotube Field-Effect Transistors, *Nano Lett.* 5: 1497–1502.

Cui, Y., Wei, Q., Park, H. & Lieber, C. M. (2001). Nanowire Nanosensors for Highly Sensitive and Selective Detection of Biological and Chemical Species, *Science* 293: 1289–1292.

Dan, Y., Lu, Y., Kybert, N. J., Luo, Z. & Johnson, A. T. C. (2009). Intrinsic Response of Graphene Vapor Sensors, *Nano Lett.* 9: 1472–1475.

Das, A., Pisana, S., Chakraborty, B., Piscanec, S., Saha, S. K., Waghmare, U. V., Novoselov, K. S., Krishnamurthy, H. R., Geim, A. K., Ferrari, A. C. & Sood, A. K. (2008). Monitoring dopants by Raman scattering in an electrochemically top-gated graphene transistor, *Nat. Nanotechnol.* 3: 210–215.

Ferrari, A. C., Meyer, J. C., Scardaci, V., Casiraghi, C., Lazzeri, M., Mauri, F., Piscanec, S., Jiang, D., Novoselov, K. S., Roth, S. & Geim, A. K. (2006). Raman Spectrum of Graphene and Graphene Layers, *Phys. Rev. Lett.* 97: 187401.

Fowler, J. D., Allen, M. J., Tung, V. C., Yang, Y., Kaner, R. B. & Weiller, B. H. (2009). Practical Chemical Sensors from Chemically Derived Graphene, *ACS Nano* 3: 301–306.

Geim, A. K. & Novoselov, K. S. (2007). The rise of graphene, *Nat. Mater.* 6: 183–191.

Gelmont, B., Shur, M. S. & Mattauch, R. J. (1995). Disk and Stripe Capacitances, *Solid-State Elec.* 38: 731–734.

Ishikawa, F. N., Chang, H.-K., Curreli, M., Liao, H.-I., Olson, C. A., Chen, P.-C., Zhang, R., Roberts, R. W., Sun, R., Cote, R. J., Thompson, M. E. & Zhou, C. (2010). Label-Free, Electrical Detection of the sars Virus N-Protein with Nanowire Biosensors Utilizing Antibody Mimics as Capture Probes, *ACS nano* 3: 1219–1224.

Lee, K., Kwon, J. H., Cho, S. M. W. S., Ju, B. K. & Lee, Y. H. (2007). pH sensitive multiwalled carbon nanotubes, *Mat. Lett.* 61: 3201–3203.

Li, C., Curreli, M., Lin, H., Lei, B., Ishikawa, F. N., Datar, R., Cote, R. J., Thompson, M. E. & Zhou, C. (2005). Complementary Detection of Prostate-Specific Antigen Using In_2O_3 Nanowires and Carbon Nanotubes, *J. Am. Chem. Soc.* 127: 12484–12485.

Lu, G., Ocola, L. E. & Chen, J. (2009). Gas detection using low-temperature reduced graphene oxide sheets, *Appl. Phys. Lett.* 94: 083111.

Lu, J., Drzal, L. T., Worden, R. M. & Lee, I. (2007). Simple Fabrication of a Highly Sensitive Glucose Biosensor Using Enzymes Immobilized in Exfoliated Graphite Nanoplatelets Nafion Membrane, *Chem. Mater.* 19: 6240–6246.

Maehashi, K., Katsura, T., Kerman, K., Takamura, Y., Matsumoto, K. & Tamiya, E. (2007). Label-Free Protein Biosensor Based on Aptamer-Modified Carbon Nanotube Field-Effect Transistors, *Anal. Chem.* 79: 782–787.

Maehashi, K., Matsumoto, K., Kerman, K., Takamura, Y. & Tamiya, E. (2004). Ultrasensitive Detection of DNA Hybridization Using Carbon Nanotube Field-Effect Transistors, *Jpn. J. Appl. Phys.* 43: L1558–L1560.

Maehashi, K., Matsumoto, K., Takamura, Y. & Tamiya, E. (2009). Aptamer-Based Label-Free Immunosensors Using Carbon Nanotube Field-Effect Transistors, *Electroanalysis* 21: 1285–1290.

Maehashi, K., Ohno, Y., Inoue, K. & Matsumoto, K. (2004). Chirality selection of single-walled carbon nanotubes by laser resonance chirality selection method, *Appl. Phys. Lett.* 85: 858–860.

Martnez, M. T., Tseng, Y.-C., Ormategui, N., Loinaz, I., Eritja, R. & Bokor, J. (2009). Label-Free dna Biosensors Based on Functionalized Carbon Nanotube Field Effect Transistors, *Nano Lett.* 9: 530–536.

Meric, I., han, M. Y., Young, A. F., Ozyilmaz, B., Kim, P. & Shepard, K. L. (2008). Current saturation in zero-bandgap, topg-ated graphene field-effect transistors, *Nat. Nanotechnol.* 3: 654–659.

Minot, E. D., Janssens, A. M., Heller, I., Dekker, H. A. H. C. & Lemay, S. G. (2007). Carbon nanotube biosensors: The critical role of the reference electrode, *Appl. Phys. Lett.* 91: 093507.

Mohanty, N. & Berry, V. (2008). Graphene-Based Single-Bacterium Resolution Biodevice and Dna Transistor: Interfacing Graphene Derivatives with Nanoscale and Microscale Biocomponents, *Nano Lett.* 8: 4469–4476.

Nagashio, K., Nishimura, T., Kita, K. & Toriumi, A. (2008). Mobility Variations in Mono- and Multi-Layer Graphene Films, *Appl. Phys. Express* 2: 025003.

Novoselov, K. S., Geim, A. K., Morozov, S. V., Jiang, D., Zhang, Y., Dubonos, S. V., Grigorieva, I. V. & Firsov, A. A. (2004). Electric Field Effect in Atomically Thin Carbon Films, *Science* 306: 666–669.

Ohno, Y., Maehashi, K., Yamashiro, Y. & Matsumoto, K. (2009). Electrolyte-gated graphene field-effect transistors for detecting ph and protein adsorption, *Nano Lett.* 9: 3318–3322.

Palaniappan, A., Goh, W. H., Tey, J. N., Wijaya, I. P. M., Moochhala, S. M., Liedberg, B. & Mhaisalkar, S. G. (2010). Aligned carbon nanotubes on quartz substrate for liquid gated biosensing, *Biosens. Bioelectron.* 25: 1989–1993.

Pan, H., Feng, Y. P. & Lin, J. Y. (2004). Ab initio study of OH-functionalized single-wall carbon nanotubes, *Phys. Rev. B* 70: 245425.

Qazi, M., Vogt, T. & Koley, G. (2007). Trace gas detection using nanostructured graphite layers, *Appl. Phys. Lett.* 91: 233101.

Robinson, J. T., Perkins, F. K., Snow, E. S., Wei, Z. & Sheehan, P. E. (2008). Reduced Graphene Oxide Molecular Sensors, *Nano Lett.* 8: 3137–3140.

Rosenblatt, S., Yaish, Y., Park, J., Gore, J., Sazonova, V. & McEuen, P. L. (2002). High Performance Electrolyte Gated Carbon Nanotube Transistors, *Nano Lett.* 2: 869–872.

Saito, R., Dresselhaus, G. & Dresselhaus, M. S. (1998). *Physical Properties of Carbon Nanotubes*, Imperial College Press, London.

Schedin, F., Geim, A. K., Morozov, S. V., Hill, E. W., Blake, P., Katsnelson, M. I. & Novoselov, K. S. (2007). Detection of individual gas molecules adsorbed on graphene, *Nat. Mater.* 6: 652–655.

Shan, C., Yang, H., Song, J., Han, D., Ivaska, A. & Niu, L. (2009). Direct Electrochemistry of Glucose Oxidase and Biosensing for Glucose Based on Graphene, *Anal. Chem.* 81: 2378–2382.

Shang, N. G., Papakonstantinou, P., McMullan, M., Chu, M., Stamboulis, A., Potenza, A., Dhesi, S. S. & Marchetto, H. (2008). Catalyst-Free Efficient Growth, Orientation and Biosensing Properties of Multilayer Graphene Nanoflake Films with Sharp Edge Planes, *Adv. Funct. Mater.* 18: 3506–3514.

Star, A., Gabriel, J.-C. P., Bradley, K. & Grüner, G. (2003). Electronic Detection of Specific Protein Binding Using Nanotube fet Devices, *Nano Lett.* 3: 459–463.

Wu, H., Wang, J., Kang, X., Wang, C., Wang, D., Liu, J., Aksay, I. A. & Lin, Y. (2009). Glucose biosensor based on immobilization of glucose oxidase in platinum nanoparticles/graphene/chitosan nanocomposite film, *Talanta* 80: 403–406.

Yamamoto, Y., Ohno, Y., Maehashi, K. & Matsumoto, K. (2009). Noise Reduction of Carbon Nanotube Field-Effect Transistor Biosensors by Alternating Current Measurement, *Jpn. J. Appl. Phys.* 48: 06FJ01.

Yamamoto, Y., Ohno, Y., Maehashi, K. & Matsumoto, K. (2010). Electrical Detection of Negatively Charged Proteins Using n-Type Carbon Nanotube Field-Effect Transistor Biosensors, *Jpn. J. Appl. Phys.* 49: 02BD10.

Yang, W., Ratinac, K. R., Ringer, S. P., Thordarson, P., Gooding, J. J. & Braet, F. (2010). Carbon Nanomaterials in Biosensors: Should You Use Nanotubes or Graphene?, *Angew. Chem., Int. Ed.* 49: 2114–2138.

Zheng, G., Patolsky, F., Cui, Y., Wang, W. U. & Lieber, C. M. (2005). Multiplexed electrical detection of cancer markers with nanowire sensor arrays, *Nat. Biotechnol.* 23: 1294–1301.

10

Non-Volatile Resistive Switching in Graphene Oxide Thin Films

Fei Zhuge, Run-Wei Li,* Congli He, Zhaoping Liu and Xufeng Zhou
Ningbo Institute of Materials Technology and Engineering,
Chinese Academy of Sciences,
People's Republic of China

1. Introduction

The semiconductor industry has long been seeking a high-density, high-speed, and low-power memory technology that retains its data even when the power is interrupted (Meijer, 2008). Though static random access memory (SRAM) and dynamic random access memory (DRAM) are very fast, both of them are volatile, which is a huge disadvantage, costing energy and additional periphery circuitry. Si-based Flash memory devices represent the most prominent nonvolatile data memory (NVM) because of their high density and low fabrication costs. However, Flash suffers from low endurance, low write speed, and high voltages required for the write operations. In addition, further scaling, i.e., a continuation in increasing the density of Flash is expected to run into physical limits in the near future. Ferroelectric random access memory (FeRAM) and magnetoresistive random access memory (MRAM) cover niche markets for special applications. One reason among several others is that FeRAM as well as conventional MRAM exhibit technological and inherent problems in the scalability, i.e., in achieving the same density as Flash today. In this circumstance, a renewed nonvolatile memory concept called resistance-switching random access memory (RRAM), which is based on resistance change modulated by electrical stimulus, has recently inspired scientific and commercial interests due to its high operation speed, high scalability, and multibit storage potential (Beck et al., 2000; Lu & Lieber, 2007; Dong et al., 2008). The reading of resistance states is nondestructive, and the memory devices can be operated without transistors in every cell (Lee et al., 2007; Waser & Aono, 2007), thus making a cross-bar structure feasible. A large variety of solid-state materials have been found to show these resistive switching characteristics, including solid electrolytes such as GeSe and Ag_2S (Waser & Aono, 2007), perovskites such as $SrZrO_3$ (Beck et al., 2000), $Pr_{0.7}Ca_{0.3}MnO_3$ (Liu et al., 2000; Odagawa et al., 2004; Liao et al., 2009), and $BiFeO_3$ (Yang et al., 2009; Yin et al., 2010; Li et al., 2010), binary transition metal oxides such as NiO (Seo et al., 2004; Kim et al., 2006; Son & Shin, 2008), TiO_2 (Kim et al., 2007; Jeong et al., 2009; Kwon et al., 2010), ZrO_2 (Wu et al., 2007; Guan et al., 2008; Liu et al., 2009), and ZnO (Chang et al., 2008; Kim et al., 2009; Yang et al., 2009), organic materials (Stewart et al., 2004), amorphous silicon (a-Si) (Jo and Lu, 2008; Jo et al., 2009), and amorphous carbon (a-C) (Sinitskii & Tour, 2009; Zhuge et al., 2010).
In last decades, carbon-based materials have been studied intensively as a potential candidate to overcome the scientific and technological limitations of traditional

semiconductor devices (Rueckes et al., 2000; Novoselov et al., 2004; Avouris et al., 2007). It is worthy mentioning that most of the work on carbon-based electronic devices has been focused on field-effect transistors (Wang et al., 2008; Burghard et al., 2009). Thus, it would be of great interest if nonvolatile memory can also be realized in carbon so that logic and memory devices can be integrated on a same carbon-based platform. Graphene oxide (GO) with an ultrathin thickness (~1 nm) is attractive due to its unique physical-chemical properties. A GO layer can be considered as a graphene sheet with epoxide, hydroxyl, and/or carboxyl groups attached to both sides. GO can be readily obtained through oxidizing graphite in mixtures of strong oxidants, followed by an exfoliation process. Due to its water solubility, GO can be transferred onto any substrates uniformly using simple methods such as drop-casting, spin coating, Langmuir-Blodgett (LB) deposition and vacuum filtration. The as-deposited GO thin films can be further processed into functional devices using standard lithography processes without degrading the film properties (Eda et al., 2008; Cote et al., 2009). Furthermore, the band structure and electronic properties of GO can be modulated by changing the quantity of chemical functionalities attached to the surface. Therefore, GO is potentially useful for microelectronics production. Considering that although a large variety of solid-state materials have been found to show resistive switching characteristics, none of them can fully meet the requirements of RRAM applications, exploration of new storage media is still a key project for the development of RRAM. Therefore, in this chapter, the resistive switching effect in GO thin films has been investigated and the possibility of GO used as a potential storage medium for RRAM has been discussed.

2. Experiments

The oxidation of graphite was carried out following the Hummer's method (Hummers & Offeman, 1958). In a typical procedure, KNO_3 (1.2 g) and graphite (1.0 g) was added into 46 mL of concentrated H_2SO_4 (98%) under stirring. After 10 mins 6.0 g of $KMnO_4$ was added slowly. The mixture was then heated to 35 °C and stirred for 6 hours. Subsequently, 80 mL of water was added dropwise under vigorous stirring, resulting in a quick rise of the temperature to ~80 °C. The slurry was further stirred at this temperature for another 30 mins. Afterwards 200 mL of water and 6 mL of H_2O_2 solution were added in sequence to dissolve insoluble manganese species. The resulting graphite oxide suspension was washed repeatedly by a large amount of water until the solution pH reached a constant value at ~4.0, and finally the suspension was diluted to 600 mL with water. 200 mL of the diluted graphite oxide suspension was transferred into a conical beaker with a volume of 500 mL. Then the suspension was gently shaken in a mechanical shaker at a speed of 160 rpm for ~6 hours. To remove the small amount of unexfoliated particles, the resulting viscous suspension was centrifuged at 2,000 rpm for 10 mins, producing a brown, homogeneous colloidal suspension of GO sheets. The colloidal suspension could be further concentrated by centrifugation at 8,000 rpm. Then GO thin films of ~30 nm in thickness were prepared at room temperature by the vacuum filtration method. 50 g of GO suspension with a concentration of 6 mg/L was filtered through a cellulose ester membrane to achieve uniform GO thin films. The film thickness could be well controlled by tuning the GO concentration or filtration volume. The as-filtered GO flakes were then transferred from the filter membrane onto commercial $Pt/Ti/SiO_2/Si$ substrates.

Atomic force microscopy (AFM) characterization was conducted by a Veeco Dimension 3100V scanning probe microscope at ambient conditions using a tapping or conductive-

AFM (CAFM) mode. Scanning electron microscopy (SEM) images were taken by a Hitachi S-4800 Field Emission Scanning Electron Microscope. Specially, a low accelerating voltage of 1.0 kV and high current of 20 μA were applied for the graphene oxide sheets on silicon substrates. In order to measure the electrical properties, metal (Cu, Ag, and Au etc.) top electrodes with a thickness of 200 nm and diameter of 100 μm were deposited at room temperature by electron beam evaporation with an in-situ metal shadow mask. The *I-V* characteristics of metal/GO/Pt structures were measured at room temperature by Keithley 4200 semiconductor characterization system with voltage sweeping mode. During the measurement, a bias voltage was applied between the top (Cu, Ag, and Au etc.) and bottom (Pt) electrodes with the latter being grounded. The schematic structure of the sandwiched devices is depicted in Fig. 1(a). In the local leakage current measurement by CAFM, the conductive tip was grounded and directly touched the GO films, as shown in Fig. 1(b). The resistances in the low resistance state and high resistance state of the metal/GO/Pt structures were measured as a function of temperature by a physical property measurement system (PPMS, Quantum Design).

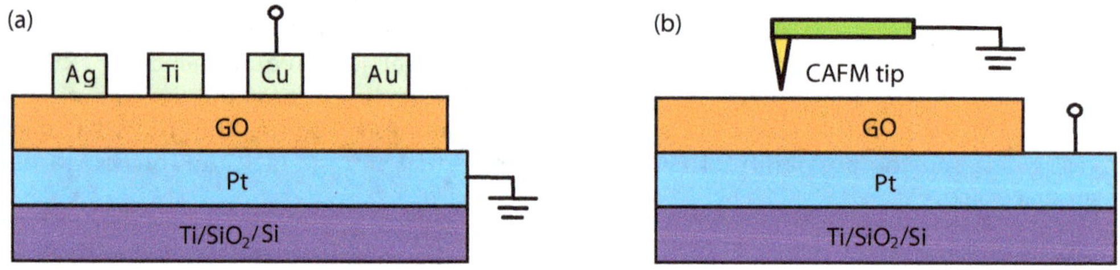

Fig. 1. Schematic configuration (a) of the metal/GO/Pt sandwiched structures, and (b) for the conductive-AFM measurements.

3. Results and discussions

3.1 Morphology of GO sheets and thin films

A typical SEM image of the GO sheets is shown in Fig. 2(a). As can be seen, the sheets have a lateral size, dominantly, in the order of 100 μm, and some of them even possess a size as large as 200 μm. To the best of our knowledge, there have been no reports on GO sheets with such a huge size. Besides the large sheets, fragments with sizes of only a few micrometres can also be observed in a small amount. These fragments might be derived from micrometre-sized graphite flakes or by inevitable breakage during the delamination process. Figure 2(b) displays an AFM image of a GO sheet which has a size far beyond the AFM scan range. The height profile denoted in Fig. 2(b) demonstrates a mean thickness of ~1 nm, indicative of a unilamellar nature of the sheet. It is worth mentioning that the sizes of the exfoliated GO sheets are closely associated with those of the starting graphite flakes. For example, if the graphite flakes of 500 mesh size (~25 μm in mean size) were used, the exfoliated GO sheets had a lateral size around 15 μm. This result indicates that the sheet sizes can be easily controlled through the size selection of the graphite material. Figure 2(c) shows an AFM image of a GO thin film of ~30 nm in thickness deposited at room temperature by the vacuum filtration method. From Fig. 2(c), we see that the as-grown GO film is not smooth and there are a large amount of protuberances on the film surface. These observed protuberances are most likely to be the gauffers of GO sheets formed during the film preparation process (Zhou & Liu, 2010).

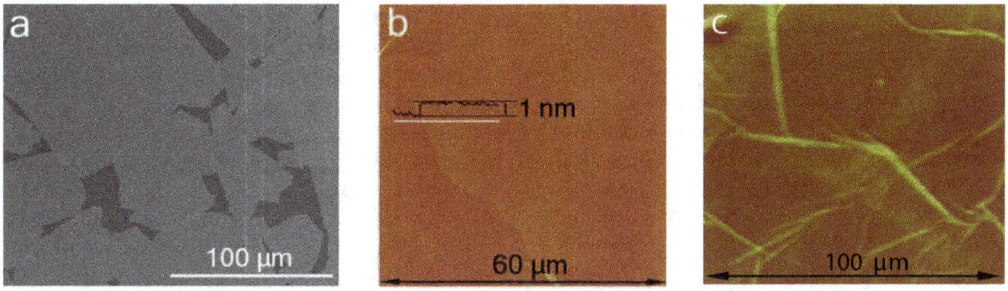

Fig. 2. (a) SEM and (b) AFM images of GO sheets that were deposited on a Si substrate by Langmuir-Blodgett assembly at a constant surface pressure of 10 mN m^{-1}. (c) AFM image of GO thin films of ~30 nm in thickness prepared at room temperature by the vacuum filtration method

3.2 Nonvolatile resistive switching in GO thin films

Metal/GO/Pt sandwiched structures show the bipolar resistive switching behavior (He et al., 2009). The *I-V* characteristics of the Cu/GO/Pt memory cell are studied by dc voltage sweep measurements to evaluate the memory effects of the obtained devices. Figure 3(a) plots a typical *I-V* curve of a Cu/GO/Pt cell. Figure 3(b) shows the same *I-V* curve in a semilogarithmic scale. During the measurement, the voltage is swept in a sequence of 0 V → 1 V → 0 V → -1 V → 0 V with a sweeping step of 0.01 V. While increasing the positive voltage steadily, the current jumps abruptly at a voltage value of about 0.8 V. The device switches from the high resistance state (HRS or Off state) to a low resistance state (LRS or On state), which is called the "Set" process. A current compliance (10 mA in this work) is usually needed during the Set process to prevent the sample from a permanent breakdown. By sweeping the voltage from 1 V to -0.4 V, the device holds on the LRS, and starts switching from the LRS to the HRS ("Reset" process) from -0.4 V. At the voltage of -0.75 V, the cell recovers to the HRS, and holds until next Set process. The *I-V* characteristics exhibit a typical bipolar switching behavior. The device yield of the Cu/GO/Pt structure is more than 50%.

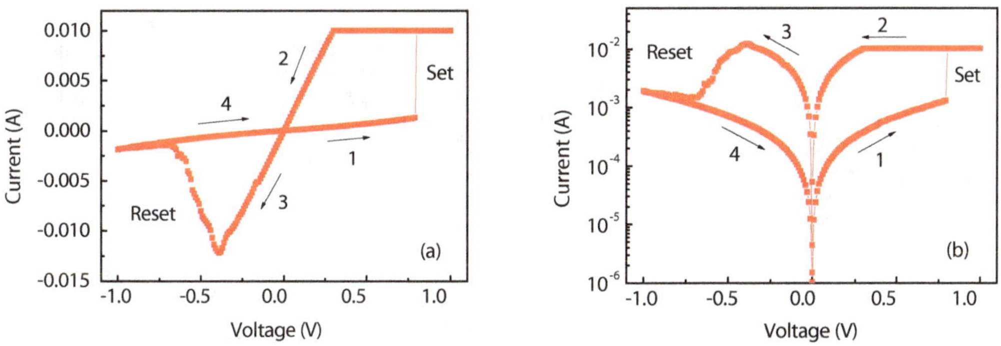

Fig. 3. Typical *I-V* characteristics of the Cu/GO/Pt sandwiched structure shown in (a) linear scale, and (b) semilogarithmic scale. The voltage is swept in the direction as follows: 0 V → 1 V → 0 V → -1 V → 0 V

In order to investigate the endurance performance of the Cu/GO/Pt memory device, cyclic switching operations are conducted. Figure 4(a) shows the evolution of resistance of the two well-resolved states in 100 cycles. The statistical distribution of the resistances in HRS and LRS is shown in Fig. 4(b). The resistance values are read out at -0.1 V in each dc sweep. Although the resistance values of both HRS and LRS show some fluctuations, the On/Off ratios are about 20 without any obvious degradation within 100 cycles. The endurance measurements ensured that the switching between On and Off states is controllable, reversible, and reproducible. After the device was switched On or Off, no electrical power was needed to maintain the resistance within a given state. The distribution of the programming voltages (V_{Set} and V_{Reset}), another critical parameter to evaluate the memory device, is shown in Fig. 4(c). Figure 4(d) shows The statistical distribution of the programming voltages. As can be seen, V_{Set} and V_{Reset} distribute in a range of 0.3 V to 1 V and -0.3 V to -0.9 V, respectively. The switching threshold voltages of the Cu/GO/Pt sandwiched structure are lower than those of most reported RRAM devices (Szot et al., 2006; Tsubouchi et al., 2007; Guan et al., 2007; Fujiwara et al., 2008; Li et al., 2008).

The retention performance of the Cu/GO/Pt memory cell at room temperature is shown in Fig. 5. The device is switched On or Off by dc voltage sweeping. Then the resistance of LRS or HRS is read out at a reading voltage of 0.1 V. The reading of the resistance state is nondestructive, and no electrical power is needed to maintain the resistance within a given state (LRS or HRS). As can be seen in Fig. 5, the LRS and HRS resistances are kept stable for more than 10^4 s, indicating that the memory device is nonvolatile and stable at room temperature.

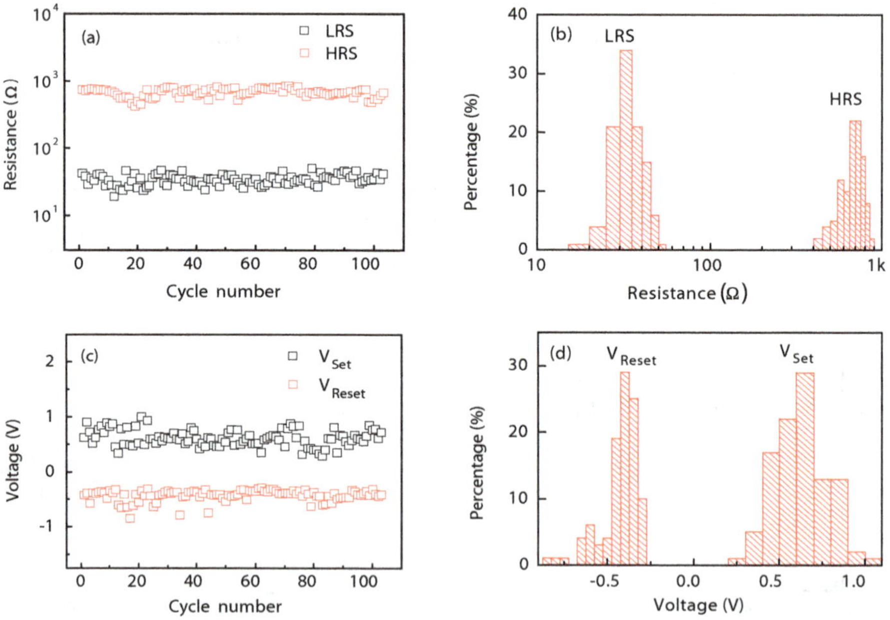

Fig. 4. (a) Endurance performance of the Cu/GO/Pt memory cell. The resistance values are read out at 0.1 V in each sweep. (b) Statistical distribution of the resistances in HRS and LRS. (c) Evolution of V_{Set} and V_{Reset} of the Cu/GO/Pt device as a function of the switching cycle. (d) Statistical distribution of V_{Set} and V_{Reset}

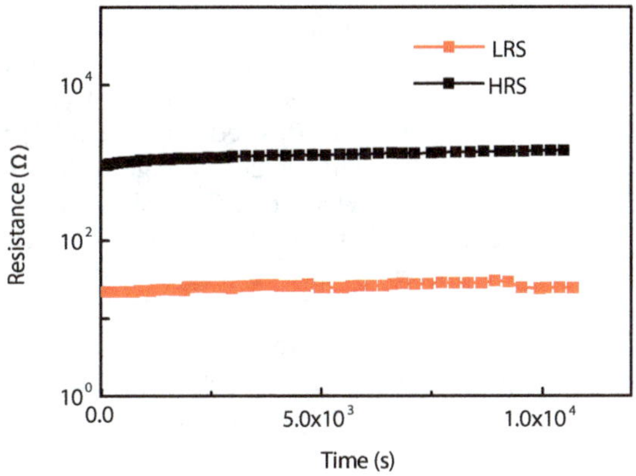

Fig. 5. Retention performance for the LRS and HRS of the Cu/GO/Pt memory cell at room temperature

To understand the conduction mechanisms of the Cu/GO/Pt memory cell, the *I-V* curves are replotted in a log-log scale. Figures 6(a) and (b) show the logarithmic plot and linear fitting of the previous *I-V* curve for the positive and negative voltage sweep regions, respectively. As shown in Fig. 6, the *I-V* curve of LRS exhibits a linearly Ohmic behavior with a slope of ~1. However, the conduction mechanisms of HRS are much more complicated. Fitting results for HRS suggest that the charge transport behavior is in good agreement with a trap-controlled space charge limited current (SCLC), which consists of two portions: the Ohmic region ($I \propto V$) and the Child's law region ($I \propto V^2$) (Strukov et al., 2008; Kim et al., 2007; Son & Shin, 2008; Chen et al., 2008). The different conduction behaviors in HRS and LRS suggest that the high conductivity in On-state cell is likely to be a confined effect rather than a homogenously distributed one (Zhuge et al., 2010).

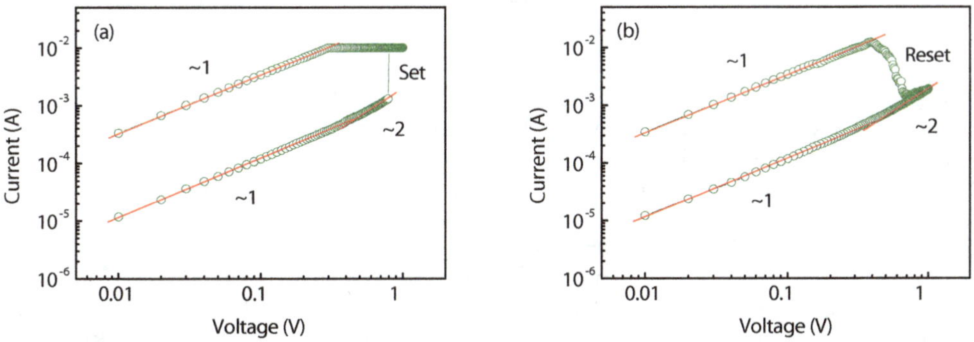

Fig. 6. *I-V* curves of the the Cu/GO/Pt memory cell plotted in log-log scale and the linear fitting results in both LRS and HRS: (a) positive voltage region and (b) negative voltage region. Also shown are the corresponding slopes for different portions

3.3 Top electrode material dependence of resistive switching of metal/GO/Pt sandwiches

Various metals (Cu, Ag, Au, and Ti) as top electrodes are deposited on GO films and the *I-V* characteristics of the corresponding metal/GO/Pt memory cells are investigated. The similar bipolar switching behaviors are observed in these four types of metal/GO/Pt sandwiched structures. Actually, a Forming process, which is the first transition from fresh state to low resistance state, is always needed before such sandwiches show good resistive switching characteristics. However, this is not desired in practical applications because the Forming voltage is usually much higher than the Set voltage of the device. Most materials, for example, NiO, TiO_2, ZnO, CuO, and $SrTiO_3$ etc. (Kinoshita et al., 2006; Do et al., 2008; Chang et al., 2008; Fujiwara et al., 2008; Oligschlaeger et al., 2006), have already been reported to be in need of Forming process. It is necessary to find some solutions to overcome this difficulty. Kinoshita et al. developed a NiO_y film (y=0.88±0.05) free from Forming process, where a grain structure was found with conductive atomic force microscopy and most current flows through grain boundaries to obviate the requirement of Forming (Kinoshita et al., 2006). Herein, the effect of top electrode materials on the Forming process in GO thin films is investigated.

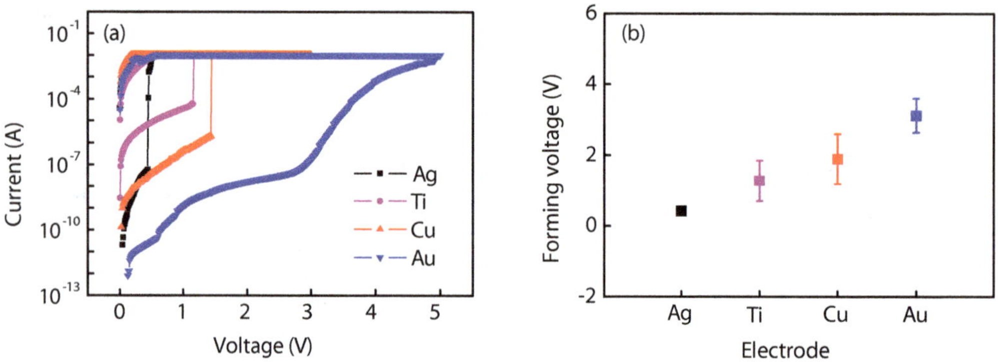

Fig. 7. (a) *I-V* curves of Forming processes of GO-based memory cells using different metals (Ag, Ti, Cu, and Au) as top electrodes. (b) Forming voltages of different metal/GO/Pt sandwiches

Figure 7(a) shows the *I-V* properties of Forming processes of Ag/GO/Pt, Ti/GO/Pt, Cu/GO/Pt, and Au/GO/Pt sandwiched structures. The Forming voltages of different metal/GO/Pt memory cells are shown in Fig. 7(b). As can be seen from Fig. 7, the Ag/GO/Pt structures have the lowest Forming voltages of about 0.4 V, while the Au/GO/Pt sandwiches show the highest Forming voltages of about 3 V. The Forming voltage of the metal/GO/Pt memory cell varies as follows: $V_{Ag}<V_{Ti}<V_{Cu}<V_{Au}$. It has been reported that solid electrolyte memory devices based on silver ions generally show low switching voltages (Kund et al., 2005; Schindler et al., 2007; Tsunoda et al., 2007). Banno et al. further clarified that the switching voltage for turn-on in RRAM devices based on solid electrolyte such as $Cu_{2-a}S$ and Ta_2O_5 could be strongly affected by the ion flux which is proportional to the product of the diffusion coefficient and the concentration of ions (Banno et al., 2008). Therefore, we can infer that a large ion diffusion coefficient can lead to a large ion flux and thus a low switching voltage. Govindaraj et al. have reported that Ag has a large diffusion coefficient in oxide films (Govindaraj et al., 2001). The Ag/ZrO_2/Au structure shows low switching voltages of <1 V without any Forming process (Li et al., 2010). In our case, it is likely that Ag also has a large diffusion coefficient in GO thin films, resulting in low Forming voltages for Ag/GO/Pt memory cells. The high Forming voltages for Au/GO/Pt structures are likely due to gold's chemical inertness. Compared to other top electrode metals (i.e. Ag, Ti, and Cu), Gold is very difficult to be oxidized to the ions. High voltages are necessary for the activation of Au/GO/Pt memory cells.

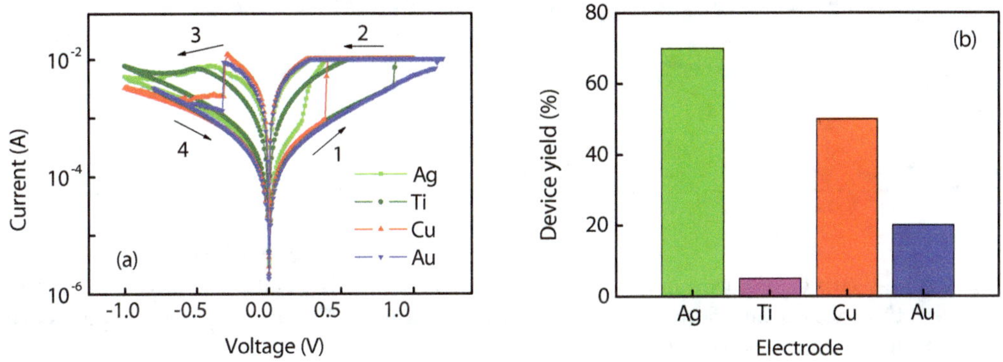

Fig. 8. (a) Typical *I-V* characteristics of GO-based memory cells using different metals (Ag, Ti, Cu, and Au) as top electrodes shown in semilogarithmic scale. (b) Device yields of different metal/GO/Pt sandwiches

Figure 8(a) shows the typical *I-V* characteristics of Ag/GO/Pt, Ti/GO/Pt, Cu/GO/Pt, and Au/GO/Pt sandwiched structures after Forming processes. Similar to the Forming processes, the Ag/GO/Pt structures have the lowest Set voltages of about 0.25 V, and the Au/GO/Pt sandwiches show the highest Set voltages of about 1.2 V. Figure 8(b) shows the top electrode dependence of device yields of metal/GO/Pt memory cells. As can be seen from Fig. 8(b), the device yield (Y) of the metal/GO/Pt memory cell varies as follows: $Y_{Ag}>Y_{Cu}>Y_{Au}>Y_{Ti}$. The Ag/GO/Pt structure has the highest device yield of about 70%. The strong top electrode material dependence of the switching characteristics is helpful for speculating about the resistive switching mechanism in GO thin films. It will be discussed later.

3.4 GO film thickness dependence of resistive switching of metal/GO/Pt sandwiches

Figure 9(a) shows the *I-V* characteristics of Forming processes of Cu/GO/Pt memory cells using GO films with different thicknesses, i.e. 15 nm, 30 nm, 60 nm, and 90 nm. The GO film thickness dependence of Forming voltages of Cu/GO/Pt memory cells is shown in Fig. 9(b). Obviously, the voltage required for the initial Forming process increases with the thickness of GO, indicating that the electric field inside the bulk GO is the controlling factor of the Forming. The similar experiment results have been also obtained in some oxides, for example, Fe_2O_3 (Inoue et al., 2008). Therefore, to reduce the Forming voltage or even entirely remove the Forming process of the metal/GO/Pt memory cell, an appropriate thickness of the GO thin film should be considered.

The resistances of the high resistive state and low resistive state after the Forming process for Cu/GO/Pt memory cells are shown in Fig. 10, as a function of the thickness of GO. No obvious variation can be observed for both HRS resistance and LRS resistance as the GO thickness changes from 15 nm to 90 nm. The thickness-insensitive property of HRS resistance and LRS resistance indicates that once the Forming process is completed, it is unlikely that the bulk region of GO contributes to the resistive switching effectively. This also underpins that the switching phenomenon occurs at the interface rather than the bulk region.

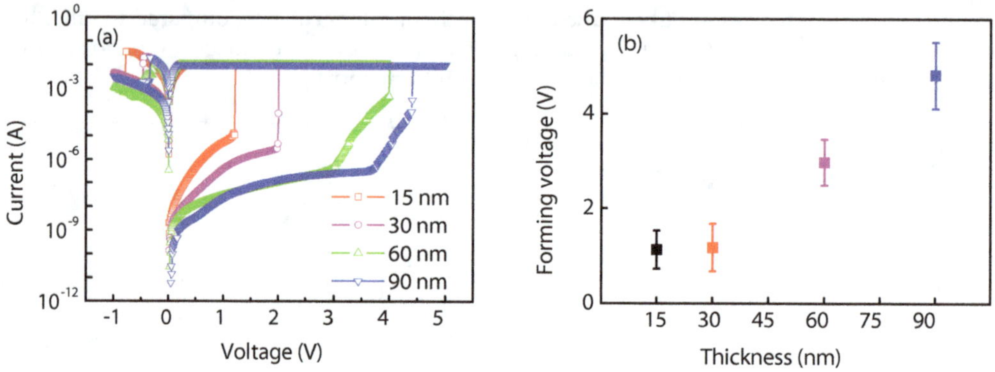

Fig. 9. (a) *I-V* curves of Forming processes of Cu/GO/Pt memory cells using GO films with different thicknesses, i.e. 15 nm, 30 nm, 60 nm, and 90 nm. (b) GO film thickness dependence of Forming voltages of Cu/GO/Pt memory cells

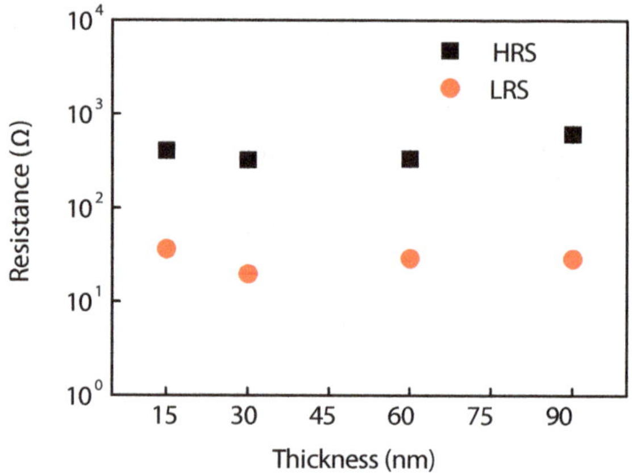

Fig. 10. Resistances of the high resistive state and low resistive state for Cu/GO/Pt memory cells, plotted against the thickness of GO

Fig. 11. Gas bubbles under the electrodes formed during the Forming processes of (a) the Pt/single crystalline $SrTiO_3$/Pt memory cell (Szot et al., 2006), (b) Cu/GO/Pt memory cell with a GO thickness of 90 nm, and (c) Au/GO/Pt memory cell with a GO thickness of 30 nm

Szot et al. investigated the resistive switching of the Pt/single crystalline $SrTiO_3$/Pt memory cell and found gas bubbles under the anode. This phenomenon was explained by the polyfilamentary models proposed in the context of amorphous-oxide films (Dearnaley et al., 1970; Rakhshani et al., 1976; Ray et al., 1990), namely the preferential diffusion of oxygen and solid-state electrolytic processes along a discrete set of filaments, which give rise to a located decomposition and subsequent accumulation of oxygen gas under the anode. In our case, such gas bubbles are also observed under the top electrodes for the Cu/GO/Pt memory cell with a GO thickness of 90 nm and Au/GO/Pt memory cell with a GO thickness of 30 nm. From Fig. 7, we see that for the Au/GO/Pt cell, the Forming voltage is more than 4 V. As can be seen from Fig. 9, the Cu/GO/Pt cell with a GO thickness of 90 nm needs a high Forming voltage of about 5V. Therefore, we infer that as the high Forming voltage is applied to the GO-based memory cell, a large amount of heat is generated in a confined region, resulting in an increase of temperature. Then, the high temperature in the confined region may cause the decomposition of epoxide, hydroxyl, and/or carboxyl groups in GO, resulting in water steam and CO_2 (Zhou et al., 2006).

3.5 Current compliance dependence of resistive switching of metal/GO/Pt sandwiches

A current compliance is usually needed during the Forming and Set processes to prevent the sample from a permanent breakdown. We find that the current compliance is another critical parameter to determine the resistive switching characteristics of metal/GO/Pt memory cells. Figure 12(a) shows the typical *I*-*V* characteristics of the Au/GO/Pt device with different current compliance (0.02 A, 0.04 A, and 0.07 A). Clearly, the Reset current (maximum current level before the Reset process) increases with increasing compliance current. As mentioned above, the different conduction behaviors in high and low resistance states suggest that the high conductivity in On-state cell is likely to be a confined effect rather than a homogenously distributed one. It is likely due to the formation of conducting filaments in GO (Yang et al., 2009). It is considered that stronger filaments with a higher density are formed at a larger current compliance (Rohde et al., 2005). The Reset process is also considered to be due to the rupture of the filaments due to the heat generated by the large current flow (Rohde et al., 2005). Thus, it can be imagined that a larger Reset current is necessary for the Reset process at the larger current compliance. The dependence of the Reset current and resistance in the low resistance state on current compliance applied in the Set process is shown in Fig. 12(b). We see that the Reset current has an approximate linear relationship to the current compliance. Thus, the power needed for the WRITE and ERASE operations in RRAM can be modulated by controlling the current compliance. As can be seen from Fig. 12(b), the resistance in the low resistance state decreases with increasing current compliance. Therefore, different resistance states can be achieved using different current compliances in the program process, showing good potential for multilevel storage (Wang et al., 2010). Figure 12(c) shows the current compliance dependence of the Set and Reset voltages. The Set voltage shows weak dependence on current compliance, whereas the Reset voltage increases obviously with increasing current compliance applied in the Set process.

3.6 Possible mechanisms of resistive switching in GO thin films

i. Filamentary model for metal/GO/Pt sandwiches

As mentioned above, the strong top electrode material dependence of the switching characteristics is helpful for speculating about the resistive switching mechanism in GO thin

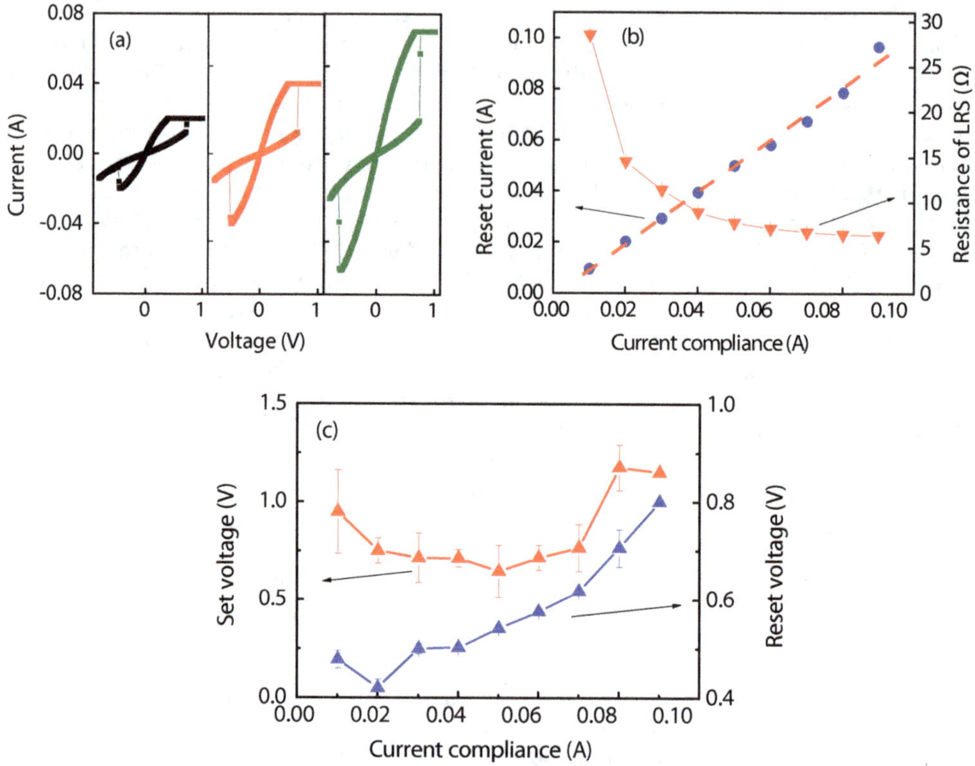

Fig. 12. (a) Typical *I-V* characteristics of the Au/GO/Pt device with different current compliance (0.02 A, 0.04 A, and 0.07 A). (b) Current compliance dependence of the Reset current and resistance of the low resistance state of the Au/GO/Pt device. (c) Current compliance dependence of the Set and Reset voltages of the Au/GO/Pt device.

films. Tokunaga et al. investigated the RS effect at the interface between metal electrodes (Pt, Au, Ag, Al, Ti, and Mg) and atomically flat cleaved (001) surfaces of $La_{1-x}Sr_{1+x}MnO_4$ (x=0−1.0) single crystals (Tokunaga et al., 2006). Hysteretic *I-V* characteristics were observed in the junctions for Mg, Al, and Ti, which have relatively shallow work functions (Φ). In our case, Ti, Cu, and Ag have a lower Φ of 3.98 eV, 4.35 eV, and 4.26 eV, respectively, whereas Au has a higher one of 5.1 eV. It indicates that the occurrence of the resistive switching in metal/GO/Pt structures is irrelevant to Φ of the top electrodes. Liao and co-workers studied the resistive switching characteristics of metal/$Pr_{0.7}Ca_{0.3}MnO_3$ (PCMO)/Pt devices and found that devices with top electrode made of Al, Ti, and Ta exhibit a bipolar resistive switching, but those with top electrode made of Pt, Ag, Au, and Cu do not (Liao et al., 2009). The resistive switching was attributed to a thin metal-oxide layer formed at the interface between the former group of top electrode and PCMO. In our work, the top electrode material includes both reactive metals such as Ti and inert metals such as Au. Therefore, the observed resistive switching is not due to the formation of a metal-oxide layer at the interface between the top electrode and GO. The strong top electrode material dependence of the resistive switching characteristics of GO thin films suggests that the resistive switching effect may be attributed to the formation/rupture of metal filaments due to the diffusion of the top electrodes under a bias voltage. Briefly, a positive voltage (>V_{Set}) on the top electrodes generates a high electric field that drives metal (e.g., Ag) ions into the GO matrix and form conducting filaments inside the GO layer, and the device reaches the On state. After the Set process, the device retains the On state unless a sufficient voltage of opposite polarity (<V_{Reset}) is applied and the electrochemical dissolution of the metal filaments Resets the device, and the Off state is finally reached again (Jo et al., 2008; Zhuge et al., 2010).

To verify the physical nature of the resistive switching effect of the metal/GO/Pt structures, resistances in the low and high resistance states of the Cu/GO/Pt device are measured as a function of temperature. Figure 13(a) shows the typical metallic behavior of the resistance in the On state. In contrast, the resistance in the Off state shows a semiconducting behavior as displayed in the inset of Fig. 13(a). The metallic conducting behavior in the low resistance state indicates the formation of conducting filaments in the GO films. The temperature dependence of metallic resistance can be written as $R(T) = R_0[1 + a(T - T_0)]$, where R_0 is the resistance at temperature T_0, and a is the temperature coefficient of resistance. By choosing T_0 as 300 K, the temperature coefficient of resistance of the filaments is calculated to be 1.7×10^{-3} K^{-1}, which is similar to the value 2.5×10^{-3} K^{-1} for high-purity copper nanowires of diameter ≥15 nm (Bid et al., 2006), confirming that the filaments are composed of Cu in metallic states due to the diffusion of the top electrode under a bias voltage. The discrepancy of the temperature coefficient of resistance is attributed to inevitable defects in the Cu filaments, since the presence of defects can reduce the temperature coefficient of resistance by shortening the mean free path of electrons (Bid et al., 2006).

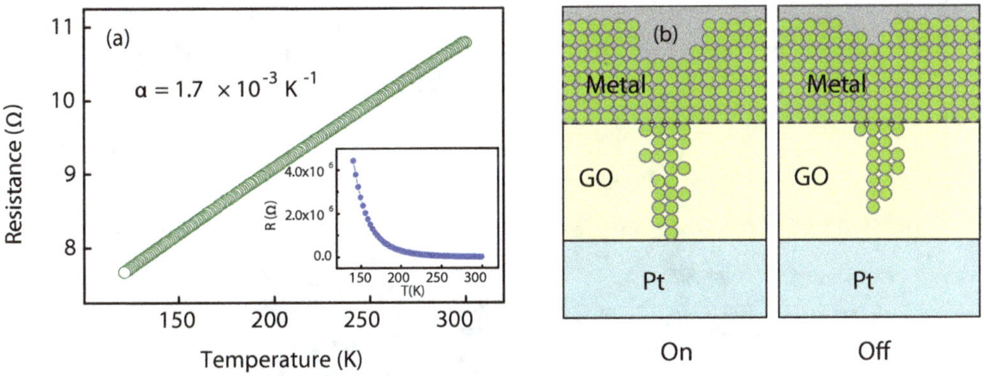

Fig. 13. (a) Temperature dependence of the resistance in the low resistance sate of the Cu/GO/Pt device. The inset shows the temperature dependence of the resistance in the high resistance state. Also shown is the temperature coefficient of resistance (a) of the conducting filaments. (b) A schematic diagram for the mechanism of resistive switching effects in metal/GO/Pt devices.

ii. Desorption/absorption of oxygen-related group on the GO sheets

Figure 14(a) shows the typical local I-V characteristics of GO films by the conductive atomic force microscopy measurements in which the AFM tip is in perpetual contact with the sample. The obvious bipolar resistive switching effect is also observed though the threshold voltages are higher probably due to the nonideal contact between the AFM probe and GO films as well as very small contact area compared to the conventional *I-V* characterization (Yin et al., 2010). In this case, we intend to attribute the observed resistive switching to the desorption/absorption of oxygen-related groups on the GO sheets. As shown in Fig. 14(c), a GO layer can be considered as a graphene sheet with epoxide, hydroxyl, and/or carboxyl groups attached to both sides, and physical properties of GO can be modulated by those chemical functionalities on the surface (Yan et al., 2009). When there are an amount of epoxide, hydroxyl, and carboxyl groups on the GO surface, the conductance of GO is assumed to be low due to sp^3 bonding feature. As a negative voltage bias is applied on the bottom electrode (i.e., Pt), some oxygen-related functional groups inside the GO layers are removed, resulting in an amount of sp^2 bonds. The conductance of GO becomes higher due to an increase in the concentration of interlayer π electrons. As a result, the sample is switched to the low resistance state (On state). While sweeping the voltage to certain positive values, the oxygen-related groups diffuse toward GO sheets and attach to them again. Correspondingly, the sample returns to the high resistance state (Off state). Therefore, as schematically shown in Fig. 14(b), the desorption and absorption of the oxygen-related functional groups on the GO sheets correspond to the low and high resistive states, respectively.

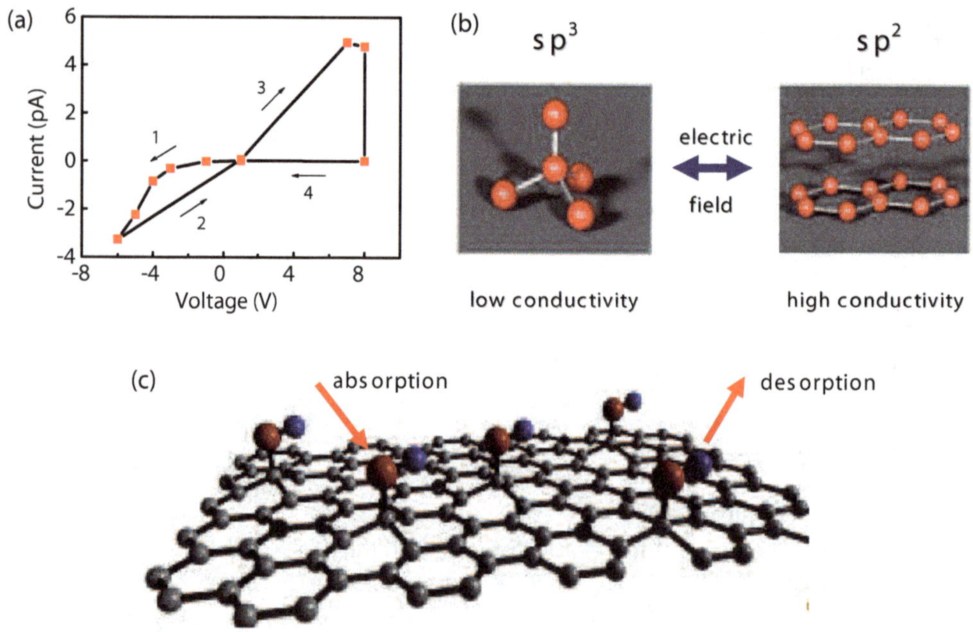

Fig. 14. (a) Typical local *I-V* characteristics of GO films measured by conductive atomic force microscopy. The voltage is swept in the direction as follows: 0 V → -6 V → 0 V → 8 V → 0 V. (b) Insulating carbon can be switched to conductive sp^2-rich carbon and conductive carbon can be switched to sp^3-rich carbon by applying appropriate electric fields (Kreupl et al., 2008). (c) Schematic of the desorption and aborption of hydroxyl or similar groups on the GO sheets (Echtermeyer et al., 2008).

4. Conclusion

Reliable and reproducible resistive switching behaviors are observed in GO thin films. The metal/GO/Pt structures exhibit an On/Off ratio of about 20, switching threshold voltages of less than 1 V, good retention and endurance performances. The Ag/GO/Pt structure has a low Forming voltage likely due to a large Ag ion diffusion coefficient in GO, whereas the Au/GO/Pt structure shows a high Forming voltage likely due to gold's chemical inertness. The Forming voltage increases with the thickness of GO, indicating that the electric field inside the bulk GO is the controlling factor of the Forming. The resistances of the high and low resistance states show weak dependence on the film thickness, demonstrating that the resistive switching occurs in a confined region. It is found that the resistance of the On state decreases with current compliance, whereas the Reset current and Reset voltage increase with current compliance. Therefore, by adjusting current compliance, the power consumption of RRAM can be modulated and multilevel storage can be realized. The resistive switching is also observed in micro-regions by conducting atomic force microscope. The dominant conduction mechanisms of low and high resistance states are Ohmic behavior and trap-controlled space charge limited current, respectively. Based on the experimental results, the possible physical mechanisms of resistive switching behaviors in GO thin films can be understood by considering the desorption/absorption of oxygen-related groups on the GO sheets as well as the formation/rupture of conducting filaments.

This work was supported by State Key Research Program of China (973 Program), National Natural Science Foundation of China, Zhejiang and Ningbo Natural Science Foundations, Chinese Academy of Sciences (CAS), Zhejiang Qianjiang Talent Project, and State Key Lab of Silicon Materials (China).

5. References

*Corresponding author. Electronic mail: runweili@nimte.ac.cn.

Avouris P.; Chen Z. H.; & Perebeinos V. (2007). Carbon-based electronics. *Nature Nanotechnology*, Vol. 2, No. 10, 605-615, ISSN: 1748-3387

Banno N.; Sakamoto T.; Iguchi N.; Sunamura H.; Terabe K.; Hasegawa T. & Aono M. (2008). Diffusivity of Cu ions in solid electrolyte and its effect on the performance of nanometer-scale switch. *IEEE Transactions on Electron Devices*, Vol. 55, No. 11, 3283–3287, ISSN: 0018-9383

Beck A.; Bednorz J. G.; Gerber Ch.; Rosse C.L. & Widmer D. (2000). Reproducible switching effect in thin oxide films for memory applications. *Applied Physics Letters*, Vol. 77, No. 1, 139-141, ISSN: 0003-6951

Bid A.; Bora A. & Raychaudhuri A. (2006). Temperature dependence of the resistance of metallic nanowires of diameter ≥15 nm: Applicability of Bloch-Gruneisen theorem. Physical Review B, Vol. 74, No. 3, 035426, ISSN: 1098-0121

Burghard M.; Klauk H.; & Kern K. (2009). Carbon-Based Field-Effect Transistors for Nanoelectronics. *Advanced Materials*. Vol. 21, No. 25-26, 2586-2600, ISSN: 0935-9648

Chang W. Y.; Lai Y. C.; Wu T. B.; Wang S. F.; Chen F. & Tsai M. J. (2008). Unipolar resistive switching characteristics of ZnO thin films for nonvolatile memory applications. *Applied Physics Letters*. Vol. 92, No. 2,022110, ISSN: 0003-6951

Chen X. M.; Wu G. H. & Bao D. H. (2008). Resistive switching behavior of Pt/$Mg_{0.2}Zn_{0.8}O$/Pt devices for nonvolatile memory applications. *Applied Physics Letters*, Vol. 93, No. 9, 093501, ISSN: 0003-6951

Cote L. J.; Kim F. & Huang J. X. (2009). Langmuir−Blodgett Assembly of Graphite Oxide Single Layers. *Journal of the American Chemical Society,* Vol. 131, No. 3, 1043-1049, ISSN: 0002-7863

Dearnaley, G.; Morgan, D. V. & Stoneham, A. M. (1970). A model for filament growth and switching in amorphous oxide films. *Journal of Non-Crystalline Solids,* Vol. 4, 593–612, ISSN: 0022-3093

Dearnaley, G.; Stoneham, A. M. & Morgan, D. V. (1970). Electrical phenomena in amorphous oxide films. *Reports on Progress in Physics,* Vol. 33, No. 3, 1129–1191, ISSN: 0034-4885

Do Y. H.; Kwak J. S.; Hong J. P.; Jung K. & Im H. (2008). Al electrode dependent transition to bipolar resistive switching characteristics in pure TiO_2 films. *Journal of Applied physics,* Vol. 104, No. 11, 114512, ISSN: 0021-8979

Dong Y.; Yu, M G.; McAlpine C.; Lu W. & Lieber C. M. (2008). Si/α-Si core/shell nanowires as nonvolatile crossbar switches. *Nano Letters,* Vol. 8, No. 2, 386-391, ISSN: 1530-6984

Echtermeyer T. J.; Lemme M. C.; M. Baus et al. (2007). A grapheme-based electrochemical switch. arXiv:0712.2026v1

Eda G.; Fanchini G. & Chhowalla M. (2008). Large-area ultrathin films of reduced graphene oxide as a transparent and flexible electronic material. *Nature Nanotechnology,* Vol. 3, No. 5, 270-274, ISSN: 1748-3387

Fujiwara K.; Nemoto T.; Rozenberg M. J.; Nakamura Y. & Takagi H. (2008). Resistance Switching and Formation of a Conductive Bridge in Metal/Binary Oxide/Metal Structure for Memory Devices. *Japanese Journal of Applied Physics,* Vol. 47, No. 8, 6266-6271, ISSN: 0021-4922

Govindaraj R.; Kesavamoorthy R.; Mythili R. & Viswanathan B. (2001). The formation and characterization of silver clusters in zirconia. *Journal of Applied Physics,* Vol. 90, No. 2, 958–963, ISSN : 0021-4922

Guan W. H.; Long S. B.; Jia R.& Liu M. (2007). Nonvolatile resistive switching memory utilizing gold nanocrystals embedded in zirconium oxide. *Applied Physics Letters,* Vol. 91, No. 6, 062111, ISSN: 0003-6951

Guan W. H.; Liu M.; Long S. B.; Liu Q. & Wang W. (2008). On the resistive switching mechanisms of Cu/ZrO_2:Cu/Pt. *Applied Physics Letters* , Vol. 93, No. 22, 223506, ISSN: 0003-6951

He C. L.; Zhuge F.; Zhou X. F.; Li M.; Zhou G. C.; Liu Y. W.; Wang J. Z.; Chen B.; Su W. J.; Liu Z. P.; Wu Y. H.; Cui P. & Li R. W. (2009). Nonvolatile resistive switching in grapheme oxide thin films. *Applied Physics Letters,* Vol. 95, No. 23, 232101, ISSN: 0003-6951

Hummers W. S. & Offeman R. E. (1958). Preparation of Graphitic Oxide. *Journal of the American Chemical Society,* Vol. 80, No. 6,1339, ISSN: 0002-7863

Inoue I. H.; Yasuda S.; Akinaga H. & Takagi H. (2008). Nonpolar resistance switching of metal/binary-transition-metal oxides/metal sandwiches: Homogeneous/inhomogeneous transtion of current distribution. *Physical Review B,* Vol. 77, No. 3, 035105, ISSN: 1098-0121

Jeong D. S.; Schroeder H. & Waser R. (2009). Abnormal bipolar-like resistance change behavior induced by symmetric electroforming in $Pt/TiO_2/Pt$ resistive switching cells. *Nanotechnology,* Vol. 20, No. 37, 375201, ISSN: 0957-4484

Jo S. H. & Lu W. (2008). CMOS Compatible Nanoscale Nonvolatile Resistance Switching Memory. *Nano Letters,* Vol. 8, No. 2, 392-397, ISSN: 1476-1122

Jo S. H.; Kim K. H. & Lu W. (2009). Programmable Resistance Switching in Nanoscale Two-Terminal Devices. *Nano Letters,* Vol. 9, No. 1, 496-500, ISSN: 1476-1122

Jo, S. H.; Kim K. H. & Lu W. (2009). High-Density Crossbar Arrays Based on a Si Memristive System. *Nano Letters*, Vol. 9, No. 2, 870-874 , ISSN: 1476-1122

Kim D. C.; Lee M. J.; Ahn S. E.; Seo S.; Park J. C.; Yoo I. K.; Baek I. G.; Kim H. J.; Yim E. K.; Lee J. E.; Park S. O.; Kim H. S.; In Chung U.; Moon J. T. & Ryu B. I. (2006). Improvement of resistive memory switching in NiO using IrO_2. *Applied Physics Letters*, Vol. 88, No. 23, 232106, ISSN: 0003-6951

Kim K. M.; Choi B. J.; Shin Y. C.; Choi S. & Hwang C. S. (2007). Anode-interface localized filamentary mechanism in resistive switching of TiO_2 thin films. *Applied Physics Letters*, Vol. 91, No. 1, 012907, ISSN: 0003-6951

Kim S.; Moon H.; Gupta D.; Yoo S. & Choi Y. K. (2009). Resistive Switching Characteristics of Sol–Gel Zinc Oxide Films for Flexible Memory Applications. *IEEE Transactions on Electron Devices*, Vol. 56, No. 4, 696-699 , ISSN: 0018-9383

Kinoshita K.; Tamura T.; Aoki M.; Sugiyama Y. & Tanaka H. (2006). Bias polarity dependent data retention of resistive random access memory consisting of binary transition metal oxide. *Applied Physics Letters*, Vol. 89, No. 10, 103509, ISSN: 0003-6951

Kinoshita K.; Yoshida C.; Aso H.; Aoki M. & Sugiyama Y. (2006). Thermal properties of NiO_y resistor practically free from the "forming" process, *Proceedings of International Conference on Solid State Devices and Materials*, pp. 570-571, Pacifico Yokohama Yokohama, Japan, Sept. 2006

Kreupl F.; Bruchhaus R.; Majewski P. et al. (2008). Carbon-based resistive memory. *Proceeding of IEEE Electron Devices Meeting*, pp. 521-524

Kund M.; Beitel G.; Pinnow C.U.; Röhr T.; Schumann J.; Symanczyk R.; Ufert K.D. & Müller G. (2005). Conductive bridging RAM (CBRAM): An emerging non-volatile memory technology scalable to sub 20 nm, *Electron Devices Meeting, 2005, IEDM Technical Digest, IEEE International*, pp. 754–757. ISBN: 0-7803-9268-x, Washington, DC, Dec. 2005, IEEE

Kwon D. H.; Kim K. M.; Jang J. H.; Jeon J. M.; Lee M. H.; Kim G. H.; Li X. S.; Park G. S.; Lee B.; Han S.; Kim M. & Hwang C. S. (2010). Atomic structure of conducting nanofilaments in TiO_2 resistive switching memory. *Nature Nanotechnology*, Vol. 5, No. 2, 148-153, ISSN: 1748-3387

Lee M. J.; Park Y.; Suh D. S.; Lee E. H.; Seo S.; Kim D. C.; Jung R.; Kang B. S.; Ahn S. E.; Lee C. B.; Seo D. H.; Cha Y. K.; Yoo I. K.; Kim, J. S. & Park B. H. (2007). Two Series Oxide Resistors Applicable to High Speed and High Density Nonvolatile Memory. *Advanced Materials*, Vol. 19, No. 22, 3919-3923, ISSN: 0935-9648

Li Y. B.; Sintskii A. & Tour J. M. (2008). Electronic two-terminal bistable graphitic memories. *Nature Materials*, Vol. 7, No. 12, 966 -971, ISSN: 1476-1122

Li M.; Zhuge F.; Zhu X. J.; Yin K. B.; Wang J. Z.; Liu Y. W.; He C. L.; Chen B. & Li R. W. (2010). Nonvolatile resistive switching in metal/La-doped $BiFeO_3$/Pt sandwiches. *Nanotechnology*, Vol. 21, No. 42, 425202, ISSN: 0957-4484

Li Y. T.; Long S. B.; Zhang M. H.; Liu Q.; Shao L. B.; Zhang S.; Wang Y.; Zuo Q. Y.; Liu S. & Liu M. (2010). Resistive switching properties of Au/ZrO_2/Ag structure for low-voltage nonvolatile memory applications. *IEEE Electron Device Letters*, Vol. 31, No. 2, 117-119, ISSN: 0741-3106

Liao Z. L.; Wang Z. Z.; Meng Y.; Liu Z. Y.; Gao P.; Gang J. L.; Zhao H. W.; Liang X. J.; Bai X. D. & Chen D. M. (2009). Categorization of resistive switching of metal-$Pr_{0.7}Ca_{0.3}MnO_3$-metal devices. *Applied Physics Letters*, Vol. 94, No. 25, 253503, ISSN: 0003-6951

Liu S. Q.; Wu N. J. & Ignatiev A. (2000). Electric-pulse-induced reversible resistance change effect in magnetoresistive films. *Applied Physics Letters*, Vol. 76, No. 19, 2749-2751, ISSN: 0003-695

Liu Q.; Long S. B.; Wang W.; Zuo Q. Y.; Zhang S.; Chen J. N. & Liu M. (2009). Improvement of Resistive Switching Properties in ZrO_2 -Based ReRAM With Implanted Ti Ions. *IEEE Electron Device Letters*, Vol. 30, No. 12, 1335-1337, ISSN: 0741-3106

Lu W. & Lieber C. M. (2007). Nanoelectronics from the bottom up. *Nature Materials*, Vol. 6, No. 11, 841-850, ISSN: 1476-1122

Meijer G. I. (2008). MATERIALS SCIENCE: Who Wins the Nonvolatile Memory Race?. *Science*, Vol. 319, No. 5870, 1625-1626, ISSN: 0036-8075

Novoselov K. S.; Geim A. K.; Morozov S. V.; Jiang D.; Zhang Y.; Dubonos S. V.; Grigorieva I. V.; & Firsov A. A. (2004). Electric field effect in atomically thin carbon films. *Science*, Vol. 306, No. 5696, 666-669, ISSN: 0036-8075

Odagawa A.; Sato H.; Inoue, I. H.; Akoh H.; Kawasaki M.& Tokura Y. (2004). Colossal electroresistance of a $Pr_{0.7}Ca_{0.3}MnO_3$ thin film at room temperature. *Physical Review B*, Vol. 70, No. 22, 224403, ISSN: 1098-0121

Oligschlaeger R.; Waser R.; Meyer R.; Karthauser S. & Dittmann R. (2006). Resistive switching and data reliability of epitaxial $(Ba,Sr)TiO_3$ thin films. *Applied Physics Letters*, Vol. 88, No. 4, 042901, ISSN: 0003-6951

Rakhshani, A. E.; Hogarth, C. A. & Abidi, A. A. (1976). Observations of local defects caused by electrical conduction through thin sandwich structures of Ag–SiO/BaO–Ag. *Journal of Non-Crystalline Solids*, Vol. 20, No. 1, 25–42, ISSN: 0022-3093

Ray, A. K. & Hogarth, C. A. (1990). Recent advances in the polyfilamentary model for electronic conduction in electro-formed insulating films. *International Journal of Electronics*, Vol. 69, No. 1, 97–107, ISSN: 0020-7217

Rohde C.; Choi B. J.; Jeong D. S.; Choi S.; Zhao J. S. & Hwang C. S. (2005). Identification of a determining parameter for resistive switching of TiO_2 thin films. *Applied Physics Letters*, Vol. 86, No. 26, 262907, ISSN: 0003-6951

Rueckes T.; Kim K.; Joselevich E.; Tseng G. Y.; Cheung C. L.; & Lieber C. M. (2000). Carbon nanotube-based nonvolatile random access memory for molecular computing. *Science*, Vol. 289, No. 5476, 94-97, ISSN: 0036-8075

Schindler C.; Meier M. & Waser R. (2007). Resistive switching in Ag-Ge-Se with extremely low write currents, *Proceedings of Non-Volatile Memory Technology Symposium*, pp. 82–85, Print ISBN: 978-1-4244-1361-4, Albuquerque, NM, Nov. 2007, IEEE

Seo S.; Lee M. J.; Seo D. H.; Jeoung E. J.; Suh D.S.; Joung Y. S.; Yoo I. K.; Hwang I. R.; Kim S. H.; Byun I. S.; Kim J.S.; Choi J. S. & Park B. H. (2004). Reproducible resistance switching in polycrystalline NiO films. *Applied Physics Letters*, Vol. 85, No. 23, 5655-5657, ISSN: 0003-6951

Son J. Y. & Shin Y. H. (2008). Direct observation of conducting filaments on resistive switching of NiO thin films. *Applied Physics Letters*, Vol. 92, No. 22, 222106, ISSN: 0003-6951

Stewart D. R.; Ohlberg D. A. A.; Beck P. A.; Chen Y.; Williams R. S.; Jeppesen J. O.; Nielsen K. A. & Stoddart J. F. (2004). Molecule-Independent Electrical Switching in Pt/Organic Monolayer/Ti Devices. *Nano Letters*, Vol. 4, No. 1, 133-136, ISSN: 1476-1122

Strukov D. B.; Snider G. S.; Stewart D. R. & Williams R. S. (2008). The missing memristor found. *Nature*, Vol. 453, No. 7191, 80 -83, ISSN: 0028-0836

Szot K.; Speier W.; Bihlmayer G. & Waser R. (2006). Switching the electrical resistance of individual dislocations in single-crystalline $SrTiO_3$. *Nature Materials* , Vol. 5, No. 4, 312-320, ISSN: 1476-1122

Tokunaga Y.; Kaneko Y.; He J. P.; Arima T.; Sawa A.; Fuji T.; Kawasaki M. et al. (2006). Colossal electroresistance effect at metal electrode/$La_{1-x}Sr_{1+x}MnO_4$ interfaces. *Applied Physics Letters*, Vol. 88, No. 22, 223507, ISSN: 0003-6951

Tsubouchi K.; Ohkubo I.; Kumigashira H.; Oshima M.; Matsumoto Y.; Itaka K.; Ohnishi T. et al. (2007). High-Throughput Characterization of Metal Electrode Performance for Electric-Field-Induced Resistance Switching in Metal/$Pr_{0.7}Ca_{0.3}MnO_3$/Metal Structures. *Advanced Materials*, Vol. 19, No. 13, 1711-1713, ISSN: 1476-1122

Tsunoda K.; Fukuzumi Y.; Jameson J. R.; Wang Z.; Griffin P. B. & Nishi Y. (2007). Bipolar resistive switching in polycrystalline TiO2 films. *Applied Physics Letters,* Vol. 90, No. 11, 113501, ISSN: 0003-6951

Wang X. R.; Ouyang Y. J.; Li X. L.; Wang H. L.; Guo J.; & Dai H. J. (2008). Room-Temperature All-Semiconducting Sub-10-nm Graphene Nanoribbon Field-Effect Transistors. *Physical Review Letters,* Vol. 100, No. 20, 206803, ISSN: 0031-9007

Waser R. & Aono M. (2007). Nanoionics-based resistive switching memories. *Nature Materials,* Vol. 6, No. 11, 833-840, ISSN: 1476-1122

Wu X.; Zhou P.; Li J; Chen, L. Y.; Lv H. B.; Lin Y. Y. & Tang T. A. (2007). Reproducible unipolar resistance switching in stoichiometric ZrO_2 films. *Applied Physics Letters,* Vol. 90, No. 18 ,183507, ISSN: 0003-6951

Yan J. A.; Xian L. & Chou M. Y. (2009). Structural and electronic properties of oxidized graphene, *Physical Review Letters,* Vol. 103, No. 8, 086802, ISSN: 0031-9007

Yang C. H.; Seidel J.; Kim S. Y.; Rossen P. B.; Yu P.; Gajek M.; Chu Y. H.; Martin L. W.; Holcomb M. B.; He Q.; Maksymovych P.; Balke N.; Kalinin S. V.; Baddorf A. P.; Basu S. R. et al. (2009). Electric modulation of conduction in multiferroic Ca-doped $BiFeO_3$ films. *Nature Materials,* Vol. 8, No. 6, 485-493, ISSN: 1476-1122

Yang Y. C.; Pan F.; Liu Q.; Liu M. & Zeng F. (2009). Fully Room-Temperature-Fabricated Nonvolatile Resistive Memory for Ultrafast and High-Density Memory Application. *Nano Letters,* Vol. 9, No. 4, 1636-1643, ISSN: 1476-1122

Yin K. B.; Li M.; Liu Y. W.; He C. L.; Zhuge F.; Chen B.; Lu W.; Pan X. Q. & Li R. W. (2010). Resistance switching in polycrystalline $BiFeO_3$ thin films. *Applied Physics Letters,* Vol. 97, No. 4, 042101, ISSN: 0003-6951

Zhou X. F. & Liu Z. P. (2010). A Scalable, Solution-phase Processing Route to Ultralarge Graphene Sheets. *Chemical Communications,* Vol. 46, No. 15, 2611-2613, ISSN: 1359-7345

Zhou Y. H.; Liu H. B.; Fu L.; Li B. & Chen Z. Z. (2006). Influence of pyrolytic temperature on structures and properties of graphite oxide. *Journal of the Chinese Ceramic Society,* Vol. 34, No. 3, 318-323, ISSN: 0454-5648

Zhuge F.; Dai W.; He C. L.; Wang A. Y.; Liu Y. W.; Li M.; Wu Y. H.; Cui P. & Li R. W. (2010). Nonvolatile resistive switching memory based on amorphous carbon. *Applied Physics Letters,* Vol. 96, No. 16,163505, ISSN: 0003-6951

11

Scanning Transmission Electron Microscopy and Spectroscopy of Suspended Graphene

Ursel Bangert[1], Mhairi Gass[2], Recep Zan[1] and Cheng Ta Pan[1]
[1]*The University of Manchester,*
[2]*SuperSTEM Laboratories, CCLRC Daresbury Laboratory,*
UK

1. Introduction

Transmission electron microscopy is the only technique by which freely suspended graphene can be scrutinised for nano-scale topography and chemistry occurring in micromechanically exfoliated graphene or having been introduced by production methods, e.g., by growth on Ni-subtsrates, or as a result of functionalistaion, e.g., upon hydrogenation. Investigations on this scale require atomic resolution, which can be catered for by electron microscopy; indeed several impressive studies have been conducted. In this chapter we summarise reports on scanning transmission electron microscopy and spectroscopy investigations, conducted with aberration corrected probes, especially recent results of atomic resolution high angle dark field microscopy, by which chemical information about atoms in graphene that are heavier then carbon can be obtained. By combining this with ultra-high spatially resolved electron energy loss spectroscopy to reveal the nature of elements, particularly those with lower atomic number (e.g., boron and hydrogen), which do not show up in high angle dark field images, one obtains an invaluable and powerful tool for graphene reserach; added to this is the information about the electronic bandstructure that can be obtained on a highly spatially resolved scale through monitoring the plasmon behaviour of graphene.

Paragraph 2 briefly describes the scanning transmission electron microscopy (STEM) technique employed for imaging and chemical analysis, and how free-standing membranes of graphene, required for these measurements, are obtained. Paragraph 3 deals with high resolution imaging of the atomic structure and configuration of intrinsic defects and paragraph 4 with stabilisation issues of single and bi-layer membranes: rippling effects in the graphene lattice are revealed. In free-standing CVD-grown graphene the membranes are an intricate patchwork of mono- and bi-layers, the latter being AB stacked or rotated at various angles to each other. In paragraph 5 atomic resolution HAADF and ultra-high spatially resolved electron energy loss spectroscopy are used to identify atomic sites and chemical nature of metal impurity atoms Au, Ni and Fe, as well as of B, intentionally introduced into graphene as electronic dopant via ion implantation. It is highly desirable to be able to reveal the presence and distribution of hydrogen in order to assess the success of H-dosing for the purpose of bandgap tailoring; we investigate the applicability of energy loss spectrum imaging for this purpose. Paragraph 6 scrutinises plasmon characteristics of graphene. Plasmon behaviour is sensitive to electron densities and electronic structure of solids; the graphitic σ-plasmon undergoes typical changes in few-layer structures and provides an unambiguous fingerprint for single layer graphene, whilst the graphitic π-plasmon can give valuable information about changes in the π-electron system upon doping and dosing. In combination with DFT calculations we test the scope of plasmon spectroscopy in assessing, especially, H-induced graphene morphologies.

2. Microscopy methods and TEM sample preparation techniques

2.1 Microscopy methods

Electron microscopy studies of graphene so far have been conducted in a transmission electron microscope (TEM), using a stationary electron beam. Here we use an aberration corrected (AC) transmission electron microscope (STEM) (Goodhew & Bleloch, 2003) with a cold field emission gun and added, analytical facilities. In a STEM a focussed electron probe is scanned in a pixelated raster over a pre-defined sample area. The benefits of this microscope are the very bright (100 pA) and very fine (≤0.1 nm diameter) probe, which, due to the small depth of focus (8nm) provides an image- (pixel-) resolution of better then 0.1 nm, allowing atomic-scale imaging to be carried out in bright field (BF) and, very importantly, in high angle annular dark field (HAADF) mode. A further benefit is the variable operating voltage from 100 down to 60 kV, ensuring measurements below the displacement threshold of C (at ~80 keV).

In BF mode electrons from the central diffraction disk are collected over an angular range of 0-6 mrad. Diffraction disks (rather then spots) are formed because of the high beam convergence angle (~24 mrad). The occurrence of atomic lattice images at high resolution arises from coherence (phase) effects in overlapping diffraction disks. In HAADF mode electrons scattered to high angles are collected in an annular detector with an angular range of 70-210 mrad. These electrons have undergone incoherent elastic scattering (Rutherford scattering), the strength (or probability) of which is approximately proportional to the square of the atomic number of the scattering element (Z-contrast), and therefore this imaging method can be used for chemical identification (Pennycook & Boatner, 1988). At high magnifications, due to the fine probe and employing sub-Å scan point separation, individual atoms can be 'lined-out', giving rise to atomic resolution HAADF images. Such images can be interpreted rather straight forwardly and much more easily than BF phase contrast images.

Chemical/elemental analysis using electron energy loss spectroscopy (EELS) (Brydson, 2003; Egerton, 1986) can be conducted with high sensitivity at the same spatial resolution as obtained in images. Inelastic scattering events can be detected with collection angles of up to 19 mrad over an energy range of 0-2000 eV, covering energy losses from valence band to K-shell core loss excitations of elements spanning the periodic table from H to noble metals (using an 'Enfina' energy loss spectrometer). The combination of the ~0.3 eV energy resolution and energy dispersions of up to 0.01 eV/channel provides access to absorption events in the 1 eV loss regime, thus enabling electronic bandstructure studies. Energy loss spectra can be obtained as individual spectra in single locations on the sample or as collective data sets along predefined lines (line scans) or in predefined image areas (spectrum images (Hunt & Williams, 1991)). The small probe size enables EELS to be carried out even on single atoms (Browning & Pennycook 1994). Image acquisition and processing as well as EELS measurements and evaluation were conducted with the Gatan Digital Micrograph software and its various 'plug-ins'; data analysis using special programs is described in the respective paragraphs.

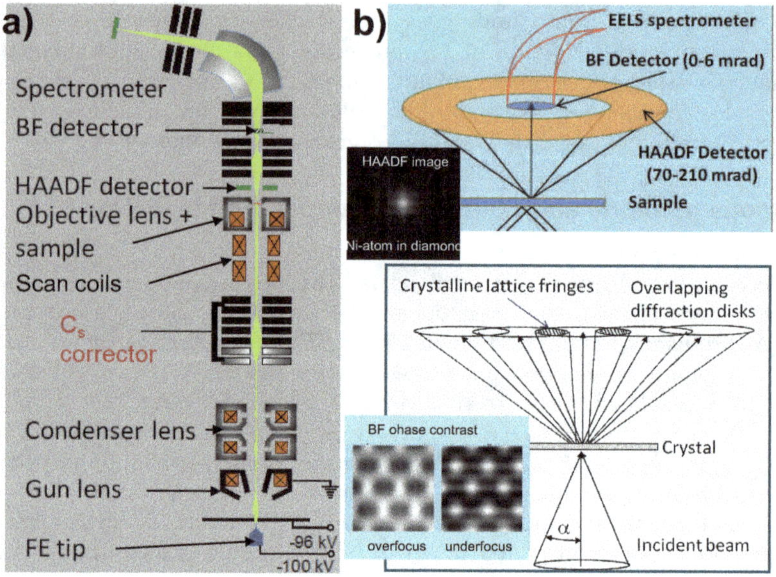

Fig. 1. a) Schematics of the STEM, b) enlarged detail of detector configuration with schematic HAADF image inset (top panel) and schematics of origin of phase contrast on the BF detector with inset showing lattice images in overfocus and underfocus (bottom panel).

The Daresbury SuperSTEMs (the UK's first aberration corrected STEMs) were used for STEM imaging and analysis. TEM results, which are also presented in this chapter, were obtained with a Tecnai F30.

2.2 TEM sample preparation

There are several methods to acquire suspended graphene layers, initially such layers were obtained by micro-mechanical cleaving and exfoliation (Booth et al, 2008). After repeated cleaving of graphite by using adhesive tape, thin flakes are transferred to an oxidised Si wafer on which few-layer graphene flakes can be identified by their colour using a light microscope. Selected flakes are then made to adhere to a TEM sample grid; this is facilitated by using, e.g., sticky grids (Quantifoil), and employing the surface tension of a drop of evaporating IPA, dripped onto the graphene flakes.

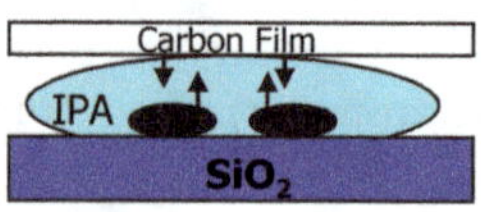

Another frequently used method employs embedding of graphene flakes in PMMA or CAB (Cellulose Acetate Butyrate) for protection, and subsequent lift off the substrate (e.g., by etching the substrate), and floating the PMMA (or CBA) containing the graphene flakes off. The details of this procedure can vary; the steps outlined in the sketch below entail for example applying a sticky tape (blue) with a window (step 2) to make a 'frame' for easier handling of the PMMA/graphene film during the transfer to the TEM grid. This sticky tape is later removed by scribing the PMMA around the target area (step 5)

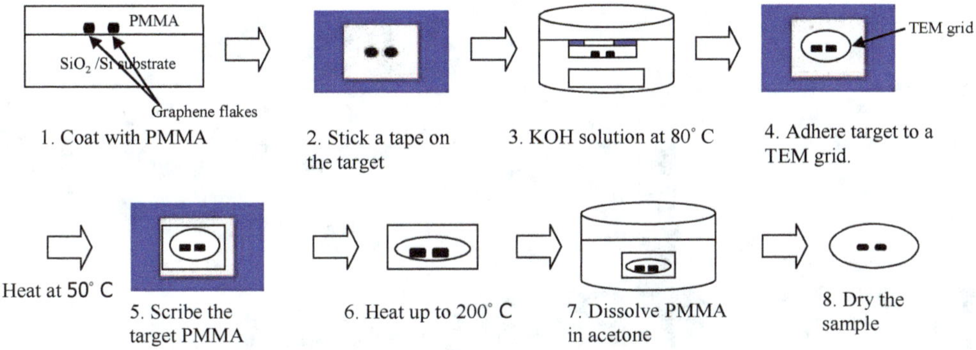

Figure 2a) shows an optical micrograph of graphene flakes obtained by micromechanical cleaving and deposition across a crack (induced by cleaving) in an oxidised Si wafer. The darkish blue graphene areas are identified as single or bi-layers. The graphene is freely suspended, where it lies across the gap and hence could be directly imaged in a TEM using the Si as sample support. Figures 2b-d are TEM BF images of the suspended flake at increasing magnification. Figure 2c shows that the graphene surface has significant contamination; 2d) shows the atomic lattice in an uncontaminated patch. Figure 2e) is a HAADF STEM image at low magnification with contamination, most likely hydrocarbon deposits, presenting them selves as whitish contrast. These deposits are interrupted by dark patches, representing uncontaminated parts of single layers. In the following paragraph it is the atomic scale topography of uncontaminated patches that we will concern ourselves with. Figures 2e&f are views at increased magnification. Figure 2g), although noisy, as it represents scattering of one single atomic layer, shows the graphene lattice. Much of the work in the following paragraphs will focus on HAADF studies, which we have carried out in the SuperSTEM; a number of papers have dealt with transmission electron microscopy (TEM) of graphene (Meyer et. al., 2007a; Meyer et. al., 2008a; Meyer et. al., 2008b) however HAADF studies have not been amongst these.

3. Atomic scale topography: defects and ad-atoms

3.1 Defects

In the case of graphene with the electron beam focussed onto the sheet, the rare occasions occurs, where electron microscopy lattice images can be seen as a direct depiction of the ball-and-stick model of an atomic lattice structure. These images are directly interpretable: bright contrast corresponds to atoms and dark contrast to the gaps in between. Effects of atomic structure extending in direction of the electron beam, resulting in de-channelling of the beam on atomic columns and hence inducing contrast changes etc, cannot occur in mono-layer structures. Moreover, the atomic arrangements of atomic-scale defects can be straight forwardly deduced. HAADF can furthermore reveal the atomic nature of species at single-atom level through the approximate Z^2-dependence of the scattering probability. It is thus feasible for detection of atom species on/in graphene of Z equal to and larger then carbon. Graphene has provided the basis for elegant theoretical calculations and predictions of defect structures alongside their effects on the physical, chemical and electronic properties (Duplock, Scheffer & Lindan, 2004; Crespi, Scheffler & Rubio, 1997; Charlier, 2002). Structural observations on suspended graphene, on the other hand, are still relatively scarce, and, as

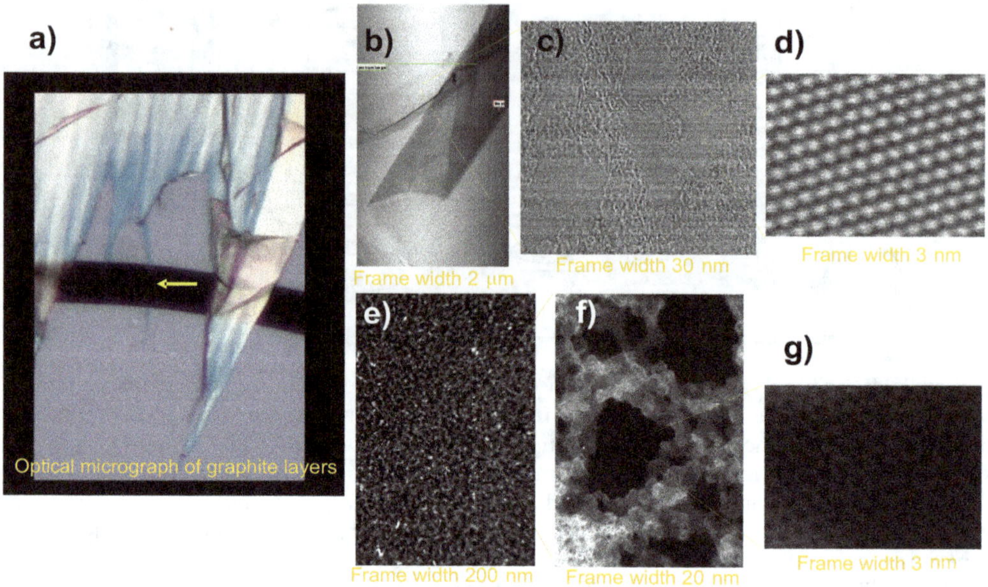

Fig. 2. a) optical micrograph of graphene flake deposited across a gap in an oxidised Si-wafer, b), c) and d) BF TEM images and e), f) and g) HAADF STEM images of the flake at increasing magnification.

noted earlier, HAADF studies are non-existent. Here we reveal directly atomic scale topological features of free standing graphene.

Electron beam energies of 100 keV, used for some of the following studies are just around the displacement threshold for atoms in carbon structures. Hence it is difficult to judge, whether defects occurring during the first image scan are radiation induced, or whether they were already present. However, changes in the films occurring over multiple scans were carefully monitored (and captured in movies), and on the basis of this it appears that discernible changes took place only after ~2 minutes of repeated scanning, and that the defect numbers during the first few scans are very small and stay constant, with the defects well separated (although some movement / reconstruction was observed even between initial scans). It is of interest to follow in-situ defect formation in the STEM, captured in sequential scans, each scan lasting no longer then 10 s. Figure 3 shows the 1st, 17th, 18th and 19th image of a 20-images series of STEM high resolution BF images. Only in the later scans, after ~2 min irradiation, does e-beam defect formation become a distinctive feature. At that stage a large number of single point defects- seen here are vacancies- suddenly appears (fig. 3c, red circle). These aggregate to form larger clusters (fig. 3d, red circle) and, eventually, holes. We emphasise we are dealing with BF-phase contrast images; these are not as directly interpretable as their HAADF counterparts. Nevertheless, since the images here were taken at ~1nm overfocus, the centre of the 6-rings, and also vacancies, will appear darker then the atoms, similar to dark field images. The delay in the apparent setoff of the beam damage may indicate that a small number of defects are already present in graphene sheets, and almost all of these are single vacancies. However, this contradicts STM studies of graphene obtained by micro-cleavage, in which the authors were unable to detect one single vacancy

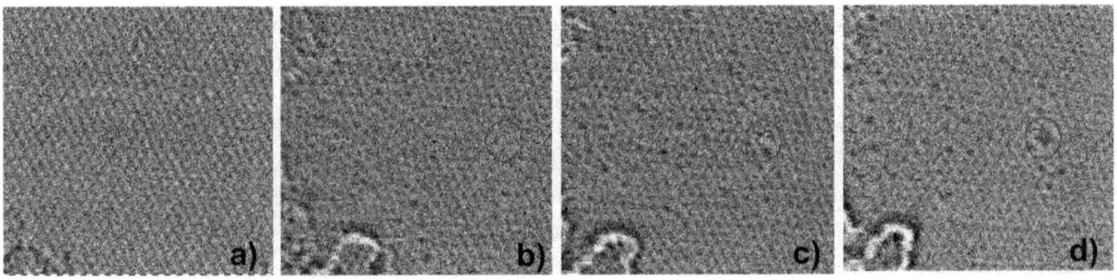

Fig. 3. Snapshots of a BF image series demonstrating development of radiation induced defects in the STEM. The darks spots are vacancy related defects. a) frame 1, b) frame 17 c) frame 18, d) frame 19; the scan time per frame was ~10 s.

(Stolyarova, 2007). Alternatively, the electron beam can radicalize molecules absorbed on the surface, which react with carbon leading to broken bonds.

Figure 4a) depicts a single-layer area imaged in HAADF. Although the lattice is visible in the raw HAADF images, it is feasible to reduce the noise by applying either a deconvolution routine based on maximum entropy image reconstruction (HREM Research Inc., 2009) or a low pass Fourier filtering method. The former sharpens features by deconvolving the images with a simulated electron probe function having a Gaussian or Lorentzian profile (or a combination of both), thus constituting a software C_s corrector, the latter improves visibility of spatial frequencies by applying a mask to the Fourier transformed (FFT) image, which includes the spatial frequencies right up to the lowest order *a*-plane diffraction spots, as well as the *a*-plane ([100], [010] and [-110]) reflections themselves (inset top left), and then obtaining the inverse Fourier transform (IFFT).

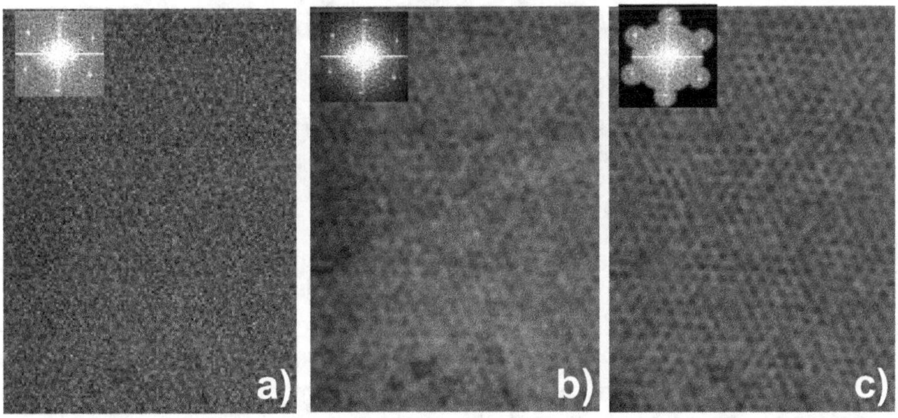

Fig. 4. a) raw HAADF image of single graphene sheet with FFT as inset, b) same HAADF image after deconvolution; the FFT in the inset is clearer as a result of noise reduction, c) IFFT of area in (a) having applied the band pass filter in the inset.

Noise reduction using deconvolution is shown in fig.4b. The low pass filter (fig.4c) achieves noise reduction by cutting out high frequencies; it preserves the information of all direct space distances larger then the a-lattice plane spacing and additionally superimposes a 'directional' component as frequencies between diffraction spots are masked out, thus enhancing the graphene lattice. In the following this latter filter is used. In order to

investigate possible artefacts introduced into the IFFT, i.e., into the end image, by the mask edges and the background noise, we subtracted a filtered Fourier transformed vacuum image from the filtered graphene Fourier space image, before we applied the inverse transform. However, this showed insignificant change in the case of this particular filter.
The vast majority of defects are vacancies. The HAADF lattice images in figures 5a and 5b are filtered in the described fashion, showing enlarged lattice structure surrounding a mono- and a di-vacancy with overlaid model structure (Gass et al., 2008). These are easy to discern, whereas Stone-Wales defects (Terrones & Terrones, 1996), constituting a reconfigured, rotated bond have not been identified so far.

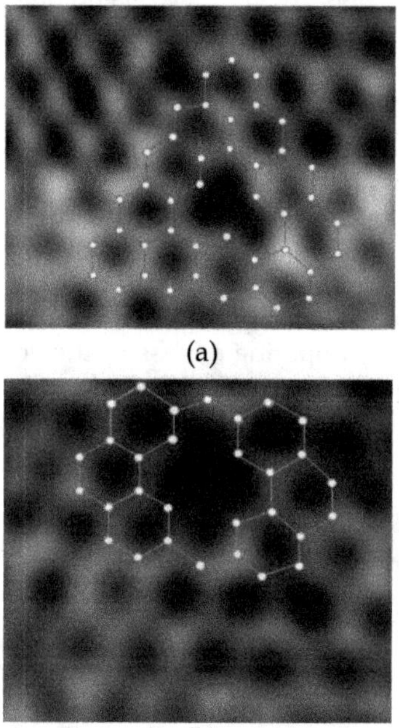

(a)

(b)

Fig. 5. a) and b): enlarged areas in HAADF images after low pass filtering showing a single vacancy and a di-vacancy with overlaid ball and stick models

There is a wealth of literature dealing with defects in HPO graphite [e.g. Hahn & Kang, 1999; Bourelle, Konno & Inagaki, 1999, Telling & Heggie 2007; Krasheninnikov et. al. 2001]. Isolated point defects on the surface of graphite are exlusively reported in connection with irradiation damage or ion bombardment, this might indicate that even the defects seen in the very first scan occur due to electron bombardment.
However, we also observe a second type of defect, related to the termination of a row of hexagons, and hence representing the 2-D or mono-layer equivalent of a dislocation (Gass et al., 2008). There are two species of unit c-axis edge dislocation in graphitic materials [Jenkins, 1969], depending on whether their glide plane cuts between atoms closely spaced in this direction or between atoms separated by a full bond length (defined as 'shuffle' and 'glide' plane respectively [Ewels, 2002]). After bond reconstruction, these two dislocation cores consist of either a pentagon-heptagon pair (5-7 defect, 'glide plane' dislocation core),

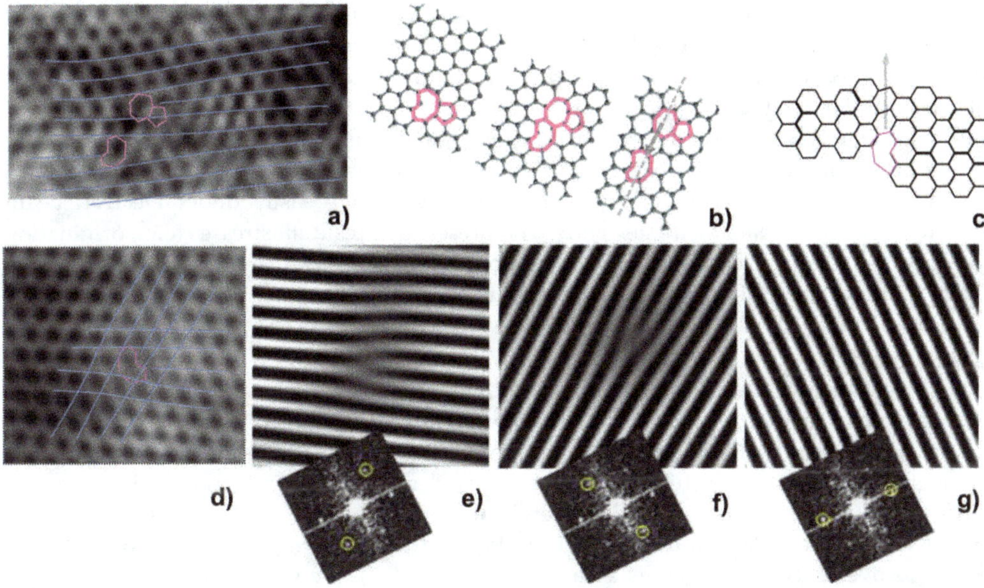

Fig. 6. a) HAADF image of dissociated shuffle-glide dipole (purple) with lattice 'planes' indicated in blue, b) model of an undissociated dipole (purple), formed by only one vacancy (leftmost panel), and subsequent separation into glide and shuffle defect via glide of the shuffle segment along the plane indicated by the dashed grey line (rightmost panel), c) formation of a shuffle defect by bond reconstruction at the edge of a graphene sheet; glide inwards along the grey arrow might be achieved by shear stress, leading as in (b) to the termination of a row of atoms, d) HAADF image of a single defect, which could have formed according to (c), e), f) and g) IFFTs obtained by masking the respective diffraction spots in the FFTs of (d), revealing termination of two different rows of atoms at the defect.

or a pentagon-heptagon pair with a 2-fold coordinated ad-atom carbon in the central bond ('shuffle' plane dislocation core; see figure 6b, e.g., 3rd schematic). These two dislocations were shown to migrate with radically different activation barriers, namely, 7.64eV for the glide core, 2.22eV for the shuffle core [Ewels, 2002]. This difference was explained in terms of the mechanism of motion. Motion occurs through a 90° rotation of a C-C bond along one side of the heptagon. For the glide dislocation core this requires breaking two C-C bonds, whereas in the shuffle dislocation core the same effect can be achieved through breaking and formation of just one bond. Thus the additional atom at the core of the shuffle dislocation is able to catalyse the motion of the dislocation [Ewels, 2002]. The glide dislocation core is essentially immobile. A single vacancy can be considered a dislocation dipole, with the two dislocations of opposite sign facing each other on neighbouring glide planes, as shown in fig. 6b (1st schematic). The 2nd and 3rd schematic in fig. 6b show dissociation of the dipole along the glide plane. The 3rd schematic depicts indeed the situation in the HAADF image of such a dissociated shuffle-glide dipole in fig. 6a.

Another possible route to create a shuffle core is through reconstruction of the edge of a graphene layer, which exhibits a distortion, as shown in fig. 6c (see row of sheared hexagons). Here an 8-atom ring defect bridges a corner. Under shear stress along the grey arrow, the defect could glide into the layer. This requires neither formation of a dipole nor vacancies, and each glide step, i.e., moving the defect by one hexagon along the grey line, only requires reconstruction of one bond. The HAADF image in figure 6d shows a single shuffle dislocation core with a clearly visible under-coordinated carbon atom in the defect core. We have demonstrated in fig. 3 that vacancies move easily under electron beam irradiation, so the e-beam facilitates glide. The presence of residual stresses (e.g., originating from micromechanical cleaving and exfoliation of the graphene) might be an additional source for glide of 8-atom ring defects formed at graphene edges.

3.2 Adatoms

Lattice resolution HAADF images of uncontaminated patches in single layer graphene films show contrast variations in the benzene rings. This can be observed in the previous figures. The area in figure 7a) is part of a clean patch and has been Fourier filtered, using a low pass filter (like this in figure 4). It becomes clear that the graphene lattice is not of uniform contrast; some of the bridges between atoms are significantly brighter. The ball and stick model, overlaid in the top right indicates the bright regions as ad-atoms. To the left of the overlaid model there is a vacancy (marked 'V'). We suggest that the regions of enhanced brightness found in fair numbers are due to C-adatoms, forming bridges above the in-plane C-C bonds. We derive this from a quantitative evaluation of the HAADF contrast (Bangert et al., 2009b). Examples of the quantification are shown in fig 7b, which includes intensity profiles (blue curves) that had been taken in *raw*, i.e., unfiltered HAADF images; the scan lines are marked in 7a) to indicate the corresponding scan locations in the filtered image. The red curves in the two panels are intensity scans across a vacuum image. They show the counts arising from camera shot noise and electron background when the beam is not on the specimen. The base line in the graphene is higher then in the vacuum image due to scattering from a single atomic layer; the high frequency variations in the intensity in the curves are of statistical nature, but it has to be noted that a lower frequency pattern with troughs and peaks modulates the noise around the base line intensity in the blue curves; this pattern reflects the periodicity of the atom array, as the 1-Å probe causes higher intensities on then between atoms. Distinctively above the graphene layer intensity are 'spikes' (orange arrows) of just below twice the intensity. These correlate with positions of the brighter 'bridges', indicated by corresponding orange arrows in the low pass filtered image. The HAADF contrast scales near linearly with the material thickness, i.e., with the number of atomic layers traversed by the e-beam. It scales, however, approximately with the square of the atomic number: measured ratios of HAADF signals from C-atoms in graphene and single atoms of Si or Cu contaminants, the nature of which was confirmed by EELS, revealed a power dependence of slightly less then 2. The HAADF intensity on the ad-atoms increases by a factor of just less then 2, whereas it should increase by 2½ or 3 times if these atoms were N or O; such an increase would lie outside the statistical error interval. In addition, highly localised core loss spectra (not shown) encompassing the range from C to O showed no sign of N and O.

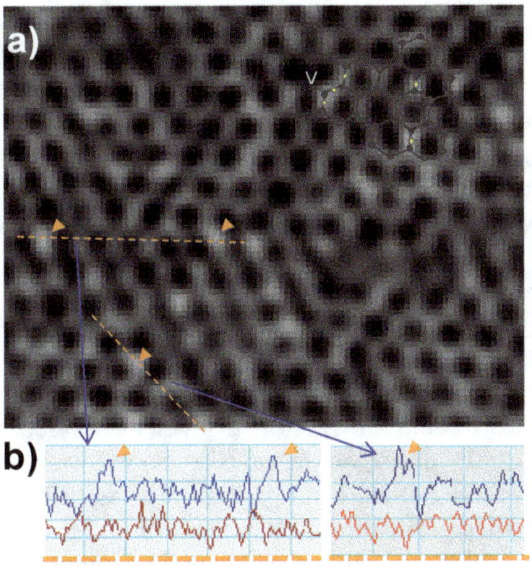

Fig. 7. a) Enlarged region from the filtered HAADF image in figure 4c with ball-and-stick model overlaid in the top right corner and a vacancy V indicated. Shown also are lines along which the intensity profiles displayed in b), blue curves, were taken. Intensity line-profiles from vacuum images are in red. Spikes of ~2x the graphene intensity are indicated with orange arrows. Enhanced intensity in the corresponding locations is emphasised and arrowed in (a). All intensity profiles were taken on raw, unprocessed images.

Carbon ad-atoms on graphene have previously been suggested (Meyer et. al. 2008b; Nordlund, Keinonen & Mattila, 1996). However, from HAADF contrast it cannot be excluded that H-atoms, too, are bound to C-adatoms or, indeed, directly to graphene C-atoms. Their Z-contrast would be lost in the noise.

4. Nano-scale topography: ripples in monolayers and turbostratic arrangements in bi-layers

4.1 Ripples

The stability of extended two-dimensional (2D) structures has been the subject of a long-standing theoretical debate, with previous suggestions that 2D films embedded in three-dimensional 3D space are crinkled. It was then countered that crinkles can be suppressed by anharmonic coupling between bending and stretching modes, such that a 2D membrane can exist but will nevertheless exhibit height fluctuations (Nelson, Piran & Weinberg, 2004; Gass et. al. 2008; Meyer et. al. 2007a&b;). The mechanical behavior of graphene can have profound impact on its extraordinary electronic properties. Recent observations suggest that suspended graphene is not perfectly flat, but rather exhibits microscopic corrugations (ripples) which can be not only dynamic (that is, through flexural phonons) but also static (Gass et. al., 2008; Booth et. al. 2008; Meyer et. al. 2007b). In those observations, large-scale ripples (>15 nm) were visualized directly (Nordlund, Keinonen & Mattila, 1996) whereas ripples on a nanoscopic scale (<15 nm) were only identified from the broadening of diffraction spots (Meyer et. al., 2007b) which prevents static bending from being distinguished from dynamic bending. Furthermore, it was argued that scaffolds supporting graphene crystals and contamination by adsorbed hydrocarbons, can induce an external compression resulting in ripples especially on the large scale, or films remember the initial non-flat configuration induced by a silicon oxide substrate used for sample preparation. In either instance, the observed ripples would not be intrinsic. The difficulty with directly imaging ripples in a transmission electron microscope arises from their small amplitude: the defocus, even with aberration-correction, is not precise enough to reliably detail changes in contrast which would arise from height differences less than a few nanometers, let alone one nanometer.

Observing perpendicularly onto an undulating plane of atoms, the bond inclination will cause variations in the projected bondlength. For example, inclinations of ~5° from the horizontal flat sheet will give rise to a change in the projected C-C bond lengths of ~1 %. Ripples possessing a wavelength of 5 nm, thus have an amplitude of 0.25 nm (this value being derived from electron beam diffraction (Meyer et al., 2007a)). However, such bond length changes are not easily observed in raw or even low-pass filtered HAADF lattice images, as seen using the filter detailed in fig. 4b and c. As FFTs of nano-scale areas, like in our case, are embedded in significant noise, the error in positioning the diffraction spot maxima gives rise to a bond length error of ~±2% in BF images, and slightly less in HAADF images. Applying a narrow annular band pass filter, with a band width corresponding to ~0.04 Å, i.e., imposing a narrow ring mask on the Fourier transform (FFT) of the raw images, passes spatial frequencies within ~2% of a selected frequency corresponding to a real space distance. The inverse transform (IFFT) reveals locations of atoms with such spacings, although with possible rotations, through intense lattice fringes. For graphene, the color coding is selected such that atoms possessing the correct bond length appear orange. Should the projected bond length change by as little as 2% owing to out-of plane bending of the atoms, the lattice periodicity will become less visible in the IFFT, and blurred patches will occur at such locations. Such regions appear blue. An example of this procedure shows ring filter and the resultant IFFT, respectively, in fig. 8a) and 8b). The latter is the projected bond length visualization; atomic scale detail can be seen as a fine raster superimposed on the colors. To better reveal the sheet rippling effect, the atomic detail has been removed by a Gaussian blurring function in the image in fig. 8d). In the Gaussian filtered images again, orange (blue) regions correspond to areas of small (large) deviations from the graphitic ***a***-plane distance of pristine and perfectly flat graphene sheets. Any spatial frequencies outside an annular mask diameter corresponding to the graphitic ***a***-plane spacing, will possess significantly weaker intensity; the color-coded IFFT resulting from both, a smaller or larger band pass would therefore be expected to exhibit more bluish colors. Since flanks of undulations reduce the projected bond length, they become visible as bluish regions in IFFTs obtained with a narrow ring mask.

We have obtained IFFTs of a number graphene images, using a series of filter diameters in each case, with increments corresponding to 0.02 Å (such an increment would arise from an inclination of ~5°), and with the above bandwidth of 0.04 Å. Although the ring mask increment lies within the error of measurement, it is still possible to follow gradual changes in contrast. In such a series of IFFT images, the spacing range covers ±10% variation of the graphitic ***a***-plane distance. In each IFFT image, a different projected lattice spacing is passed by the respective filter and comes 'into focus'; this is equivalent to observing the lattice in slices 'cut' through the crumpled 3D graphene sheet at different distances parallel to the flat sheet reference plane. In figure 9a and b, selected images from a series of IFFTs using ring mask diameter increments as described are detailed, the masks were applied to raw, single

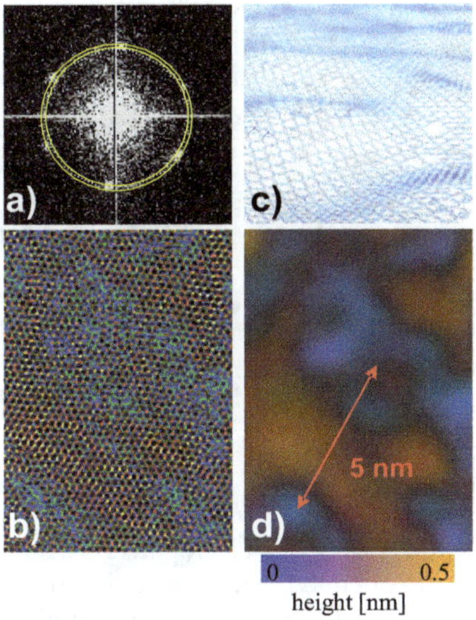

Fig. 8. a) A ring mask typically used to construct the IFFTs in (b), which is the inverse FFT of a raw HAADF image. c) Model of graphene, with undulations and defects. d) Gaussian-blurred image of (b) with the atomic detail removed, to highlight larger scale structure (flanks of undulations appear blue, approximate height scale given underneath in colour with flank inflection points set at zero).

layer images, i.e. allowing spatial frequencies both below and above the first order diffraction spot distance to pass. Fig. 9 (uppermost and middle rows) detail identical features imaged in BF and HAADF, with their associated IFFTs. BF IFFTs for the same band pass are very similar in their overall pattern to those of HAADF images, taking into account that the color scheme is slightly different in both cases. This lends credibility to the method being applied to HAADF images with weak intensities. The strongest intensity variations (orange to blue) can be observed in Fig. 9 columns iv-vii, where the ***a***-plane frequency has filtered through, whilst deviations from this band pass result in images of overall lower contrast (bluish colors, columns ii, iii and viii). In row c, the same method is applied to a 5-layer graphene sample, (the number of graphene layers was determined using the procedures described in chapter 6 and also in (Eberlein et. al., 2008). The breadth of images exhibiting the brown-orange colors (columns iv-vii) suggests flatter sheets, but also reflects the accuracy limit of the measurement; however, the radius of the ring filter used in column v (blue arrow) corresponds most closely to the ***a***-plane spacing. We note that pattern changes with band passes larger than the graphitic *a*-plane frequency (right of arrow, i.e. in columns vi and vii.) arise from a reduction in the bond length projection. For example, the circled locations mark contrast inversions, where the lattice at the flank of a ripple comes into focus, and then goes out of focus. In a simplistic picture, the peaks / troughs of ripples should be 'in focus' at the same band pass as the flat sheet, and the flanks should then be 'out of focus'. The increment in the annular filter radius required to change the color from brown to blue (images in rows a and b, columns vi and vii, circled area), allows one to deduce the inclination angle as ~12°. The ripple width of ~5 nm yields a height of ~0.5 nm. Although this is slightly greater than the value reported (Meyer et. al., 2007b; Fasolino, Los

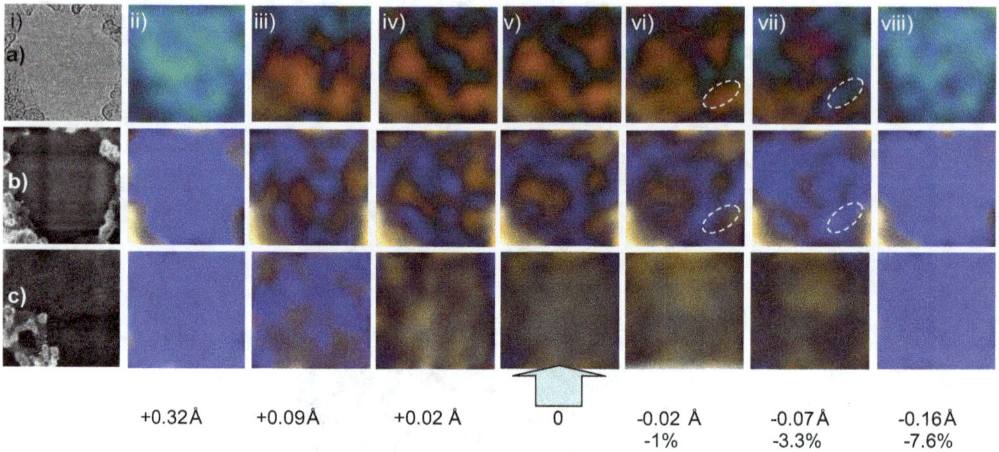

Fig. 9. In a) image (i) is a BF lattice image of a clean patch in a single graphene sheet, from which the IFFTs in (ii) to (viii) are obtained. All IFFT images have undergone Gaussian blurring. The arrow in column (v) marks the image corresponding to the correct graphitic lattice spacing. Bandpass deviations from the graphitic a-plane spacing, and hence from the C-C bond length are, (ii) +15%, (iii) +4.3%, (iv) +1%, (v) none, (vi) -1%, (vii) -3.3% and (viii) -7.6%. In b) (i) represents the HAADF lattice image of the same area as in (a), and in c) (i) is the HAADF image of a 5-layer graphene patch. (ii) to (viii) in (b) and (c) represent IFFTs of the respective images in (i) with the same band pass filters as detailed for (a) above. The frame width of panels is 15 nm.

& Katnelson, 2007), ripples of this height are frequent in occurrence and pronounced in our samples. Ripples of lesser height are most likely present too, but they are concealed by larger undulations and are also close to the resolution limit of the present method: the smallest increment in Fig. 9 corresponds to 1% change in spatial frequency equivalent to ~6° inclination. The contrast variations reflect the complex and intricate sheet buckling, which is quite severe in certain regions. Additionally, topography on a larger scale arising from thin hydrocarbon deposits, as can be seen bordering 'clean' areas' in all lower magnification images of graphene, also affects the IFFT image: such deposits introduce some amorphicity, hence many spatial frequencies are present in such locations, ranging from the lattice frequency down to low frequencies representing local, large scale disturbances. This accounts for the pink coloration occurring at such locations at all band passes.

Although undulations are observed in every clean graphene 'patch' (graphene area free from hydro-carbon deposit that is) in the absence of visible topological defects, we observe that the ripple patterns can, however, be influenced by the latter. Further results on this are given in Bangert et al. (2009a).

4.2 Turbostratic graphene

All results shown so far were obtained of micro-mechanically cleaved and exfoliated graphene. Electron microscopy investigations were also carried out on graphene grown by chemical vapour deposition (CVD) on nickel and copper substrates and lifted off these substrates via aforementioned procedures. We note that hydrocarbon contamination is generally more extensive and more finely dispersed on all such samples.

The images in figure 10, in contrast to the STEM images of the previous figures, are high resolution phase contrast images, obtained in a transmission electron microscope, a Tecnai F30. Ni-grown samples consisted of large films (10's to 100's of μm), partially consisting of single layer, but in the main of few-and bi-layers, either in graphitic AB stacking or in turbostratic arrangements (de-coupled layers, rotated at various angles with respect to each other). Figure 10a) depicts suspended CVD-graphene extracted from a Ni substrate. The small omni-present worm-like contrast arises from molecular hydrocarbons, presumably in form of chains, on the surface of bi- or triple-layer graphene. This contrast is present on all such CVD sample surfaces, uncontaminated surface areas are only of the order of few nm; upon closer inspection (fig. 10a might have to be viewed with slight enlargement) the underlying graphene lattice becomes obvious. The inset in the left hand top corner is an electron diffraction pattern of the image area, revealing the turbostratic nature by the two hexagonal spot patterns rotated with respect to each other (12 spots instead of 6). The right hand bottom inset is the FFT of the image, showing the same orientation relationship of the sheets as the diffraction pattern. Areas with apparent AB stacking can also be found (not shown here). This might, however, arise from differences in the rotation angle across the film: fig 10c) shows a schematics of graphene sheet rotated with respect to each other with different rotation angles, leading to varying Moiré patterns and different atomic lattice appearances within.

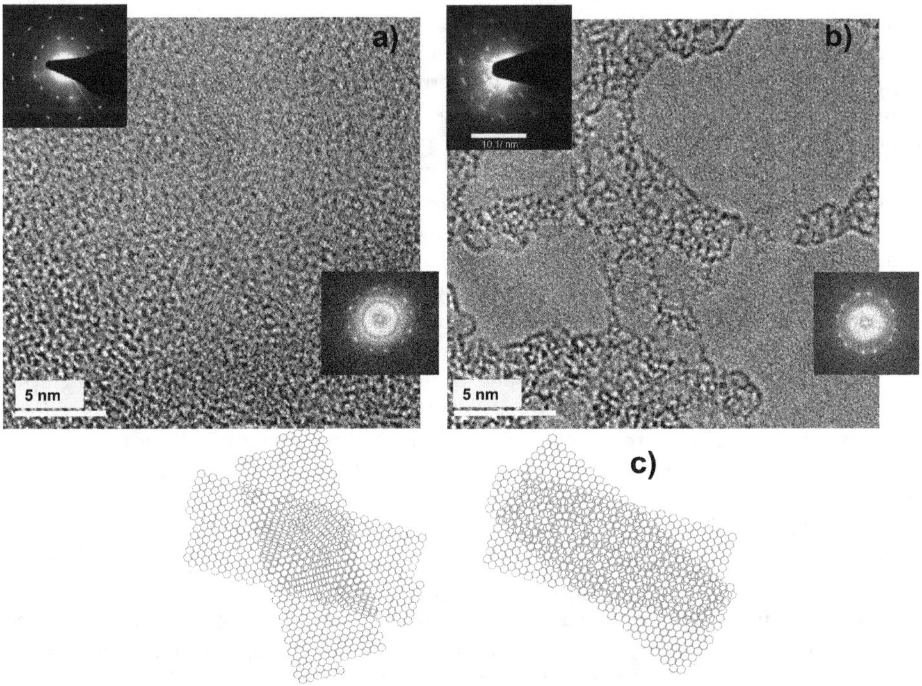

Fig. 10. HREM phase contrast images of suspended graphene grown by CVD after lift-off from a) a nickel and b) a copper substrate. Insets in both cases are the experimental diffraction pattern (top left in each image) and the Fast Fourier Transform (FFT; bottom right) of the image areas shown; c) schematics of turbostratic arrangements leading to different appearances of the atomic-scale structure.

Figure 10b) shows an image of suspended CVD-graphene grown on copper. Regions of larger uncontaminated areas (some 10s of nm) are found in these samples, however, the films consist in large, like in the Ni-substrate case, of turbostratic graphene. The insets show again the electron diffraction pattern (left top) and the FFT (right bottom) of the image area. Close inspection of the lattice structure reveals a Moiré pattern similar to the right-hand schematics in fig.10c).

Hence micromechanically exfoliated and CVD-grown graphene have quite different topographies, larger single-layer areas appear to be rare in the latter.

5. Atomic scale chemical information: impurities- native, dosed and doped

5.1 Impurities with Z ≥ 6

We have explained and demonstrated in paragraphs 1-3 that HAADF is the prime tool for pinpointing impurities with Z ≥6 on graphene. We cannot identify hydrogen attached to C-atoms via HAADF, but we will show other means of detecting H in paragraphs 5 and 6. We can identify extensive C-adatom coverage (paragraph 3), and although HAADF contrast evaluation gives strong evidence that the ad-atoms are indeed C (the HAADF signal strength in respective locations is twice that of graphene), due to the error in quantifying weak HAADF signals, we do not want to exclude oxygen atoms as being part of the ad-atom population (although the signal strength in locations of O-adatoms should be ~3x that of pure graphene), nor do we want to exclude, at this stage, the existence of OH-groups. However, we should be able to identify fluorine with a Z number of 9, giving a theoretical HAADF signal ratio F:C of 81:36 and hence an expected HAADF signal strength, in places of F-adatoms, of ~4x that of pure graphene). We have dosed graphene with F (details not given here), however, we could not detect any F retention on pristine graphene patches.

Extensive HAADF investigations, in which we have scrutinised many clean single-layer graphene patches for foreign atoms have revealed it is free of impurities with Z≥9. Graphene investigated for residual impurities was exfoliated by micromechanical cleavage as well as from CVD grown layers on Ni and Cu. Included were also attempts to dope/dose graphene with impurities via metal deposition (Au and Cr).

It should be emphasised that the above said applies strictly to clean, single layer graphene; we have never observed impurity atoms of any kind attached to such sheets. However, we observed many types of impurities in the hydrocarbon contamination on single and multiple sheets, and we did also observe impurity atoms on clean areas in multi-layer graphene.

Transition, noble, and group IV ad-atoms are predicted to modify the graphene electronic states to a large degree indicating covalent bonding with strong hybridization between ad-atoms and graphene. Metal-graphene interactions have been much studied by means of Density Functional Theory (DFT) calculations. H-sites (see fig.12e) are predicted as preferred location for most metals (e.g., Ti, Fe), T-sites for Sb, Sn and Ni, whereas Pd, Cr and Pt are expected to bind strongly to B sites (Suarez-Martinez et al, 2009; Uchoa, Lin & Neto, 2008; Chan, Neaton & Cohen, 2008; Mao, Juan & Zhong, 2008; Sevincli et al, 2008). Experimental exploration of the metal-graphene systems is still limited, especially TEM observations are practically non-existent. Such observation would be of great interest, as calculations have lead to discrepancies in the site of metal atoms: the two common approximation for the calculations, local density approximation and generalized gradient approximation -the latter does not represent Van-der-Waals forces very well (Leenaerts, Partoens & Peeters, 2008)-, lead to different binding energies and thereby to different sites for Au atoms, T-sites for the former and B-sites for the latter method (Aktürk & Tomak., 2009). In addition, arbitrary variables used in the calculation, such as cut-off energy (Varns & Strange, 2008) and size of the supercell (Leenaerts, Partoens & Peeters, 2009) can affect the result of DFT calculations. It was furthermore found, theoretically and experimentally, that point defects in graphene provide nucleation sites and even further substitutional incorporation of metals (Charlier et al., 2009; Boukhvalov & Katsnelson, 2009; Gan., Sun & Banhart, 2008; Krasheninnikov et al., 2009; Zhou, Gao & Goodman, 2010).

Gold atoms and its dimers on the graphene surface have been studied by using first-principles calculation with the local density approximation (Varns & Strange, 3008). The results show that the gold–gold interaction is significantly stronger than the gold–graphene interaction (for this reason gold is highly mobile on graphene) as is experimentally confirmed by Gan., Sun & Banhart (2008) with observation of gold clusters -rather then atoms- on the graphene sheet. For a single gold atom the favourable energy configuration is found to be directly above a carbon atom. We have evaporated 0.1 nm of gold onto graphene. Figures 11a) and b) show results of such deposition: indeed on single or de-coupled (turbostratic) layers gold atoms do not remain dispersed, but cluster into nano-crystals, and notably, are exclusively located in the hydrocarbon contamination. This demonstrates the extreme mobility of gold on pure graphene. Only in stacked few-layer films (here: exfoliated by cleaving) can single gold atoms be observed: it appears that atoms in sub-surface layers are needed to contribute to the bonding with surface gold atoms in order to prohibit dissociation and diffusion. Figure 11c) is a BF STEM image, and represents phase contrast (like an HREM image). The beam is focussed on the exit surface of the sample; this is the surface, on which the gold is evaporated. So the electron beam proceeds through the graphene layers before it encounters the gold atoms. The aggregate of gold atoms is so thin, that it is invisible in phase contrast, however, the graphene in the BF image exhibits strong contrast and relatively little noise, and shows the lattice periodicities well. This means that for BF analysis the raw (unprocessed) image can be used. Figure 11d) is the corresponding HAADF image, and is obtained simultaneously with the BF image. The image intensity arising from incoherent elastic scattering of few atom layers is very much lower than in BF, and the noise is so high that the benzene rings are barely recognisable. However, the gold atoms are clearly visible. They have formed a loose aggregate with single atoms dotted around the aggregate, some of which we have numbered. We note that comparing identical positions in images (c) and (d), the benzene ring centres, known to be black in HAADF images, correspond to bright patches in the BF images; this can be seen from the schematic benzene rings (red) overlaid on the HAADF image and positioned in identical locations in the BF image. Hence the C-atoms surrounding the black centres in HAADF images, constituting bright contrast, correspond to dark contrast in BF images. Due to the weak contrast and the noise of the raw HAADF image, although gold atoms can be located rather accurately, it would be near impossible to assign them exact sites on the benzene rings. However, the benzene ring shape is clearly discernible in the BF image, and by locating the exact corresponding position here (circled), the sites can be identified. It thus occurs that all gold atoms sit on T-sites. This is in agreement with theoretical predictions.

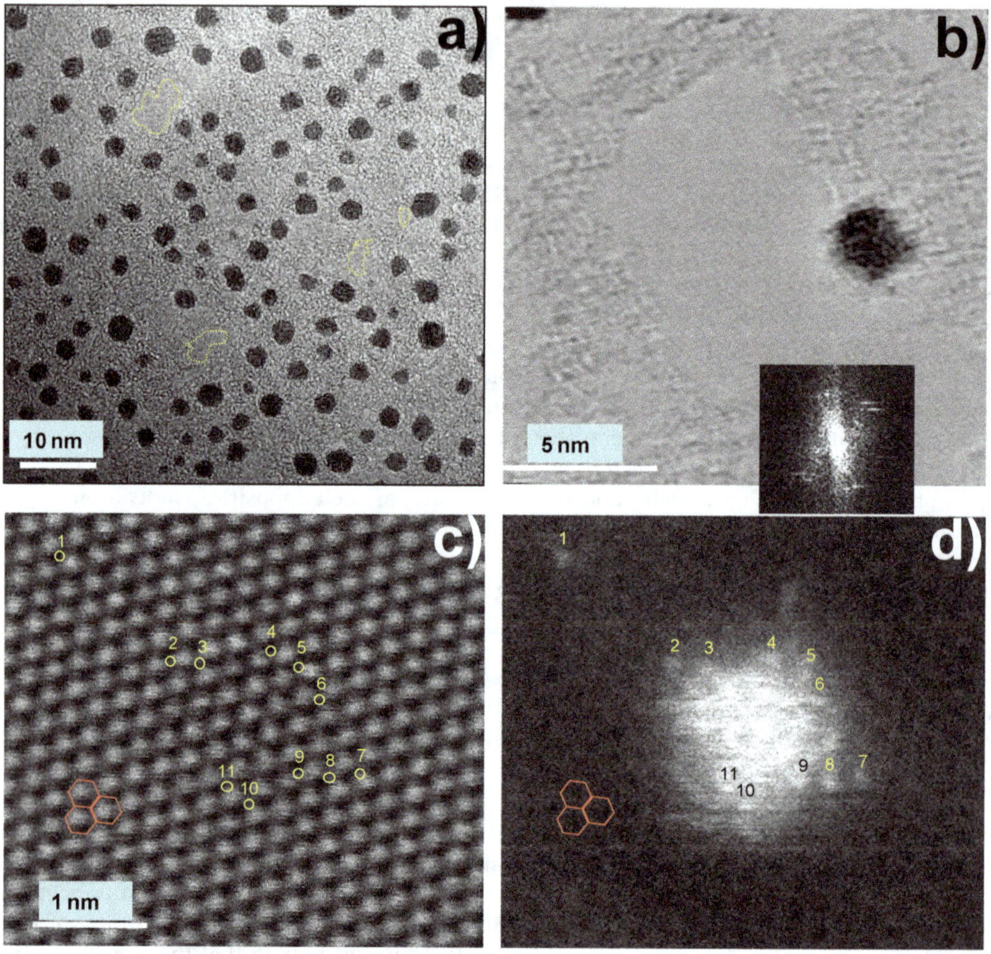

Fig. 11. a) TEM image of 0.1 nm gold evaporated on mono-layer graphene, b) BF STEM image of 0.1 nm gold evaporated on 2-layer turbostratic graphene with FFT of the image as inset to show, by the double spots, that the two layers are rotated with respect to each other, c) and d) BF and HAADF STEM lattice image of the same area in few-layer graphene. Here individual gold atoms can bee seen separated from the small cluster in the middle. Identical positions of the benzene rings are marked in red, showing that bright contrast in the BF image corresponds to the dark, hollow centres of benzene rings in the HAADF image. Single atoms 1-11 are marked with the numbers occurring just above the atoms in the HAADF image; identical places are marked by yellow circles in the BF image, showing that Au-atoms sit on T-sites (see fig. 12e) on the sample surface. All images represent raw, unfiltered data.

Iron-atoms have not been deliberately deposited, but reside as impurities on many samples, presumably as a result of chemical processing. The graphene sample in the here presented results was obtained by micromechanical exfoliation. Figure 12a) shows again that on uncontaminated single layer areas (black patches) the sticking probability is very small: all Fe clusters and single atoms (white spots) are located on hydrocarbon contamination (greyish areas). On the surfaces of multi-layer graphene on the other hand individual atoms can be seen. Figures 12c) and d) are HAADF and BF images of 1-4-layer staggered graphene sheets; with the sheet number increasing from the right to the left (i.e., on the right edge of the image is one sheet). Fe-atoms are clearly visible in (c) and corresponding positions have been circled in (d). The spectrum on the right of fig. 12c) is an energy loss spectrum taken on an individual Fe-atom, showing the characteristic $L_{2,3}$ absorption edge of Fe at ~708 eV. In fig.12b) the areas around two Fe-atoms are enlarged with a model of the AB-stacked graphene lattice overlaid. From this it becomes clear that the Fe-atoms in both cases sit on B-sites. This contradicts calculations, which predict H-sites.

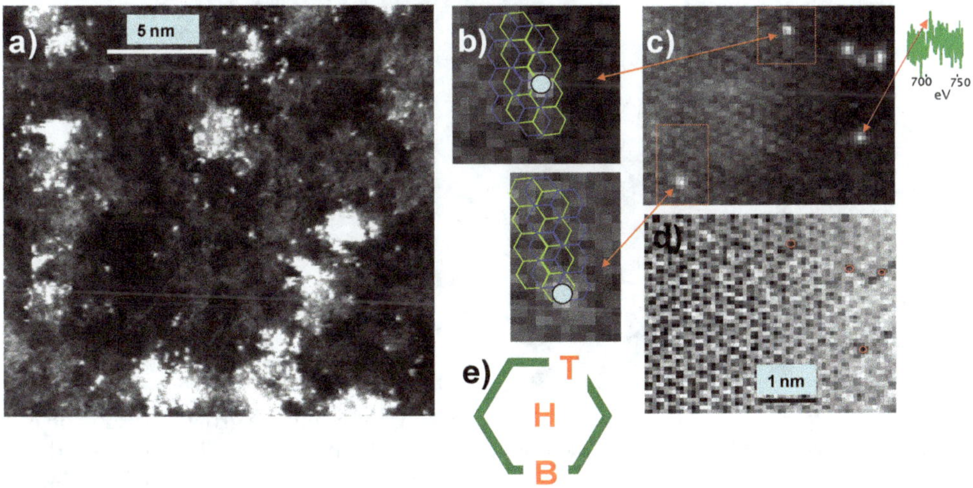

Fig. 12. a) Fe-atom contamination on single layer graphene, showing that all Fe- impurities sit on hydrocarbon contamination and not on clean graphene patches, b) enlargement of the red framed areas in the atomic resolution HAADF image of few-layer AB-stacked graphene in (c). The model of the lattice and positions of the Fe-impurity atoms are overlaid, c) HAADF image of larger area showing individual Fe-atoms on few-layer AB-stacked graphene. The inset on the right hand side of (c) shows an EEL spectrum with Fe $L_{2,3}$ absorption peak, taken on the arrowed atom. d) BF STEM lattice image of the same area as in (c); positions identical to those of atoms in the HAADF image in (c) are circled in red. The graphene sheets are staggered starting with a single sheet with on the right, followed by a second, third and fourth sheet towards the left with edges running parallel to the sides of the frame. e) model of benzene ring with sites of impurity atoms.

Nickel-atom contamination is shown in figure 13. The graphene film in this case was grown by CVD on Si with a sacrificial Ni-layer on top of the Si-substrate. Ni-atoms have remained 'stuck' to the film after lift-off. The image in this example is a low-pass filtered HREM phase contrast image taken in a TEM. The large underlying image shows mono-layer Ni-atom islands or – rafts on top of graphene, which itself consists of a patchwork of 1 and 2 layers and more.

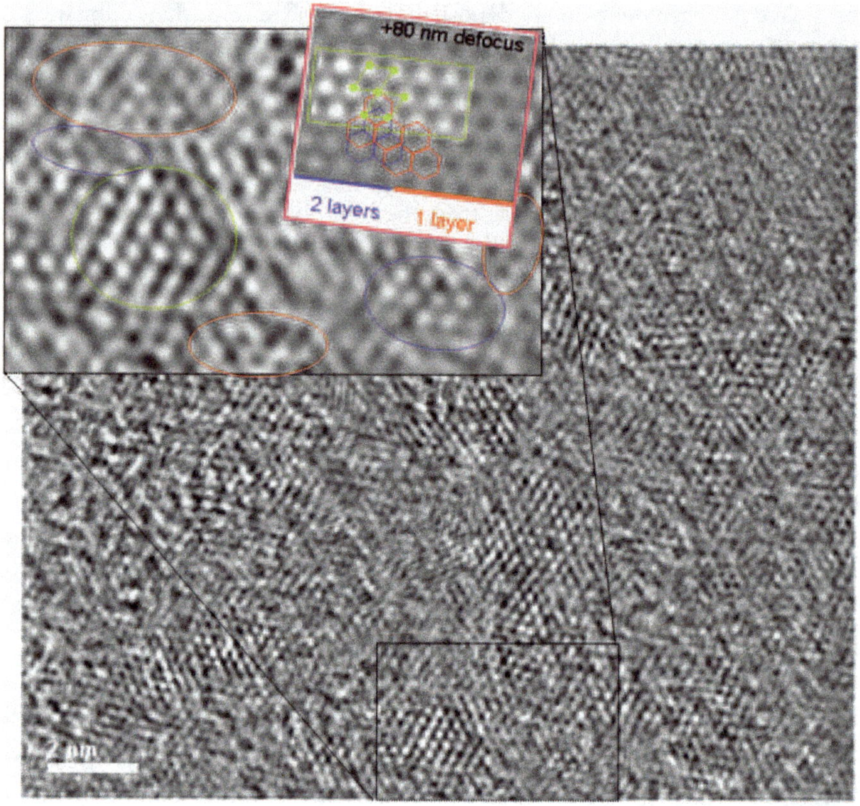

Fig. 13. Low pass filtered HREM image, taken in a TEM, of graphene grown by CVD on a sacrificial Ni-layer on Si. The area in the image consists of patches of mainly mono- and bi-layers, as well as of mono-layer Ni-islands. The larger inset is an enlargement of the black-framed area, the smaller inset is an image simulation, using the TEMSIM program, of mono- and bi-layer graphene with a mono-layer island of Ni stretching over both. Features with similar contrast in the experimental image in the larger inset are circled with the respective colours. The focus value of +80 nm achieves strongest contrast of Ni and graphene simultaneously.

The contrast in phase contrast images is not straight-forwardly interpretable and depends on sample lens aberrations, sample thickness and defocus, and it is common practice to run image simulation., if information other then that about lattice periodicities and perfection is to be extracted. We have used the TEMSIM code (Kirkland, 2005). The small inset shows simulated HREM contrast for 1 and 2 layers and a Ni-atom island covering parts of both, the single and the bi-layer. The model lattice for the graphene layers (red and blue) and a Ni-atom raft (green) is overlaid. The simulations were run for a series of focus values; we have depicted the simulation for 80 nm defocus, as this focus condition achieved the strongest contrast (best visibility) in the simulated images, so this value would be 'automatically' tuned to in the experiment. It should be noted that in overfocus conditions Ni-atoms have bright contrast similar to the centres of benzene rings in 2-layer graphene, whereas in 1-layer graphene the centres of benzene rings are dark. Comparing this to the larger inset, which in turn is an enlargement of the black famed area in the underlying image, we can then identify single layer (circled in red) and bi-layer (circled in blue) patches. Atoms constituting a Ni-island (circled in green) can be recognised by their strong, bright contrast (like in the image simulation). In relationship to the C-hexagons the position of these Ni-atoms can be assigned T-sites in benzene ring in agreement with above mentioned calculations.

5.2 Impurities with Z≤ 6

Interesting potential electronic dopants in graphene, such as nitrogen and ***boron***, due to the similarity in atomic number with carbon, cannot easily be detected in HAADF images. However, their presence, and even local bonding environment, should be accessible via core loss EELS. With probe aberration corrected STEM it is possible to detect EEL signals and positions of single atom (we have already shown an example of this in fig. 12c). Boron was ion implanted at 100 eV to a level giving on average one atom per nm^2 ($10^{14}cm^{-2}$) (Bangert et al., 2010a) Figure 14a) is an atomic resolution STEM BF image of few-layer graphene. A spectrum image was taken over the whole image area, and pixels, in which B K-edge signals were detected, are filled in yellow. Although the dopants appear dispersed the retention is less then 10%. We do presently not have an explanation as to why the retention is so small, other then that the implantation energy might not have been large enough to incorporate B in the lattice. This might have been the case, since Pomoell et al. (2003) estimate that ions ranging from carbon to xenon loose on average 300 eV when penetrating a single walled CNT, i.e., 150 eV per graphene sheet. Figure 14f) shows EEL spectra of the B K-edge, extracted from the yellow pixels. Although noisy, the spectra exhibit a sharp peak at 192 eV, which very much resembles the π^*-peak spectra of boron-doped carbon nanotubes measured by Fuentes et al. (2003), indicative for substitutional boron.

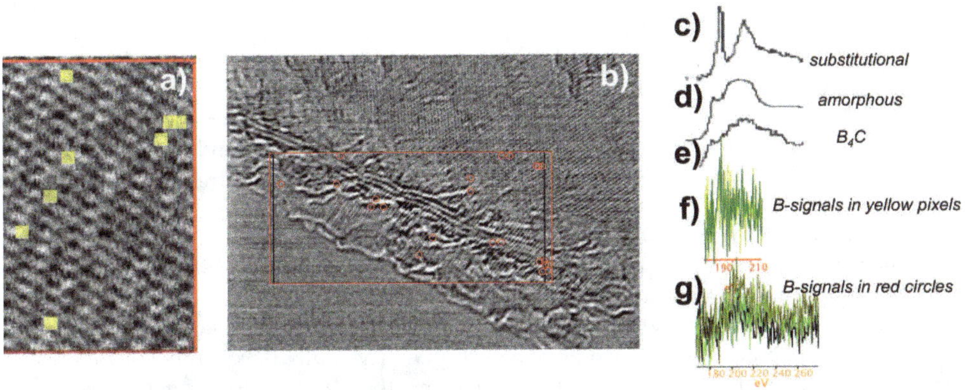

Fig. 14. a) STEM BF image of few-layer graphene and b) of staggered graphene sheets, starting with a single sheet, whose edge runs diagonally from left side to bottom right, implanted with B to a dose of 10^{14} atoms cm^{-2} at an energy of 100 eV. Positions were B K-edge EEL signals were recorded in spectrum images (area indicated by red frames) are indicated by yellow squares (a) or red circle (b), c) literature reference spectrum of BN (Gatan EELS atlas), d) of amorphous B, e) of B_4C, f) spectra extracted from yellow pixels in (a), g) spectra extracted from red circles in (b).The spectra in (f) and (g) indicate that B assumes substitutional positions in multilayers, and attaches along graphene edges in form of few-atom clusters.

Interestingly, the rather strong B-signal obtained from staggered graphene sheets in fig. 14g) does not exhibit the π^*-peak. Signal positions are indicated by red circles in fig.14b); they coincide in most cases with the edge of graphene sheets: the leftmost portion of the image in 14b) represents a single layer, followed (towards the right) by the 'frayed' edge of a second sheet along which a number of signal positions are marked. This edge is followed by, to a large part, straight edges of four further graphene sheets, with a cluster of boron signal positions marked at the bottom left hand corner of the image, and two further positions higher up along these edges. Typical spectra of the boron K-edge in red-circled pixels are shown in fig. 14g). Comparison with spectra of amorphous boron (fig.14d) and B_4C (fig.14e) suggests that small few-atom boron clusters have formed attached to sheet edges or mono-atomic steps. Hence, although there is indication of substitutional B-incorporation via low energy ion implantation in few and many-layer samples (as in the case of carbon nanotubes (Bangert et al., 2010a), B does not seem to substitute atoms in single, bi- or triple-layers, nor does it appear bound to the surface. As in the case of metal impurities in previous examples mobilities of single-atom species on graphene surfaces seem very high.

The possibility for ***atomic hydrogen*** to be chemi- or physisorbed on graphite surfaces (Jeloaica & Sidis, 1999) and graphene has been researched theoretically and experimentally (e.g., Ito, Nakamura & Takayama, 2009; Elias et al., 2009), and direct experimental observation has been attempted via transmission electron microscopy imaging (Meyer et al., 2008b). H-dosing appears promising for band-gap engineering of graphene. In order to carry out investigations of the effect of hydrogen on the graphene atomic and electronic structure we have dosed graphene with hydrogen by exposure to a hydrogen plasma, described in Elias et al., (2009).

H cannot be revealed in HAADF images, so in order to assess H adsorption on graphene we have carried out EELS. Figure 15a) shows low loss spectra, which were obtained of uncontaminated graphene patches, typically a few 10 nm^2 in size. The blue spectrum in (a) is a reference spectrum taken of pristine, hydrogen-free graphene. To ensure that we have hydrogen-free surface areas few-layer graphene regions were repeatedly scanned in the STEM. The scan raster was successively decreased, and HAADF images in combination with EEL measurements revealed material being peeled off layer-by-layer, until finally a hole appeared. The blue spectrum is from the last remaining layer after such a repeat-scan procedure; the Lorentzian fit curve (red dotted line) was obtained from least squares fitting, and was used as fit for the rise of the $\sigma+\pi$ plasmon. Spectra of such freshly revealed areas did not show the bump at 13 eV like in the orange spectrum above, which was obtained from a deliberately hydrogen dosed sample.

We applied the spectrum imaging method (Jeanguillaume & Colliex, 1989), and scrutinised entire EEL-spectrum images for occurrence of the ~13 eV core-level excitation signal of hydrogen. Spectrum images were obtained in the following way: a raster was defined over an area of an uncontaminated patch and in each pixel (typically $0.3x0.3nm^2$) of this raster a spectrum was taken. Intensity maps, extracted from these spectrum images are shown in figs 15b) and c). These maps were obtained by displaying in each pixel the intensity of the respective spectrum integrated over an energy window ~12-14 eV, after having subtracted the plasmon background determined by the Lorentzian fit. These maps represent the distribution and relative amount of hydrogen on the graphene surface. Since spectra from individual pixels display a great deal of noise, we carried out Principle Component Analysis (PCA) on the spectrum images prior to background subtraction (e.g., Borglund, Astrand & Csillag, 2005). This reduces the noise significantly, and clearly reveals the bump on the rise of the $\sigma+\pi$ plasmon at ~ 13.2 eV as seen in the orange curve in fig.15a).

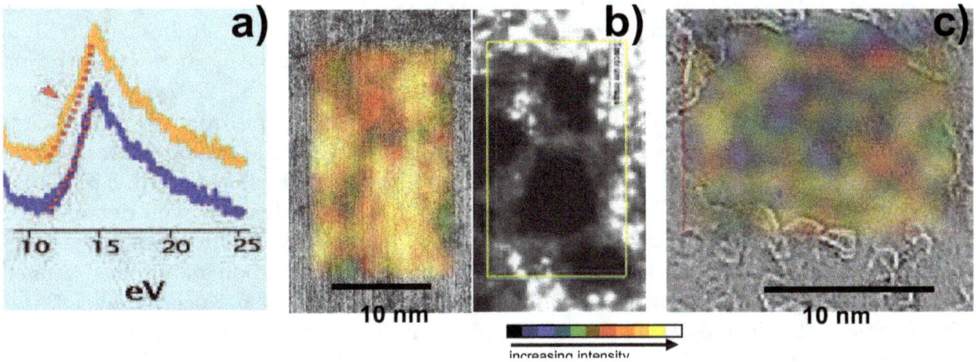

Fig. 15. a) low loss EEL spectra showing the π+σ-plasmon of ***H-free*** (electron beam 'zapped') single layer graphene with Lorentzian fit (blue spectrum) and of hydrogen-dosed graphene with the same Lorentzian fit-curve (orange spectrum), b) EEL intensity map of hydrogen-dosed graphene, showing the integrated intensity -after Lorentzian background subtraction- in the energy region 12-14 eV; the map is overlaid on the respective STEM BF image area (left); the corresponding HAADF image is shown on the right, c) EEL intensity map of pristine, i.e., undosed graphene, obtained with the same procedure as in (b), overlaid on the respective STEM BF image area.

Figure 15b) shows an EEL intensity map of the H-core loss signal of hydrogen-dosed graphene. The intensity map is overlaid on the BF STEM image area where the spectrum image was taken. The corresponding HAADF image is also shown. It reveals a hydrocarbon 'net' (greyish contrast), bordering uncontaminated single layer graphene patches (black), as well as contaminants of higher atomic number (white dots) on top of hydrocarbons. The intensity map shows highest H-signals (yellow/orange) in regions free of hydrocarbons, whereas hydrocarbon contamination shows up with lower intensity values (green). The reason for this is that hydrocarbons with longer chains have a distribution of energies of the molecular orbitals causing a spread in the H-core level energies with a shift towards lower values (Murell & Schmidt, 1972). These will not have been picked up in the energy window set for the intensity map in (b). Figure 15c) shows an intensity map obtained with the same energy window of 12-14 eV in an uncontaminated region in pristine graphene, revealing that hydrogen is present on clean graphene even without hydrogenation. Again the map is overlaid on the STEM BF image. The amount of hydrogen, however, is significantly lower (by 2-3 times) then in the dosed sample, as evidenced by the colour-scale. A detailed discussion of hydrogenation studied by EELS can be found in (Bangert et al., 2009b).
The question about coverage, detailed atomic structure and bonding arrangements of the hydrogenated areas poses itself. We can get a fair insight into this via highly spatially resolved valence band EELS, i.e., EELS in the extreme low loss regime in combination with DFT calculations; an introduction into this will be given in paragraph 6.

6. Plasmons in graphene: indicators of local electronic bandstructure

6.1 Pure graphene

From an experimental viewpoint, an essential task in the research into 2D structures is to provide evidence that they do indeed exist. This is especially important, because theory does not allow the existence of perfect crystals in 2D space. STM-, AFM- and Raman studies have given indirect proof of the existence of single-layer graphene. The most conclusive evidence for the existence of *free-standing* graphene, except for the HAADF studies reported in Bangert et al. (2009b) and Gass et al. (2008), a resumé of which has been given in paragraph 3, has been obtained from electron diffraction experiments (Meyer et al., 2007a).

By carrying out highly spatially resolved electron energy loss spectroscopy (EELS), we observe significant red-shifts in the frequency of plasmons in sample positions concomitant with single graphene sheets (Eberlein et al. 2008), very similar to those observed in SWCNTs (Stéphan et al., 2002). Surface plasmon behaviour in thin metal sheets is well documented experimentally and explained using dielectric theory (Raether, 1977; Otto, 1967). There are also numerous reports on surface plasmons in graphite and carbon nanotubes (Annet, Palmer & Willis, 1988; Laitenberger & Palmer, 1996; Perez & Que, 2006; Kociak et al., 2001).

A characteristic of thin foils is the vanishing of the bulk plasmon mode, leaving only the surface plasmon mode; the out-of-plane and in-plane contributions of the latter split in energy when the product of foil thickness d and plasmon momentum q, tends to zero; in graphite the maximum/minimum energy of the $\pi \rightarrow \pi^*$ transition tends to roughly 7 and 5 eV, and of the $\sigma \rightarrow \sigma^*$ transition to 20 and 15 eV, respectively (Stéphan et al., 2002). The ***E***-field of a fast moving particle is elongated along its direction of travel, therefore, when passing perpendicularly through a graphene foil mainly the out-of plane mode, with momentum ***q*** parallel to ***E***, should be excited. However, these modes are forbidden in a single layer (Eberlein et al. 2008), and they have weak intensity in graphite. In an EELS experiment carried out in a STEM, although the momentum transfer is close to zero, nonetheless, ***q*** has a considerable in-plane component (along ***a***, the in-plane spacing) because the collection angle is several mrad. For this reason we will observe surface and bulk plasmons excited with ***E*** (or ***q***) parallel to ***a***.

Graphene samples for the plasmon studies were obtained by exfoliation after micromechanical cleavage. An HAADF image of a region including one to up to a few graphene layers is shown in Fig. 16a), top panel; the narrow, rectangular bar shows the line along which the intensity (overlaid profile), integrated over the width of the bar, is traced. The step heights in the contrast profile are multiples of the smallest height at position 1 (to the right of which is vacuum), confirming that the contrast in this location arises from one single graphene sheet.

Plasmon spectra taken in positions near the edges of one, two, three and several sheets (fig.16a, bottom panel) show that the π-mode, at 7 eV in graphite, has shifted to 4.8 eV at position 1 relating to single graphene layer. Furthermore, the spectrum here exhibits only the $\pi+\sigma$ -surface mode at 15 eV and the 26 eV bulk mode of graphite is not present. The shape and intensity of the one-layer plasmon structure was repeatedly measured in different places on the same sample, in different samples and even in different experimental sessions. Given the same acquisition conditions (e.g., energy dispersion, electron beam current and dwell time), nearly identical spectra were obtained, which at the same time, constituted the lowest plasmon signal measured overall. For two sheets the triangular shape of the $\pi+\sigma$ resonance gives way to a plateau and the integrated intensity under the peak approximately doubles, for three layers it triples; this linear relationship carries on up to ~8 layers, but for increasing numbers of sheets, features above 15 eV start to appear and the plasmon maximum moves to higher energies accompanied by further broadening. The multi-layer plasmon structure in fig.16a) strongly resembles that of graphite. The plasmon characteristics are thus supreme indicators for the presence of single layers.

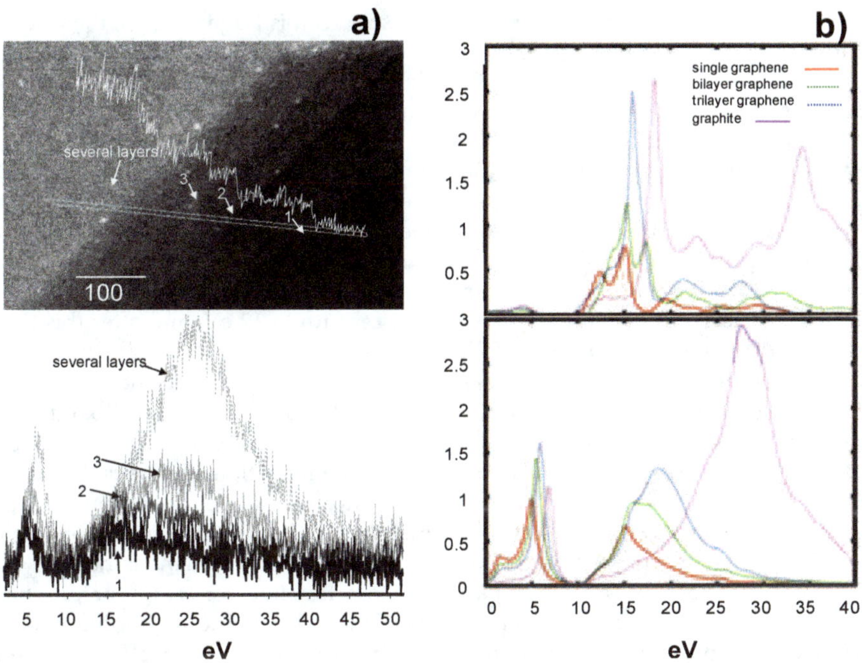

Fig. 16. a) top panel: HAADF image of staggered graphene layers with overlaid intensity profile, which was taken in the turquoise box. The arrows indicate the positions of the edges of new layers, the bottom right corner is vacuum; bottom panel: b) EEL spectra taken of one, two, three and six layers, showing the π- and the π+σ-plasmon shifting to higher energies with increasing layer number; b) AIMPRO calculations of EEL spectra of single, bi- and triple-layer graphene and of graphite for ***E***-fields (plasmon $\boldsymbol{q}$-vectors) parallel to the ***c***-axis (top panel) and parallel to the graphene sheet (bottom panel). The latter spectra show good agreement with the experimental spectra, indicating that the plasmons in graphitic structures are in-plane plasmons.

In support of the experimental results we have carried out EEL spectra calculations using the AIMPRO local density functional code (Jones et al., 2000; Fall et al., 2002) to evaluate the imaginary part of the dielectric tensor for q=0 and use the Kramers-Kronig relations to derive its real part. The loss function is then found for graphite, graphene and bi- and tri-multilayers. We have also carried out density functional calculations using the WIEN2K code (Nelhiebel et al., 1999), (results of this are shown in the last part), and, after optimising the input parameters and obtaining convergence, acquired identical results as with AIMPRO.

In modelling physical properties of extended periodic (e.g., crystalline) structures the desired process is applied to a unit cell of the structure and the 'outcome' for the entire sample obtained by assembling unit cells in a 3D repeat operation. The representation of nano-objects by repetition of a crystallographic unit cell within the nano-object is inappropriate (the object itself may only consist of few unit cells), here the usual approach is to build large 'supercells' with the entire structure of the nano-object placed in the middle surrounded by vacuum. A repeat structure of these supercells is then obtained and the calculations carried out as described above. If the supercell is large enough, the calculations of individual unit cells do not impact on each other and the modelled physical property represents that of the supercell alone. To find the required size of such a supercell is tricky, and often a series of calculations for increasing size is carried out until the 'outcome' converges.

Graphene, was modelled by expanding the lattice parameter along ***c*** (Eberlein et al., 2008); convergence occurs at around 20 lattice parameters in this direction, but qualitative changes in the low loss spectra already cease to happen at 5 *c*-parameters. Figure 16 b) shows calculated spectra for single, bi- and triple-layer graphene for plasmon propagation (momentum vectors) ***q*** along the ***c***-axis (top panel) and in-plane (bottom panel). Notably in single layers (red line) the loss for ***q*** | | ***c*** is almost zero up to 12 eV, and after this the onset occurs at a similar energy to graphite. However, the peak heights are very different from graphite. The disappearance of the 4 eV peak can be explained by a selection rule. As stated above the peak is due to a transition between occupied and unoccupied π * bands at the M (1/2,0,0) point of the Brillouin zone. (Inspection of the wave functions for these two states shows that both transform as ***pz,*** and are odd under the reflection symmetry present in the basal plane of graphene, but not AB graphite, and hence the dipole matrix element between them vanishes for transitions for which ***q*** is parallel to ***c***, although the transition is allowed for ***q*** parallel to ***a***.) This shows that the dielectric constant of graphene is not the same as graphite. Fig. 16b), bottom panel, compares the loss function for ***q*** | |***a*** for graphite and graphite. We note there are substantial red-shifts of the peaks found in graphite (Eberlein et al., 2008). The 7 eV plasmon peak has shifted downwards to about 4.8 eV while the broad peak around 27 eV has sharpened and shifted to 14 eV just as seen in the experimental spectra; in fact the latter seem to be entirely dominated by the in plane plasmon component (*q* | |*a*). The loss functions for bi-layers, triple-layers stacked as in graphite, and graphite itself are shown as green, blue and purple curves. The separation between periodically repeated multilayers was chosen as 5 times the separation in graphite. The supercell containing the trilayer for example, has three layers of graphene separated by the interlayer separation found in graphite, but the separation of these planes from similar planes in adjacent unit cells along the c-axis is now 5 times the separation in graphite. The peak positions depend on the number of layers. There is an increasing red shift of the π+σ plasmon resonance as the number of layers decreases. Markedly, there is also a red-shift to ~4.7eV in the graphitic π-plasmon, the latter only occurs for ***q*** | |***c*** (see bottom panel 16b). The relative amplitudes of the two plasmon peaks here for the different layers seem roughly consistent with experimental spectra in Fig.16a).

It is the π-plasmon, which bears significant information about electronic properties of graphene and is especially sensitive to modifications of graphene. In the last sub-paragraph we will give a glimpse of the wealth of properties of graphene that can be deduced from the π-plasmon.

6.2 π-plasmon in modified graphene

To give an example of the sensitivity of the low loss region to modification of graphene we return to the hydrogenation issue. In paragraph 5 we observed presence and distribution of hydrogen atoms by their core level excitation, but the detail of the atomic-scale structure was not revealed; questions concerning issues that might have implications for electronic properties include: Is hydrogenation complete? If not, is there ordering of H-atoms? Is H bonded on both sides? If not, does this induce strain (Xue & Xu, 2010)? Does hydrogenation induce diamond-like, corrugated atomic stucture (Sofo, Chaudhari & Barber, 2007; Elias et al. 2009)?

In all spectra, which showed the H-core loss signal, additional structure was found in the low loss region on the high energy tail of the π-plasmon. Experimental low loss spectra of pure and of hydrogenated graphene, including the π-plasmon, can be seen in figures 17a) (purple curve) and 17c) (orange curve). An extra peak at ~7eV has occurred in the latter, whereas the pre-plasmon peak at ~2eV in pure graphene has vanished. EEL intensity maps of the integrated intensity (after subtraction of a fitted plasmon tail) in the 6-8 eV region, from the same area as in fig. 15b show identical intensity distributions as seen in the latter figure. This suggests that the 7-eV feature is induced by hydrogen. More interestingly, the π-plasmon remains present throughout the spectrum image, in co-existence with the new feature, which is similar to a feature occurring on unpassivated, reconstructed diamond surfaces (Bangert et al., 2006), which have undergone reconstruction of their dangling bonds into π-bonded chains. Hydrogenation ought to passivate the π-bonds, and cause formation of sp^3 bonds instead, and hence dramatically diminish the occurrence of the π-plasmon. However, the π-plasmon in graphene exposed to a hydrogen plasma is nearly undiminished in areas, which, at the same time, show the H-ground state excitation. Co-existence of the π-plasmon and the 13.2 eV signal suggests that hydrogenation is not complete, i.e., not every C-atom is bonded to an H-atom.

In order to support the above interpretation of the EELS observations and to reproduce the EEL spectra theoretically, DFT calculations using the WIEN 2K code, have been performed. WIEN 2K is an all-electron code and uses the General Gradient Approximation with a mixed linearized augmented plane basis set of wave functions (Nelhiebel et al. 1999; see also Blaha et al. 2010). There is a fair number of bandstructure calculations of hydrogenated graphene in the recent literature, predicting the opening of a band-gap of ~3.7 eV and C atoms displaced from planar geometries by H atoms (e.g., Haberer et al., 2010), optimum corrugation angles and resulting bandstructure (Sofo, Chaudhari & Barber, 2007), and influence of hydrogenation on strain and corrugation angle (Xue & Xu, 2010).

As mentioned before WIEN2K calculations carried out on graphite and pure graphene show identical results to those from ab-initio calculations using the AIMPRO code, which reproduces measured spectra very well (Eberlein et. al. 2008) This gives us confidence in using WIEN calculations -not previously applied to graphene- on hydrogenated structures. It is beyond the scope of this chapter to go into detailed discussion of the input parameters and how they were established; it suffices to say here that all parameters (k-point number, maximum g-vector, muffin tin potential radius) were the minimum value requirements for convergence of the calculations to occur, and that a detailed discussion of the procedures and the results of WIEN calculations of hydrogenated graphene can be found in Bangert et al. 2010b.

Also shown in figure 17a) is a calculated EEL spectrum for single-layer graphene. It matches the experimental spectrum very well. Surprisingly flat and corrugated unhydrogenated graphene both produce similar π-plasmons, indicating the presence of a π-electron system even in corrugated graphene. Figure 17b) shows the evolution of the π-plasmon in fully hydrogenated graphene (also termed graphane) with increasing corrugation angle. Interestingly, flat graphane, similar to graphene, possesses a π-plasmon. However, its energy is slightly higher then experimentally observed. The π-plamon has nearly vanished

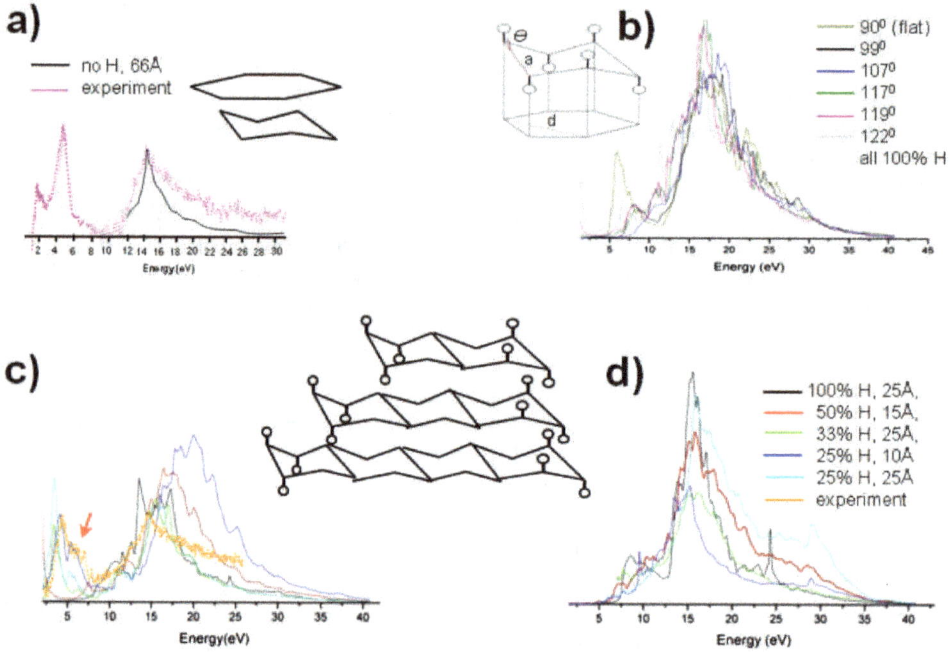

Fig. 17. WIEN 2K calculations of low loss EEL spectra for a) non hydrogenated graphene in flat and corrugated geometry for the in-plane plasmon component ($q||a$) with overlaid experimental spectrum. Note the agreement in the pre-plasmon structure; b) fully hydrogenated graphene with varying corrugation angle Θ, again, for ($q||a$). c) and d) show calculated in-plane and out-of-plane ($q||c$) plasmon components for corrugated graphene (Θ=114^0) with different hydrogen content (given in % in the legend; model structures are shown in the sketch). The orange spectrum overlaid in (c) is an experimental spectrum of a hydrocarbon-free region in H-dosed graphene. Note the disappearance of the pre-plasmon feature and the occurrence of an extra peak at ~7eV. The distances given in Å in the legends are the separation distances of the graphene sheets, or the supercell dimension along the c-axis. Further detail and fuller explanation of the input parameters (k-point number, muffin-tin radii etc) are given in Bangert et al. 2010b.

for 10^0 out-off-plane bending of bonds, and has disappeared for 22^0 bond bending. A feature at ~6-7 eV emerges at the rise of a (presumably) pure σ-plasmon, which sits at slightly higher energy then the π+σ-plasmon in graphene. Hence sp^3 character does become established in corrugated graphene, when all bonds are used. It should be noted that (a) and (b) show the in-plane plasmon component. Figures 17c) and d) show plasmon calculations of the in-plane and the out-off-plane plasmon component for different hydrogen coverages for a sheet corrugation angle of 114^0 (as deduced from diffraction experiments (Elias et al, 2009)). The overlaid experimental spectrum fits the >25% hydrogenation cases rather well. The out-off plane component does not have a π-plasmon for any hydrogen coverage. Models to describe possible structural scenarios are shown as inset. It becomes clear that corrugation is essential to produce the ~7eV feature, as is occurrence of sp^3-character. In order be preserve π-bonding, however, the corrugation angles have either to be small (as would be the case for strained graphene, if it is hydrogenated mainly on one surface (Xue & Xu, 2010)) or unhydrogenated areas have to be present,. Alternatively or simultaneously, hydrogenated patches could be interleaved with pure graphene areas. The low loss region thus provides a wealth of information, which we have only just started to explore.

7. Acknowledgements

We gratefully acknowledge our colleagues from the University of Manchester, A. Geim and K. Novoselov for stimulating discussions and R. R. Nair for provision and hydrogenation of samples. We thank Q. Ramasse from the SuperSTEM for conducting the Au-on-graphene STEM measurements and J. Vandenberg from Salford University for the low energy ion-implantations.

8. Conclusions

Combined HAADF/BF imaging in an AC STEM has proven a most powerful method to 'see' directly the atomic-scale landscapes of graphene; the former imaging mode provides chemical information and position of atoms, the latter, due to the stronger contrast, helps relate positions of ad-atom/ impurities with sites on benzene rings. Ultra-high resolution EELS has proven an invaluable add-on to identify the chemical nature of individual atoms, especially those with Z≤6, and to gain information about local bonding and electronic bandstructure.

The imaging studies revealed extensive hydrocarbon contamination on every single graphene sample, independent of growth method or fabrication procedure, as well as a fair proportion of multiple layers and -in CVD grown samples- predominance of turbostratic arrangements. Here we present facts concerning uncontaminated ('model') graphene, which exists in patches of only few 10s of nm in diameter interspersed with extended chain-like adsorbate networks. In 'model' graphene we have shown existence of nanometer-size ripples, point- and ring defects and omnipresence of C-adatoms above C-C bonds. We have identified metal impurity sites and found Au- and Ni-atoms taking up T-sites on benzene rings, and Fe-atoms to occupy B-sites; the former in agreement with and the latter contradicting theoretical predictions. H can be found on clean graphene, and to a much larger extent on hydrogen-dosed graphene. EELS in combinations with DFT WIEN calculations shows the structure of the latter is concomitant with corrugated graphene possessing partial H-coverage and small corrugation angles, or/and co-existence of unhydrogenated areas. Plasmons in graphene show unique behaviour and can be used to identify single layers.

9. References

Annet, J. F., Palmer, R. B & Willis, R. F., 'Surface dielectric response of a semimetal- electron energy loss spectroscopy of graphite', *Phys. Rev. B* 37, 2408-2414 (1988)

Aktürk, O.Ü. & Tomak, M., 'AuPt clusters adsorbed on graphene studied by first-principles calculations', *Phys. Rev. B* 80, 085417 (2009)

Bangert, U., Barnes, R., Hounsome, L. S., Jones, R., Blumenau, A. T., Briddon, P. R., Shaw, M. J. & Öberg, S., 'Electron energy loss spectroscopic studies of brown diamonds' *Phil Mag* 86 (29-30), 4757-4779 (2006)

Bangert, U., Eberlein, T., Nair, R. R., Jones, R., Gass, M. H., Bleloch, A. L., Novoselov, K. S., Geim, A., Briddon P. R., 'STEM plasmon spectroscopy of free standing graphene', *phys. stat. sol. A* 205 (9), 2265-2269 (2008)

Bangert, U., Gass, M. H., Bleloch, A. L., Nair, R. R. & Geim, A. K., 'Manifestation of ripples in free-standing graphene in lattice images obtained in an aberration-corrected scanning transmission electron microscope', *phys. stat. sol. A* 206 (6), 1117-1122 (2009a)

Bangert, U., Gass, M. H., Bleloch, A. L., Nair, R. R. & Eccles, J., 'Nanotopography of graphene' *phys. stat. sol. A,* 206 (9), 2115-2119 (2009b)

Bangert, U., Bleloch, A. L., Gass, M. H., Seepujak, A. & Vandenberg, J., 'Doping of few-layered graphene and carbon nanotubes using ion implantation', *Phys. Rev. B* 81(24), 245423 (2010a)

Bangert, U., Pan, C.-T., Nair, R. R. & Gass, M. H., 'Atomic structure and low electron loss characteristics of hydrogen-dosed graphene deduced from STEM experiments and DFT-WIEN calculations', *Appl. Phys. Lett.*, 97, 253118 (2010b)

Batson, P. E., 'Simultaneous STEM imaging and electron-energy-loss-spectroscopy with atomic-column sensitivity' *Nature* 366 (6457), 727-728 (1993)

Blaha, P., Schwarz,K., Madson, G., Kvasnicka D. & Luitz, J, *User's guide WIEN 2k_10.1* at www.wien2k.at/reg_user/textbooks/usersguide.pdf

Booth, T. J., Blake P., Nair, R. R., Jiang, D., Hill, E.W. , Bangert, U., Bleloch, A. L., Gass, M., Novoselov, K. S. M., Katsnelson , I. & Geim , A. K., 'Macroscopic graphene membranes and their extraordinary stiffness', *Nano Lett.*, 9 (4), 1433-1441 (2008)

Borglund, N., Astrand, P.-G. & Csillag, S., 'Improved background removal method using principal components analysis for spatially resolved electron energy loss spectroscopy' *Microsc. Microanal.* 11(1), 88-96 (2005)

Browning, N. & Pennycook, S., 'Atomic-resolution electron energy-loss spectroscopy in the scanning transmission electron microscope', *J. Microsc.-Oxford* 180, 230-237 (1995)

Brydson, R., *Electron Energy Loss Spectroscopy*, Garland Science, England (2003)

Boukhvalov, D.W. & Katsnelson, M.I., 'Destruction of graphene by metal adatoms', *Appl. Phys. Lett.* 95 (2), 023109 (2009)

Bourelle, E., Konno, H. & Inagaki, M., 'Structural defects created on natural graphite surface by slight treatment of oxygen plasma - STM observations', *Carbon* 37 (12), 2041-2048 (1999)

Chan, K. T., Neaton, J. B. & Cohen, N. L., 'First-principles study of metal adatom adsorption on graphene', *Phys Rev B* 77 (23), 235430 (2008)

Charlier, J.-C., 'Defects in carbon nanotubes', *Acc. Chem. Res.* 35 (12), 1063-1069 (2002)

Charlier, J.-C., Arnaud, L. , Delgado, M., Demoisson, F., Espinosa, E. H., Ewels, C. P., Felten, A., Guillot, J., Ionescu, R., Leghrib, R., Llobet, EMansour, A., Migeon, H-N, Pireaux, J-J., Reniers, F., Suarez-Martinez, I., Watson, G. E. & Zanolli, Z., 'Carbon nanotubes randomly decorated with gold clusters: from nano(2)hybrid atomic structures to gas sensing prototypes', *Nanotechnology* 20 (37), 375501 (2009)

Crespi, V. H., Cohen, M. L. & Rubio, A., 'In situ band gap engineering of carbon nanotubes', *Phys Rev Lett* 79 (11), 2093-2096 (1997)

Duplock, E. J., Scheffler, M. & Lindan, P. J. D, 'Hallmark of perfect graphene', *Phys Rev Lett* 92 (22), 225502 (2004)

Eberlein, T., Bangert, U., Nair, R. R., Jones, R., Gass, M., Bleloch, A. L., Novoselov, K. S., Geim, A. K. & Briddon, P. R., 'Plasmon spectroscopy of free-standing graphene films', *Phys Rev B* 77, 233406 (2008)

Egerton, R. F., *Electron Energy Loss Spectroscopy in the Electron Microscope'*, Plenum Press N.Y. (1986)

Elias, D. C., Nair, R. R., Mohiuddin, T. M. G., Morozov, S. V., Blake, P.,. Halsall, M. P., Ferrari, A. C., Boukhvalov, D. W., Katsnelson, M. I.,Geim, A. K. & Novoselov, K. S., 'Control of Graphene's Properties by Reversible Hydrogenation: Evidence for Graphane', *Science* 323 (5914), 610-613 (2009)

Ewels, C. P., Heggie, M. I. & Briddon, P. R., 'Adatoms and nanoengineering of carbon', *Chem. Phys. Lett.* 351, 178 (2002)

Fall, C. J., Blumenau, A. T., Jones, R., Briddon, P. R., Frauenheim, T.,Gutiérrez-Sosa, A., Bangert, U., Mora, A. E., Steeds, J. W. & Butler, P. E., 'Dislocations in diamond: Electron energy-loss spectroscopy', *Phys. Rev. B* 65 (24), 205206 (2002)

Fasolino, A., Los, J. H. & Katnelson, M. I., 'Intrinsic ripples in graphene', *Nature Materials 6* (11), 858-861 (2007)

Fuentes, G. G., Borowiak-Palen, E., Pichler, T., Liu, X., Graff, A., Behr, G., Kalenczuk, R. J., Knupfer, M. & Fink, J., 'Electronic structure of multiwall boron nitride nanotubes', *Phys. Rev. B* 67 (3), 035429 (2003)

Gan, Y., Sun, L.T, & Banhart F., 'One- and two-dimensional diffusion of metal atoms in graphene', *Small* 4 (5), 587–591 (2008)

Gass, M. H., Bangert, U., Bleloch, A. L., Wang, P., Nair R. R. & Geim, A. K., 'Free-standing graphene at atomic resolution', *Nature Nanotechnology* 3 (11), 676-681, (2008)

Goodhew, P. & Bleloch, A. L., 'SuperSTEM - the microscope with glasses', Mater. World 11 (3), 23-24 (2003)

Haberer, D., Vyalikh, D. V., Taioli, S., Dora, B., Farjam, M., Fink, J., Marchenko, D., Pichler, T., Ziegler, K., Simonucci, S., Dresselhaus, M. S., Knupfer, M., Büchner B.& Grüneis, A., 'Tunable Band Gap in Hydrogenated Quasi-Free-standing Graphene', *Nano Lett.* 10, 3360-3366 (2010)

Hahn, R. & Kang, H., 'Vacancy and interstitial defects at graphite surfaces: Scanning tunneling microscopic study of the structure, electronic property, and yield for ion-induced defect creation', *Phys. Rev. B* 60 (8), 6007-6017 (1999)

HREM Research Inc., DeConvHAADF User's Guide (2009)

Hunt, A. & Williams, D. B., 'Electron energy-loss spectrum imaging', Ultramicroscopy 38 (1), 47-73 (1991)

Ito, A., Nakamura, H. & Takayama, A., 'Molecular Dynamics Simulation of the Chemical Interaction between Hydrogen Atom and Graphene', *J. Phys. Soc. Jpn* 77 (11), 114602 (2008)

Jenkins, G. M., 'Basal plane distortion in pyrolytic carbon', *Carbon* 7, 9 (1969)

Jeanguillaume, C. & Colliex, C., 'Spectrum image- the next step in EELS digital acquisition and processing', *Ultramicroscopy* 28 (1-4), 252-257 (1989)

Leenaerts, O., Partoens, B., Peeters, F. M., 'Adsorption of H_2O, NH_3, CO, NO_2, and NO on graphene: A first-principles study', *Phys. Rev. B* 77 (12), 125416 (2008)

Leenaerts, O., Partoens, B., Peeters, F. M., 'Adsorption of small molecules on graphene', *Microelectronics Journal* 40 (4-5), 860–862 (2009)

Jeloaica, L. & Sidis, V., *Chem. Phys. Lett.* 300, 157 (1999)

Jones, R., Coomer, B. J., Goss, J. P., Hourahine, B. & Resende A., 'The interaction of hydrogen with deep level defects in silicon' in: *Special defects in semiconducting materials,* Volume 71 of *Solid State Phenomena*, Chapter 173-248. Scitech Publications Ltd., Zuerich-Uetikon, Switzerland, (2000)

Kirkland, E., *Advanced Computing in Electron Microscopy*, New York: Plenum (2005)

Kociak, M., Stéphan, O., Henrard, L., Charbois, V., Rothschild, A., Tenne, R. & Colliex, C., 'Experimental evidence of surface-plasmon coupling in anisotropic hollow nanoparticles', *Phys. Rev. Lett.* 87 (7), 077501 (2001)

Krasheninnikov, A. V., Nordlung, K., Sirviö, M., Salonen, E. & Keinonen, J., 'Formation of ion-irradiation-induced atomic-scale defects on walls of carbon nanotubes', *Phys Rev B* 63 (24), 245405 (2001)

Krasheninnikov, A. V., Lehtinen, P. O., Foster, A. S., Pyykko, P., Nieminen, R. M., 'Embedding transition-metal atoms in graphene: structure, bonding, and magnetism', *Phys. Rev. Lett.* 2 (12), 126807 (2009)

Laitenberger, P. & Palmer, R. B., 'Plasmon dispersion and damping at the surface of a semimetal', *Phys. Rev.Lett.* 76 (11), 1952-1955 (1996)

Mao, Y., Juan, J. & Zhong, J., 'Density functional calculation of transition metal adatom adsorption on graphene', *J Phys: Condens Matter* 20 (11), 115209 (2008)

Meyer, J. C., Geim, A. K., Katsnelson, M. I., Novoselov, K. S., Booth, T. J. & Roth, 'The structure of suspended graphene sheets,' *Nature* 446, 60-63 (2007a)

Meyer. J. C., Geim, A. K., Katsnelson, M. I., Novoselov, Obergfell, D., Roth, S., Girit, C. & Zettl, A., 'On the roughness of single- and bi-layer graphene membranes', *Solid State Commun.* 143, 101-109 (2007b)

Meyer, J. C., Kisielowski, C., Erni, R., Rossell, M. D., Crommie, M F & Zettl, A., 'Direct imaging of lattice atoms ond topological defects in graphene membranes' *Nano Lett.* 8(11), 3582 (2008a)

Meyer, J. C., Girit, C. O., Crommie, M. F. & Zettl, A., 'Imaging and dynamics of light molecules on graphene', *Nature* 454, 319-322 (2008b)

Murrell, J. N.& Schmidt, W., *J.* 'Photoelectron spectroscopic correlation of molecular-orbitals of methane, ethane, propane, isobutane and neopentane', *Chem. Soc.-Faraday Trans.II,* 68 (10) 1709-1718 (1972)

Nelhiebel, M., Louf, P. H., Schattschneider, P, Blaha, P, Schwarz, K.& Jouffrey, B., 'Theory of orientation-sensitive near-edge fine-structure core-level spectroscopy', *Phys. Rev. B* 59 (20), 12807-12814 (1999)

Nelson, N. D., Piran T. & S. Weinberg, (eds.) *Statistical Mechanics of Membranes and Surfaces* , World Scientific, Singapore, (2004).

Nordlund, K., Keinonen, J. & Mattila, T., 'Formation of ion irradiation induced small-scale defects on graphite surfaces', *Phys. Rev. Lett.* 77 (4), 699-702 (1996)

Otto, A., 'Theory of plasmon excitations in thin films by electrons of non-normal incidence', *phys. stat. sol.* 22 (2), 401-406 (1967)

Pennycook, S., & Boatner, L.A., 'Chemically sensitive structure-imaging with a scanning transmission electron microscope', *Nature* 336 (6199), 565-567 (1988)

Perez, R. & Que, W., 'Plasmons in isolated single-walled carbon nanotubes', *J. Phys: Condens. Matter* 18 (12), 3197-3216 (2006)

Pomoell, J., Krasheninnikov, A. V., Nordlund, K. & Keinonen, J., 'Stopping of energetic ions in carbon nanotubes', *Nucl. Instr. Meth. in Phys. Res. B* 206, 18-21 (2003).

Raether, H.,' Surface plasma oscillations and their applications' in: *Physics of Thin Films*. Vol 9, p 145ff, Academic Press, New York - London (1977)

Sevincli, H., Topsakal, M., Durgun, E. & Ciraci, S., 'Electronic and magnetic properties of 3d transition-metal atom adsorbed graphene and graphene nanoribbons *Phys Rev B* 77 (19), 195434 (2008)

Sofo, J. O., Chaudhari, A. S. & Barber, G. D., ‚Graphane: A two-dimensional hydrocarbon', *Phys. Rev.* B 75 (15), 153401 (2007)

Stéphan, O., Taverna, D., Kociak, M., Suenaga, K., Henrard, L. & Colliex, C., 'Dielectric response of isolated carbon nanotubes investigated by spatially resolved electron energy-loss spectroscopy: From multiwalled to single-walled nanotubes', *Phys. Rev. B* 66 (15), 155422 (2002)

Stolyarova, E., Rim, K.- T., Ryu, S. M., Maultzsch, J., Kim, P., Brus, L. E., Heinz, T. F., Hybertsen, M. S. & Flynn, G. W., 'High-resolution scanning tunneling microscopy imaging of mesoscopic graphene sheets on an insulating surface', Proceedings of PNAS of the U. S. A, 104 (22), 9209-9212, May 2007

Suarez-Martinez, I., Felten, A., Pireaux, J. J., Bittencourt, C. & Ewels, C. P., 'Transition Metal Deposition on Graphene and Carbon Nanotubes' *J Nanosci. Nanotechnol.* 9 (10), 6171-6175 (2009)

Telling, R. H. & Heggie, M. I., 'Radiation defects in graphite', *Phil Mag* 87 (31), 4797-4846 (2007)

Terrones, M. & Terrones, H., 'The role of defects in graphitic structures', *Fullerene Sci and Tecnol* 4 (3), 517-533 (1996)

Uchoa, B., Lin, C.-Y. & Neto, A. H. C, 'Tailoring graphene with metals on top' *Phys Rev B* 77 (3), 35420 (2008)

Varns, R. & Strange, P., 'Stability of gold atoms and dimers adsorbed on graphene', *J. Phys.: Condens. Matter* 20 (22), 225005 (2008)

Xue, K. & Xu, Z.-P., 'Strain effects on basal-plane hydrogenation of graphene: A first-principles study', *Appl. Phys. Lett.* 96 (6), 063103 (2010)

Zhou, Z.-H., Gao, F. & Goodman, D. W., 'Deposition of metal clusters on single-layer graphene/Ru (0001): Factors that govern cluster growth', *Surface Science* 604 (13-14), L31–L38 (2010)

Permissions

All chapters in this book were first published in PAGE, by InTech Open; hereby published with permission under the Creative Commons Attribution License or equivalent. Every chapter published in this book has been scrutinized by our experts. Their significance has been extensively debated. The topics covered herein carry significant findings which will fuel the growth of the discipline. They may even be implemented as practical applications or may be referred to as a beginning point for another development.

The contributors of this book come from diverse backgrounds, making this book a truly international effort. This book will bring forth new frontiers with its revolutionizing research information and detailed analysis of the nascent developments around the world.

We would like to thank all the contributing authors for lending their expertise to make the book truly unique. They have played a crucial role in the development of this book. Without their invaluable contributions this book wouldn't have been possible. They have made vital efforts to compile up to date information on the varied aspects of this subject to make this book a valuable addition to the collection of many professionals and students.

This book was conceptualized with the vision of imparting up-to-date information and advanced data in this field. To ensure the same, a matchless editorial board was set up. Every individual on the board went through rigorous rounds of assessment to prove their worth. After which they invested a large part of their time researching and compiling the most relevant data for our readers.

The editorial board has been involved in producing this book since its inception. They have spent rigorous hours researching and exploring the diverse topics which have resulted in the successful publishing of this book. They have passed on their knowledge of decades through this book. To expedite this challenging task, the publisher supported the team at every step. A small team of assistant editors was also appointed to further simplify the editing procedure and attain best results for the readers.

Apart from the editorial board, the designing team has also invested a significant amount of their time in understanding the subject and creating the most relevant covers. They scrutinized every image to scout for the most suitable representation of the subject and create an appropriate cover for the book.

The publishing team has been an ardent support to the editorial, designing and production team. Their endless efforts to recruit the best for this project, has resulted in the accomplishment of this book. They are a veteran in the field of academics and their pool of knowledge is as vast as their experience in printing. Their expertise and guidance has proved useful at every step. Their uncompromising quality standards have made this book an exceptional effort. Their encouragement from time to time has been an inspiration for everyone.

The publisher and the editorial board hope that this book will prove to be a valuable piece of knowledge for researchers, students, practitioners and scholars across the globe.

List of Contributors

Haobin Zhang, Cao Chen, Wenge Zheng and Qing Yan
Ningbo Key Laboratory of Polymer Materials, Division of Polymers and Composites, China
Ningbo Institute of Material Technology & Engineering, Chinese Academy of Sciences, China

Shunlun He
Orinko Plastics Group, China

Tonica Valla and Zhihui Pan
Condensed Matter Physics and Materials Science Department, Brookhaven National Laboratory, USA

Elena Voloshina
Physikalische und Theoretische Chemie, Freie Universität, Berlin

Yuriy Dedkov
Fritz-Haber-Institut der Max-Planck-Gesellschaft, Berlin, Germany

Kristóf Tahy, Tian Fang, Pei Zhao, Aniruddha Konar, Chuanxin Lian, Huili (Grace) Xing, Michelle Kelly and Debdeep Jena
University of Notre Dame, USA

Ado Jorio and Luiz G. Cançado
Departamento de Física, Universidade Federal de Minas Gerais , Brazil

Erlon H. Martins Ferreira and Carlos A. Achete
Instituto Nacional de Metrologia, Normalização e Qualidade Industrial, Brazil

Rodrigo B. Capaz
Instituto de Física, Universidade Federal do Rio de Janeiro, Brazil

Qianqian Liu
Department of Physics, Astronomy, and Materials Science, Missouri State University, U.S.A

Li Wang and Kezheng Chen
College of Materials Science and Engineering, Qingdao University of Science and Technology, P. R. China

Lifeng Dong
Department of Physics, Astronomy, and Materials Science, Missouri State University, U.S.A
College of Materials Science and Engineering, Qingdao University of Science and Technology, P. R. China

Filippo Giannazzo, Sushant Sonde and Vito Raineri
Consiglio Nazionale delle Ricerche, Istituto per la Microelettronica e Microsistemi, Catania, Italy

E.V.Rut'kov and N.R.Gall
Ioffe Physico-Technical Institute of Russian Academy of Sciences, Russia

Yasuhide Ohno, Kenzo Maehashi and Kazuhiko Matsumoto
The Institute of Scientific and Industrial Research, Osaka University, Japan

Fei Zhuge, Run-Wei Li, Congli He, Zhaoping Liu and Xufeng Zhou
Ningbo Institute of Materials Technology and Engineering, Chinese Academy of Sciences, People's Republic of China

Ursel Bangert, Recep Zan and Cheng Ta Pan
The University of Manchester

Mhairi Gass
SuperSTEM Laboratories, CCLRC Daresbury Laboratory, UK

Index

www.ingramcontent.com/pod-product-compliance
Lightning Source LLC
LaVergne TN
LVHW051047220926
848485LV00001BA/33
* 9 7 8 1 6 3 9 8 7 2 6 9 5 *